BEAM ACCELERATION
IN CRYSTALS AND
NANOSTRUCTURES

BEAM ACCELERATION IN CRYSTALS AND NANOSTRUCTURES

Edited by

Swapan Chattopadhyay
Northern Illinois University, USA

Gérard Mourou
École Polytechnique, France

Vladimir D. Shiltsev
Fermi National Accelerator Laboratory, USA

Toshiki Tajima
University of California, Irvine, USA

World Scientific

NEW JERSEY · LONDON · SINGAPORE · BEIJING · SHANGHAI · HONG KONG · TAIPEI · CHENNAI · TOKYO

Published by

World Scientific Publishing Co. Pte. Ltd.

5 Toh Tuck Link, Singapore 596224

USA office: 27 Warren Street, Suite 401-402, Hackensack, NJ 07601

UK office: 57 Shelton Street, Covent Garden, London WC2H 9HE

British Library Cataloguing-in-Publication Data
A catalogue record for this book is available from the British Library.

BEAM ACCELERATION IN CRYSTALS AND NANOSTRUCTURES
Proceedings of the Workshop

ISBN 978-981-121-712-8

For any available supplementary material, please visit
https://www.worldscientific.com/worldscibooks/10.1142/11742#t=suppl

Contents

Summary of the "Workshop on Beam Acceleration in Crystals and Nanostructures" (Fermilab, June 24–25, 2019)[*]

V. D. Shiltsev

*Fermi National Accelerator Laboratory, MS 312,
Batavia, IL, 60510, USA
shiltsev@fnal.gov*

T. Tajima

*Department of Physics and Astronomy, University of California,
Irvine, CA 92697-4575, USA
ttajima@uci.edu*

Here we present a short summary of the "Workshop on Beam Acceleration in Crystals and Nanostructures" which has taken place at Fermilab on June 24–25, 2019.

Keywords: Accelerators; crystals; carbon nanotubes; nanostructures.

1. General Information about the Workshop

The concept of beam acceleration in solid-state plasma of crystals or nanostructures like CNTs (or alumna honeycomb holes) has the promise of ultra-high accelerating gradients $O(1-10)$ TeV/m, continuous focusing and small emittances of, e.g., muon beams and, thus, may be of interest for future high energy physics colliders. The goal of the "Workshop on Beam Acceleration in Crystals and Nanostructures" which took place at Fermilab on June 24 and 25, 2019, was to assess the progress of the concept over the past two decades and to discuss key issues toward proof-of-principle demonstrations and next steps in theory, modeling and experiment. The Workshop was endorsed by the American Physical Society (APS) Division of Physics of Beams (DPB) and the APS Topical Group in Plasma Astrophysics (GPAP), the International Committee on Ultra-High Intensity Lasers (ICUIL) and the International Committee on Future Accelerators' Panel on Advanced and Novel Accelerators (ICFA ANA).

The Workshop had 40 participants from 6 countries, representing all relevant areas of research such as accelerators and beam physics, plasma physics, laser physics, and astrophysics. More than 20 presentations covered a broad range of topics relevant to acceleration in crystals and carbon nanotubes (CNTs), including:

1. overview of the past and present theoretical developments toward crystal acceleration, ultimate possibilities of the concept;
2. concepts and prospects of PeV colliders for HEP;
3. effective crystal wake drivers: beams, lasers, other;

---

[*]Fermi National Accelerator Laboratory is operated by Fermi Research Alliance, LLC under Contract No. DE-AC02-07CH11359 with the United States Department of Energy.

4. beam dynamics in crystal acceleration;
5. instabilities in crystal acceleration (filamentation, etc.);
6. acceleration in nanostructures (CNTs, etc.);
7. muon sources for crystal acceleration;
8. application of crystal accelerators (X-ray sources, etc.);
9. astrophysical evidence of wakefield acceleration processes;
10. steps toward "proof-of-principle": 1 GeV gain over 1 mm, open theory questions, modeling and simulations;
11. possible experiments at FACET-II, FAST, AWAKE, AWA, RHIC, LHC, CEBAF, or elsewhere.

There were many vivid discussions on these subjects. All the talks and summaries of the discussions are available at https://indico.fnal.gov/event/19478/.

2. Major Outcomes

Several interesting proposals for further explorations or experimental tests were made by Sahel Hakimi *et al.* (University of California, Irvine, on how to drive wakes in CNTs by ultimate or existing X-ray pulses from, e.g., the LCLS SASE FEL); by Aakash Sahai *et al.* (University of Colorado, on production of detectable number of muons and their subsequent acceleration either at BELLA or FACET-II facilities); by Vladimir Shiltsev *et al.* (Fermilab, on demonstration of effective micromodulation of electron beams at FAST and FACET-II and subsequent experiments with micromodulated beams sent through CNTs at FAST with kA peak current type beams and then at the FACET-II facility with up to 300 kA bunches, e.g., to demonstrate the CNT channeling or to study the electron beam filamentation phenomena in structured materials); by Gennady Stupakov (SLAC, on possibility to use 1-nm-SASE-modulated electron bunches at the end of LCLS-I undulators to excite crystals and demonstrate acceleration); by Johnathan Wheeler *et al.*, (Ecole Polytechnique, to use the APOLLO laser facility to demonstrate Peta-Watt optical pulses/single cycle pulses via thin-film-compression technique); by Valery Lebedev (FNAL, to explore effectiveness of the wake excitation in crystals or CNTs by high-Z high energy ions, e.g. by 450 GeV ion beams from the CERN SPS available at the AWAKE facility, and observation of possible acceleration of externally injected electrons).

Formation of the research teams has began and follow-up presentations are being planned for the FACET-II Annual Science Workshop (SLAC, October 29–November 1, 2019).

These Proceedings of the Workshop are co-edited by Profs. Gerard Mourou (Ecole Polytech, 2018 Nobel Prize in Physics), Toshiki Tajima (UCI), Swapan Chattopdhyay (NIU) and Vladimir Shiltsev (Fermilab).

Fig. 1. Group photo of the Workshop.

Ultimate Colliders for Particle Physics: Limits and Possibilities

V. D. Shiltsev

Fermi National Accelerator Laboratory, MS 312,
Batavia, IL 60510, USA
shiltsev@fnal.gov

The future of the world-wide HEP community critically depends on the feasibility of the concepts for the post-LHC Higgs factories and energy frontier future colliders. Here we overview the accelerator options based on traditional technologies and consider the need for plasma colliders, particularly, muon crystal circular colliders. We briefly address the ultimate energy reach of such accelerators, their advantages, disadvantages and limits in the view of perspectives for the far future of the accelerator-based particle physics and outline possible directions of R&D to address the most critical issues.

Keywords: Colliders; accelerators; plasma accelerators; muons, crystals.

1. Current Landscape of Accelerator-based Particle Physics

Colliding beam facilities which produce high-energy collisions (interactions) between particles of approximately oppositely directed beams have been on the forefront of particle physics for more than half a century and. In total, 31 colliders ever reached operational stage and six of them are operational now.[1–4] These facilities essentially shaped the modern particle physics as their energy has been on average increasing by an order of magnitude every decade until about the mid-1990s. Since then, following the demands of high energy physics (HEP), the paths of the colliders diverged: to reach record high energies in the particle reaction the Large Hadron Collider was built at CERN, while record high luminosity e^+e^- colliders called "particle factories" were focused on detailed exploration of phenomena at much lower energies. Currently, the HEP landscape is dominated by the LHC. The next generation of colliders is expected to lead the exploration of the smallest dimensions beyond the current Standard Model.

Given the cost, complexity and long construction time of the collider facilities, the international HEP community regularly goes through extensive planning exercises. For example, in the recent past and at present we have the European Strategy planning (2012–2013), the US "Snowmass"and P5 plan (2013–2014), the European Strategy Update (2018–2020), under consideration now is the ILC250 project in Japan (decision by Spring 2020) and potential CepC project in China, the next US Snowmass and P5 process is set for 2019–2022. Discussions at the most recent 2019 European Particle Physics Strategy Update symposium (EPPSU, May 2019, Granada, Spain)[5] were focused on two types of the longer term (20–50 yrs)

Table 1. Main parameters of proposed colliders for high energy particle physics research.[6]

Project	Type	Energy TeV, c.m.e.	Int.Lumi./ Oper.Time.	Power years	Cost B(unit)
ILC	e^+e^-	0.25	2 ab^{-1} / 11yrs.	129	5.3ILCU
		0.5	4 ab^{-1} / 10 yrs.	163(204)	7.8ILCU
		1		300	?
CLIC	e^+e^-	0.38	1 ab^{-1} / 8yrs.	168	5.9CHF
		1.5	2.5 ab^{-1} / 7 yrs.	370	+5.1CHF
		3	5ab^{-1} / 8 yrs.	590	+7.3CHF
CEPC	e^+e^-	0.091	16 ab^{-1} / 4 yrs.	149	5 \$
		0.24	5.6 ab^{-1} / 7yrs.	266	+?
FCC-ee	e^+e^-	0.091	150 ab^{-1}/ 4 yrs.	259	10.5CHF
		0.24	5 ab^{-1} / 3 yrs.	282	
		0.365	1.5 ab^{-1} / 4 yrs.	340	+1.1 CHF
LHeC	ep	0.06/7	1 ab^{-1} / 12 yrs.	(+100)	1.75 CHF
HE-LHC	pp	27	20 ab^{-1} / 20 yrs.	220	7.2 CHF
FCC-hh	pp	100	30 ab^{-1} / 25 yrs.	580	24 CHF
$\mu\mu$Coll.	$\mu\mu$	14	50 ab^{-1} / 15 yrs.	230	10.7* CHF

HEP facilities: Higgs Factories(HF) and the Energy Frontier (50–100 TeV pp or 6–15 TeV lepton). There are four possible concepts fof these machines: linear e^+e^- colliders, circular e^+e^- colliders, pp/ep colliders, and muon colliders. (Table 1 summarizes main parameters of the future facilities, paramters of the muon collider are given according to Ref. 7.) They all have limitations in the energy, luminosity, AC power consuption, and cost which in turn mostly depend on five basic underlying accelerator technologies: normal-conducting (NC) magnets, superconducting (SC) magnets, NC RF, SC RF and plasma. The technologies are at different level of performance and readiness, cost efficiency and required R&D.[6]

Feasibility of the future colliders depends on their energy reach, luminosity, cost, length and power efficiency. So far, the most advanced of the proposals for the energy frontier collidsers call for acceleration by wakefields in plasma which can be excited by: lasers (demonstrated electron energy gain of about 8 GeV over 20 cm of plasma with density $3 \cdot 10^{17}$ cm^{-3} at the BELLA facility in LBNL); very short electron bunches (9 GeV gain over 1.3 m of $\sim 10^{17}$ cm^{-3} plasma at FACET facility in SLAC) and by proton bunches (some 2 GeV gain over 10 m of 10^{15} cm^{-3} plasma at the AWAKE experiment at CERN). In principle, the plasma wake field acceleration (PWFA) is thought to make possible multi-TeV e^+e^- colliders. There is a number of critical issues to resolve along that path, though, such as the power efficiency of the laser/beam PWFA schemes; acceleration of positrons (which are defocused when accelerated in plasma); efficiency of staging (beam transfer and matching from one short plasma accelerator cell to another); beam emittance control in scattering media; the beamstrahlung effect that leads to the rms energy spread at IP of about 30% for 10 TeV machines and 80% for 30 TeV collider.

An attempt to assess options for ultimate future energy frontier collider facility with c.o.m. energies of 300–1000 TeV (20–100 times the LHC) was made in Ref. 8.

There it was argued that for the same reason the circular e^+e^- collider energies do not extend beyond the Higgs factory range (~ 0.25 TeV), there will be no circular proton-proton colliders beyond 100 TeV because of unacceptable synchrotron radiation power — therefore, the colliders will have to be linear. Moreover, electrons and positrons even in linear accelerators become impractical above about 3 TeV due to beam-strahlung (radiation due to interaction) at the IPs and beyond about 10 TeV due to the radiation in the focusing channel. That leaves only $\mu^+\mu^-$ or pp options for the far future colliders. If one goes further and requests such a flagship machine not to exceed ~ 10 km in length then an accelerator technology is needed to provide average accelerating gradient of over 30 GeV/m (to be compared with ~ 0.5 GeV per meter in the LHC). There is only one such option known now: super-dense plasma as in, e.g., crystals,[9] but that excludes protons because of nuclear interactions and leaves us with muons as the particles of choice. Acceleration of muons (instead of electrons or hadrons) in crystals or carbon nanotubes with charge carrier density $\sim 10^{22}$ cm^{-3} has the promise of the maximum theoretical accelerating gradients of 1–10 TeV/m allowing to envision a compact 1 PeV linear crystal muon collider.[1] High luminosity can not be expected for such a facility if the beam power P is limited (e.g., to keep the total facility site power to some affordable level of $P \sim 100$ MW). In that case, the beam current will have to go down with the particle energy as $I = P/E_p$, and, consequently, the luminosity will need to go down with energy E_p. Therefore, there is a need in the paradigm shift for the particle colliders which in the past expected the luminosity to scale as $L \propto E_p^2$.

2. Acceleration in Crystals and Nanostructures

The very first proposal to accelerate muons in crystals[9] assumed excitation of solid plasmas by short intense X-ray pulses. The density of charge carriers (conduction electrons) in solids $n_0 \sim 10^{22-24}$ cm^{-3} is significantly higher than that in gaseous plasma, and correspondingly, the longitudinal accelerating fields of upto 100–1000 GeV/cm (10–100 TV/m) are possible according to

$$E\,[\mathrm{GV/m}] \approx 100\sqrt{n_0\,[10^{18}\ \mathrm{cm}^{-3}]}. \tag{1}$$

The are several critical phenomena in the solid plasma due to intense energy radiation in high fields and increased scattering rates which result in fast pitch-angle diffusion over distances of $l_d\,[\mathrm{m}] \sim E_p\,[\mathrm{TeV}]$. The latter leads to particles escaping from the driving field; thus, it was suggested that particles (muons) have to be accelerated in solids along major crystallographic directions, which provide a channeling effect in combination with low emittance determined by an Angstrom-scale aperture of the atomic tubes.[10,11] Channeling in the nanotubes was later brought up as a promising option.[12–14] Positively charged particles are channeled more robustly, as they are repelled from ions and thus experience weaker scattering. Radiation emission due to the betatron oscillations between the atomic planes is thought to be the major source of energy dissipation, and the maximum beam energies are limited to about 0.3 TeV for positrons, 10^4 TeV for muons and 10^6 TeV for protons.[10] For energies of 1 to 10 PeV, muons offer much more attraction because they are point-like particles and, contrary to protons, do not carry an intrinsic energy spread of elementary constituents; and they can much easier propagate in solid plasma than protons which will extinct due to nuclear interactions. The muon decay becomes practically irrelevant in the proposed very fast acceleration scheme as the muon lifetime quickly grows with energy as $2.2\,\mu s \times \gamma$. Very high gradient crystal accelerators have to be disposable if the externally excited fields exceed the ionization thresholds and destroy the periodic atomic structure of the crystal (so acceleration will take place only in a short time before full dissociation of the lattice). For the fields of about $1\,\mathrm{GV/cm} = 0.1\,\mathrm{TV/m}$ or less, reusable crystal accelerators can probably be built which can survive multiple pulses. Possible conceptual scheme of a crystal linear muon collider — see Fig. 1 — includes two high brightness muon sources, two continuous crystal linacs of a total length of 1 to 10 km driven by numerous X-ray sources (or other type of drivers) to reach 1–10 PeV c.m.e. at the interaction point with a crystal funnel.[1] Initial luminosity analysis of such machine assumes the minimal overlap area of the colliding beams to the crystal lattice cell size $A \sim 1\,\text{Å}^2 = 10^{-16}\,\mathrm{cm}^{-2}$ and that the crystals in each collider arm are aligned channel to channel. The number of muons per bunch N

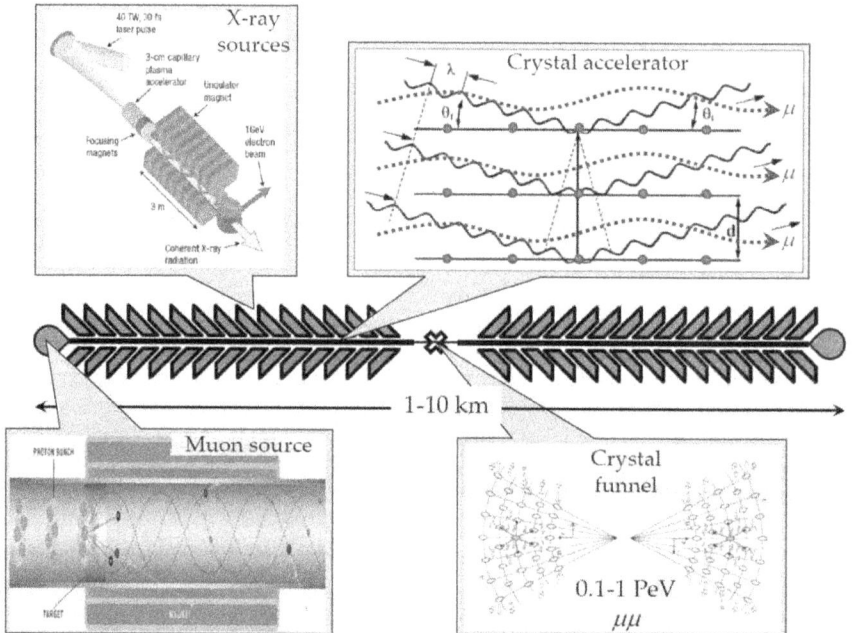

Fig. 1.   Concept of a linear X-ray crystal muon collider (adapted from Ref. 1).

also cannot be made arbitrary high due to the beam loading effect and should be $N \sim 10^3$. Excitation many parallel atomic channels n_{ch} can increase the luminosity $L = f n_{\text{ch}} N^2 / A = f \cdot 10^{16} \cdot 10^6 \cdot n_{\text{ch}} \, [\text{cm}^{-2}\text{s}^{-1}]$ which can reach $10^{30} \, \text{cm}^{-2}\text{s}^{-1}$ at, e.g., $f = 10^6$ Hz and $n_{\text{ch}} \sim 100$. Exceeding the value of the product $f n_{\text{ch}}$ beyond 10^8 Hz can be very costly as the total beam power $P = f n_{\text{ch}} N E_p$ will get beyond a practical limit of ~ 10 MW. Instead, using some kind of *crystal funnel* to bring microbeams from many channels into one can increase the luminosity by a factor of n_{ch} to some $10^{32} \, \text{cm}^{-2}\text{s}^{-1}$.

3. Challenges and Open Questions

Until now, crystals were of interest for particle accelerators because of their strong inter-planar electric fields $\sim 10 \, \text{V}/\text{Å} = 1 \, \text{GV/cm}$.[15] Given their unique radiation hardness and stability, crystals were used even in the highest energy hadron colliders like the 2 TeV Tevatron and the 14 TeV LHC for particle focusing and/or for deflection (with efficiency notably growing with the energy, e.g. better than $\sim 95\%$ in the Tevatron and over 99% in the LHC for some 4 mm bent crystals[16,17]).

Several methods can be envisioned for the wakefield excitation in the crystals[18] — see Fig. 2. Historically first was the suggestion to use ultrashort and powerful 40 keV X-ray pulses injected in the crystal at a proper angle to achieve Bormann anamalous transmission over longer distances.[9] Extreme X-ray pulse power density $O(10^{23-24} \, \text{W/cm}^2)$ can now be achieved at the SASE FELs like LCLS at SLAC, and the gradients of about $0.2 \, [\text{TV/m}] \cdot a_0^2$ are predicated in CNTs which can lead to 100s of MeV of acceleration in few micron long structures[13,14] (here $a_0 \sim O(1)$ is the normalized field intensity of a $O(1 \, \text{nm})$ wavelength laser). Further opportunities to increase the laser intensities can be offered by recently proposed ICAN and thin film compression schemes.

Bunches of charged particles can excite plasma effectively if their transverse and longitudinal sizes are comparable or shorter than the plasma wavelength of $\lambda_p \sim 0.3 \, \mu\text{m}$ for $n_0 = 10^{22} \, \text{cm}^{-3}$ and the total number of particles in that volume approaches the number of free electrons in the solid plasma $\sim n_0 \lambda_p^3$. Arguably the closest to such conditions are the electron bunches prepared for the FACET-II experiments at SLAC — at the initial stage of 3D compression they will be $8 \times 7 \times 2 \, \mu\text{m}$ that for the total charge of 2 nC results in $n_e = 6 \cdot 10^{18} \, \text{cm}^{-3}$, while at the ultimate compression $2 \times 2 \times 0.4 \, \mu\text{m}$ the density will be about $n_e = 2 \cdot 10^{20}$ cm^{-3} (and corresponding peak current of about 300 kA).

Relativistic fully stripped heavy ions can offer yet another possibility for wakefields excitation in crystals or carbon nanotubes[18] as the fields they leave behind in the media are about the ionization loss gradient of

$$E_i \approx 2 \, [\text{MeV}/(\text{g/cm}^2)] \times Z^2, \tag{2}$$

that gives $E_i \approx 2 \, \text{TV/m}$ for $Z = 70\text{–}80$ in silicon. Naturally, one can envision these ions either channeling in crystals ahead of the accelerating particles (e.g. muons) or

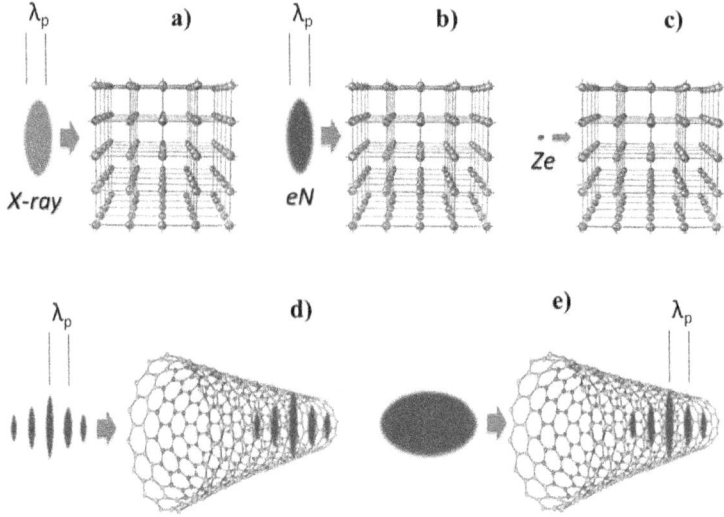

Fig. 2. Possible ways to excite plasma wakefields in crystals or/and nanostructures: (a) by short X-ray laser pulses; (b) by short high density bunches of charged particles; (c) by heavy high-Z ions; (d) by modulated high current beams; (d) by longer bunches experiencing self modulation instability in the media.

being well aligned with them so the latter are always kept in sync with accelerating wake. At present, the highest energy heavy ions are available at RHIC (100 GeV/u gold, $Z = 79$) and LHC (2.5 TeV/u lead, $Z = 82$) and the dephasing length $2\gamma_p^2\lambda_p$ can be as long as few cm to few meters.

Figures 2(d) and 2(e) conceptually depict two other possibilities to excite structured solid plasmas by either pre-modulated high density bunches of charged particles or by initially unmodulated long bunches which get microbunched while propagating in the media due to self modulation instability (SMI). In both cases it is important that the drive beam density modulation is resonant to the plasma waves, i.e., occurs at λ_p so the waves excited by individual microbunches add up coherently. The first of these methods can employ, e.g., either the nanomodulated bunches at the end of SASE (self amplified spontaneous emission) process in modern X-ray FELs or micromodulated beams obtained via slit-masking in chicanes as, e.g., it was proposed in Ref. 19. The SMI in longer proton bunches traversing low density gaseous plasma has been demonstrated in the AWAKE experiment at CERN. Of course, in the solid plasma of crystals or CNTs the SMI will compete with other phenomena, such as the Weibel or filamentation instabilities and that issue requires detail study.

In general, there are many important topics for future research on acceleration in crystals and nanostructures, including: (a) critical overview of the past and present theoretical developments toward crystal acceleration, exploration of the ultimate possibilities of the concept; (b) further development of the concepts and most optimal schemes of PeV crystal colliders for HEP; (c) theory, modeling and experiments on effective crystal wave drivers such as beams (including self-modulation instability), lasers, other schemes; (d) particle and beam dynamics in crystal acceleration channels; (e) instabilities in crystal acceleration channels, such as filamentation/Weibel instability, etc.; (f) acceleration in nanostructures (CNTs, alumna honeycomb holes, zeolites, others); (g) high brightness muon sources for crystal acceleration; (h) possible practical applications of crystal accelerators (X-ray sources, etc.); (i) comprehensive study of possible steps toward "proof-of-principle" experiment to demonstrate 1 GeV energy gain over 1 mm; (j) preparation of possible crystal acceleration experiments at FACET-II, FAST, BELLA, AWAKE, AWA, RHIC, LHC or elsewhere (including addressing open theory questions, modeling and simulation, hardware and diagnostics development, etc.).

4. Conclusions

The concept of beam acceleration in solid-state plasma of crystals or nanostructures like CNTs (or alumna honeycomb holes) has the promise of ultra-high accelerating gradients $O(1\text{–}10)$ TeV/m, continuous focusing and small emittances of, e.g., muon beams and, thus, may be of interest for future high energy physics colliders. Recent advances in the acceleration in gaseous beam- or laser-driven plasma and muon production and cooling, progress in the intense X-ray pulse generation, production of short very high peak current bunches of charged particles, development of sophisticated high-performance PIC codes to model high density plasmas — all that paves the way for comprehensive studies of the theory, corresponding modeling, and eventually experiments on the wakefield excitation in solid plasmas, acceleration of particles in crystals or CNTs, muon production and detection, etc. Some schemes of the crystal/CNT excitation can be tested at the beam test facilities such FACET-II at SLAC, FAST at Fermilab, BELLA at LBNL and AWAKE at CERN. One can also explore opportunities for proof-of-principle experimental studies with either high energy high-Z ions available at RHIC or LHC or to exploit unique properties of the self-modulated electron beams in the SASE FEL facilities, like, e.g. the LCLS-I and -II at SLAC. Past experience with crystals in high energy particle accelerators as well as available hardware might very helpful for the initial studies.

Acknowledgments

Some materials of this summary were also presented at the *European Particle Physics Strategy Update Symposium* (May 13–16, 2019, Granada, Spain). I would like to thank T. Tajima (UCI), V. Lebedev (FNAL), G. Stupakov (SLAC) and F. Zimmermann (CERN) for fruitful discussions on the subject of this presentation.

Fermi National Accelerator Laboratory is operated by Fermi Research Alliance, LLC under Contract No. DE-AC02-07CH11359 with the United States Department of Energy.

References

1. V. Shiltsev, High-energy particle colliders: Past 20 years, next 20 years, and beyond, *Physics-Uspekhi* **55**(10), 965 (2012).
2. A. Chao *et al.* (eds.) *Handbook of Accelerator Physics and Engineering* (World Scientific, 2013).
3. A. Chao and W. Chou (eds.), *Reviews of Accelerator Science and Technology. Volume 7: Colliders* (World Scientific, 2015).
4. S. Myers and O.Bruning (eds.), *Challenges and Goals for Accelerators in the XXI Century* (World Scientific, 2016).
5. *European Particle Physics Strategy Update Symposium*, May 13–16, 2019, Granada, Spain, https://indico.cern.ch/event/765096/contributions/
6. V. Shiltsev, arXiv:1907.01545.
7. D. Neuffer and V. Shiltsev, On the feasibility of a pulsed 14 TeV CME muon collider in the LHC tunnel, *JINST* **13**(10), T10003 (2018).
8. V. Shiltsev, Will there be energy frontier colliders after LHC?, *PoS(**ICHEP2016**)*, 050 (2016), also Report No. FERMILAB-CONF-16-369-APC.
9. T. Tajima and M. Cavenago, Crystal X-ray accelerator, *Phys. Rev. Lett.* **59**(13), 1440 (1987).
10. P. Chen and R. Noble, Crystal channel collider: Ultra-high energy and luminosity in the next century, *AIP Conf. Proc.* **398**(1), 273 (1997).
11. I. Dodin and N. Fisch, Charged particle acceleration in dense plasma channels, *Phys. Plasmas* **15**(10), 103105 (2008).
12. T. Tajima *et al.,* Beam transport in the crystal x-ray accelerator, *Part. Accel.* **32**, 235 (1989).
13. X. Zhang *et al.,*, Particle-in-cell simulation of x-ray wakefield acceleration and betatron radiation in nanotubes, *Phys. Rev. Accel. Beams* **19**(10), 101004 (2016).
14. Y. M. Shin, D. Still and V. Shiltsev, X-ray driven channeling acceleration in crystals and carbon nanotubes, *Phys. Plasmas* **20**(12), 123106 (2013).
15. V. Biryukov, Y. Chesnokov and V. Kotov, *Crystal Channeling and its Application at High-Energy Accelerators* (Springer, 2013).
16. N. Mokhov *et al.*, Crystal collimation studies at the Tevatron (T-980), *Int. J. Mod. Phys. A* **25**, 98 (2010).
17. W. Scandale *et al.*, Observation of channeling for 6500 GeV/c protons in the crystal assisted collimation setup for LHC, *Phys. Lett. B* **758**, 129 (2016).
18. V. Shiltsev, Presentation at *EuCARD-2 XBEAM Strategy Workshop*, 13–17 February 2017, Valencia, Spain); https://indico.cern.ch/event/587477/contributions/ and in F. Zimmermann *et al.*, *Strategy for the Future Extreme Beam Facilities*, EuCARD Monograph no. XLIV (Institute of Electronic Systems, Warsaw University of Technology, Warsaw, 2017).
19. Y. M. Shin, A. Lumpkin and R. Thurman-Keup, TeV/m nano-accelerator: Investigation on feasibility of CNT-channeling acceleration at Fermilab, *Nucl. Instr. Meth. B* **355**, 94 (2015).

Novel Laser-Plasma TeV Electron-Positron Linear Colliders

Kazuhisa Nakajima[†], Jonathan Wheeler, Gérard Mourou

International Center for Zetta-Exawatt Science and Technology,
École Polytechnique, Route de Saclay, Palaiseau, 91128, France
[†]naka115@dia-net.ne.jp

Toshiki Tajima

Department of Physics and Astronomy, University of California
Irvine, CA 92697, USA
ttajima@uci.edu

TeV center-of-mass energy electron-positron linear colliders comprising seamlessly staged capillary laser-plasma accelerators are presented. A moderate intensity laser pulse coupled with the single electromagnetic hybrid mode in a gas-filled capillary can generate plasma waves in the linear regime, where laser wakefields can accelerate equally focused electron and positron beams. In multiple stage capillary accelerators, a particle beam with respect to the laser wakefield can undergo consecutive acceleration up to TeV energies, associated with continuous transverse focusing in a beam size down to a nanometer level, being capable of a promising electron-positron linear collider with very high luminosities of the order of 10^{34} cm^{-2}s^{-1}. The transverse and longitudinal beam dynamics of beam particles in plasma wakefields with the effects of radiation reaction and multiple Coulomb scattering are investigated numerically to estimate the luminosities in beam-beam collisions with the effects of beamstrahlung radiation and bunch disruption.

Keywords: Laser-plasma accelerator; Linear collider; CAN laser.

1. Introduction

Laser-driven plasma accelerators (LPAs) [1-3] can support a wide range of potential applications requiring high-energy and high-quality electron/positron beams. In particular, field gradients, energy conversion efficiency and repetition rates are essential factors for practical applications such as compact x-ray free electron lasers [4, 5] and high energy frontier colliders [6-9]. For such applications, LPAs have some drawbacks in laser guiding and dephasing that limit energy gains despite their high accelerating gradients. In the LPA concept [1], one of the critical issues is optical guiding of relativistic laser pulses in underdense plasma to the extent of a dephasing length, which scales as $\propto n_e^{-3/2}$, where n_e is the operating plasma density of LPAs [2]. The propagation of laser pulses in plasmas is described by refractive guiding, in which the refractive index can be modified from the linear free space value mainly by relativistic self-focusing, ponderomotive channeling and a preformed plasma channel [10]. The self-guided LPA [11-13] relies only on intrinsic effects of relativistic laser-plasma interactions such as relativistic self-focusing and ponderomotive channeling. On the other hand, the channel-guided LPA exploits a plasma waveguide with a preformed density channel [14-16] or a gas-filled capillary waveguide made of metallic or dielectric materials [17]. The plasma waveguide with preformed density channel is likely to propagate a single mode laser pulse through a radially parabolic distribution of the refractive index and generate plasma waves inside the density channel, the properties of which are largely affected by a plasma density profile and laser power. In a capillary waveguide filled with gas, a laser pulse is guided via Fresnel reflection at the inner capillary wall and plasma waves are generated in initially homogeneous plasma, relying on neither laser power nor plasma density as long as the laser intensity on the capillary wall is kept below the material breakdown [18, 19].

One challenging application of LPAs is the future high-energy electron-positron linear colliders (LC) being capable of providing TeV center-of-mass (c.m.) energies at the luminosities of more than 10^{34} cm^{-2} s^{-1} [20]. It is very reasonable in conventional fashion that two-linac complex comprising successive LPA stages accelerate electron/positron beams to the high-energies required and then deliver the colliding beams focused to the minuscule dimensions at the interaction point to produce a high-enough event rate required for particle physics experiments. In each of these linacs, an independent laser is synchronously operated at a considerably high-repetition rate to drive the plasma waves of the LPA. Recent progress of multi-GeV LPA experiments [12-16] and high-average power lasers at kHz-level repetition rate [21] may make a laser-plasma-based collider concept feasible, achieving considerable reduction in a structural size and simplicity, compared to the proposed designs of the future electron-positron colliders based on both conventional technology [20] and so-called advanced accelerator concepts [22]. Despite rapid progress and growing interest in laser-plasma accelerators, there are few studies on a full-scale model to explore beam dynamics and beam-beam interactions in multistage laser-plasma-based electron-positron colliders in the multi-TeV energy range, apart from visionary parameter scaling based on single-stage, small-scale model studies [6-9]. Previously only one study [23] on the particle dynamics in multistage LPAs for a conceptual collider has been carried out by using a linear particle transport model based on a simple harmonic oscillator approximation for the betatron motion in plasma wakefields, while missing considerations on effects of synchrotron (betatron) radiation and multiple Coulomb scattering as well as the beam loading, *i.e.*, beam self-field. This study points out that increasing drift length turns out to encounter a "blow out" of the betatron amplitude because the stability condition for the trace of the particle transfer matrix will be violated [23].

For the electron-positron collider design, a most essential requirement is the capability of generating a substantial event rate of the fundamental particle reaction of interest, characterized by $Y = \mathcal{L}\sigma$ where $\mathcal{L}$ is the luminosity and σ is the collision cross section for the reaction $e^+e^- \rightarrow e^+e^-$ at a c.m. energy W_0. Since this cross section scales as $\sigma \propto W_0^{-2}$, a reasonable guideline for the luminosity may be imposed as $\mathcal{L}$ [10^{34} cm^{-2}s^{-1}] $\sim W_0^2$ [TeV] [6] such that the event rate in particle physics experiments becomes constant regardless of c.m. energies. This requirement often yields high-density bunches of colliding highly relativistic electron and positron beams that lead to two major effects, namely, beamstrahlung [24, 25] and disruption [26, 27], arising from the beam-

beam interaction with strong electromagnetic fields created in the oppositely oncoming bunch. While strong disruption contributes to enhancement of the luminosity, the associated beamstrahlung effects induce radiation loss of the particle energies and degradation of the resolution in c.m. energies.

In this paper, we present a novel scheme of laser-plasma-based electron-positron linear collider comprising multiple gas-filled capillary waveguides without coupling drift spaces for avoiding a "blow out" of betatron oscillations. Each of which is driven by a laser pulse formed from the lowest order capillary eigenmode, so-called electromagnetic hybrid mode EH_{11} [18]. The numerical model on the bunched beam dynamics in laser wakefields, based on the exact solution of single particle betatron motion, taking into account radiation reaction and multiple Coulomb scattering, shows that the normalized transverse emittance and beam radius can be consecutively reduced during continuous acceleration in the presence of optimally phased recurrence of longitudinal and transverse wakefields in the multistage LPAs, leading to three orders of magnitude smaller values. The final properties of the particle beams reached the energy objective, thus meeting the luminosity requirements without resorting to an additional focusing system. The remaining part of this paper is organized as follows. In Sec. 2, the complete descriptions on laser wakefields generated by the single electromagnetic hybrid mode with moderate intensity and the beam dynamics of a single particle (electron or positron) in laser wakefields, beam loading and self-focusing fields are provided. The numerical calculation on particle transport based on the solution of betatron motion is described, taking into account radiation reaction and multiple Coulomb scattering. In Sec. 3, the numerical results of the particle transport calculation for the multistage capillary accelerators are shown and the analytical estimates on the evolution of bunch radius, transverse normalized emittance and radiative energy loss are presented. In Sec. 4, the effects of beam-beam interactions of the colliding electron and positron beams extracted from the capillary accelerating structure on the c.m. energy and luminosity are evaluated in terms of beamstrahlung and disruption for various vacuum drift distances between the plasma accelerator and the interaction point. In Sec. 5, an embodiment of the LPA stage is envisioned by exploiting a tens kW-level high average power laser such as a coherent amplification network (CAN) consisting of fiber laser systems. In Sec. 6, design examples of the multi-TeV laser-plasma-capillary collider are discussed.

2. Beam Dynamics in a Capillary Accelerator

2.1 *Laser wakefield and acceleration driven by a single capillary mode*

Considering the electromagnetic hybrid modes EH_{1n} [18] to which the most efficient coupling of a linearly polarized laser pulse in vacuum occurs, the normalized vector potential for the eigenmode of the n-th order is written by [28]

$$a_n(r,z,t) \sim a_{n0} J_0\left(\frac{u_n r}{R_c}\right) \exp\left[-k_n^I z - \frac{(z-v_g t)^2}{2c^2\tau^2}\right] \cos(\omega_0 t - k_{zn} z), \qquad (1)$$

where a_{n0} is the amplitude of the normalized vector potential for the EH_{1n} mode, J_0 the zero-order Bessel function of the first kind, u_n the n-th zero of J_0, r the radial coordinate of the capillary in cylindrical symmetry, R_c the capillary radius, z the longitudinal coordinate, τ the pulse duration and ω_0 the laser frequency. The longitudinal wave number k_{zn}, the damping coefficient k_n^l and the group velocity of the n-th mode v_g are given by [18]

$$k_{zn} = \left(k_0^2 - \frac{u_n^2}{R_c^2}\right)^{1/2}, \quad k_n^l = \frac{u_n^2(1+\varepsilon_r)}{2k_{zn}^2 R_c^3(\varepsilon_r - 1)^{1/2}}, \quad v_g \simeq c\left(1 - \frac{u_n^2}{k_0^2 R_c^2}\right)^{1/2},\tag{2}$$

where $k_0 = \omega_0/c$ is the laser wavenumber and ε_r is the relative dielectric constant. The coupling efficiency C_n defined by an input laser energy with a spot radius r_0 and amplitude a_0 coupled to the E_{1n} mode in the capillary is calculated for a linearly polarized Gaussian beam,

$$C_n = 8\left(\frac{R_c}{r_0 J_1(u_n)}\right)^2\left[\int_0^1 x\exp(-\frac{x^2 R_c^2}{r_0^2})J_0(u_n x)dx\right]^2,\tag{3}$$

and for an Airy beam,

$$C_n = \frac{4}{J_1^2(u_n)}\left[\int_0^1 J_1(\frac{v_1 R_c x}{r_0})J_0(u_n x)dx\right]^2,\tag{4}$$

where $v_1 = 3.8317$ is the first root of the equation of $J_1(x) = 0$ [18]. When a linearly polarized Gaussian beam is focused at the capillary entrance ($z = 0$), the maximum coupling coefficient is given by $C_1 \simeq 0.981$ at $R_c/r_0 \simeq 1.55$ (or $r_0/R_c \simeq 0.65$). For an Airy beam, the maximum coupling efficiency is $C_1 \simeq 0.837$ at $R_c/r_0 \simeq 0.95$ (or $r_0/R_c \simeq 1.05$).

In the linear limit $|a| = e|A|/(m_e c^2) \ll 1$, the ponderomotive force exerted on plasma electrons by the capillary laser field a_n is given by $F_p = -m_e c^2\beta_g\nabla a_n^2/2$, where $m_e c^2$ is the electron rest energy, $\beta_g = v_g/c$ and a_n^2 is defined by averaging the nonlinear force over the laser period $2\pi/\omega_0$, i.e., using $a_{n0}^2 = C_n a_0^2$,

$$a_n^2(r,z,t) = \frac{1}{2}C_n a_0^2 J_0^2\left(\frac{u_n r}{R_c}\right)\exp\left[-2k_n^l z - \frac{(z - v_g t)^2}{c^2\tau^2}\right].\tag{5}$$

The electrostatic potential $\Phi(r,t)$ of the wake excited in plasma with electron density n_e by a laser pulse propagating the capillary is obtained from the equation [2]

$$\left(\frac{\partial^2}{\partial t^2} + \omega_p^2\right)\Phi(r,t) = \frac{\omega_p^2 m_e c^2\beta_g}{2e}a_n^2(r,t),\tag{6}$$

where $\omega_p = (4\pi e^2 n_e/m_e)^{1/2}$ is the plasma frequency. The solution of Eq. (6) is given by

$$\Phi(r,t) = \frac{\sqrt{\pi}}{8}\left(\frac{m_e c^2}{e}\right)\beta_g k_p c\tau e^{-\left(\frac{k_p c\tau}{2}\right)^2} a_0^2 C_n J_0^{\,2}\left(\frac{u_n r}{R_c}\right)e^{-2k_n^l z}$$
$$\times\left[S(z)\cos k_p(z-v_g t)+C(z)\sin k_p(z-v_g t)\right] \tag{7}$$

where $k_p = \omega_p / v_g$ is the plasma wavenumber in the capillary and

$$C(z) = -\Re\mathrm{erfc}\left(\frac{z-v_g t}{c\tau}+i\frac{k_p c\tau}{2}\right), \quad S(z) = \Im\mathrm{erf}\left(\frac{z-v_g t}{c\tau}+i\frac{k_p c\tau}{2}\right), \tag{8}$$

with the real ($\Re$) and imaginary ($\Im$) part of the error function $\mathrm{erf}\ z = (2/\sqrt{\pi})\int_0^z e^{-s^2}ds$ and $\mathrm{erfc}\ z = 1-\mathrm{erf}\ z$. For $k_n^l \ll k_p$, the longitudinal electric field can be obtained from $E_z = -\partial\Phi/\partial z$ as

$$E_z(r,z,t) = \frac{\sqrt{\pi}}{8}\left(\frac{m_e c\omega_p}{e}\right)a_0^2 C_n k_p c\tau e^{-\left(\frac{k_p c\tau}{2}\right)^2} J_0\left(\frac{u_n r}{R_c}\right)e^{-(\alpha_d a_0^2/\gamma_g^2)k_p z}$$
$$\times\left[S(z)\sin k_p(z-v_g t)-C(z)\cos k_p(z-v_g t)\right]. \tag{9}$$

The transverse focusing force is obtained from $F_r = e(E_r - B_\varphi) = -\partial\Phi/\partial r$ as

$$F_r(r,z,t) = \frac{\sqrt{\pi}}{4}\left(\frac{m_e c\omega_p}{e}\right)\frac{c\tau}{R_c}e^{-\left(\frac{k_p c\tau}{2}\right)^2} a_0^2 u_n C_n J_0\left(\frac{u_n r}{R_c}\right)J_1\left(\frac{u_n r}{R_c}\right)e^{-(\alpha_d a_0^2/\gamma_g^2)k_p z}$$
$$\times\left[C(z)\sin k_p(z-v_g t)+S(z)\cos k_p(z-v_g t)\right], \tag{10}$$

where $J_1(z) = -J_0'(z)$ is the Bessel function of the first order. Here, while propagating through plasma and generating wakefields, the laser pulse loses its energy as $\partial\mathcal{E}_L/\partial z \sim -\mathcal{E}_L/L_{pd}$ [29], where L_{pd} is the characteristic scale length of laser energy deposition into plasma wave excitation, referred to as the pump depletion length. In the quasi-linear wakefield regime, i.e., $a_{n0}^2 = C_n a_0^2 \le 1$, the pump depletion length is $k_p L_{pd} = \gamma_g^2/(\alpha_d a_0^2)$ [7, 29], where $\gamma_g = \omega_0/\omega_p$ and $\alpha_d \approx C_n/17.4$ for a Gaussian laser pulse. Assuming that the pulse duration is fixed, the laser energy evolution in the capillary can be written as $\mathcal{E}_L(z) \propto a_{n0}^2 \exp[-(\alpha_d a_0^2 e^{-k_n^l z}/\gamma_g^2)k_p z - k_n^l z]$ as indicated in Eqs. (9) and (10).

In the linear wakefields excited by the single hybrid mode, the longitudinal motion of electron ($-$) and positron ($+$) moving along the capillary axis at a normalized velocity $\beta_z = v_z/c \approx 1$ is described as [2, 10]

$$\frac{d\gamma}{dz} = \pm k_p\frac{E_{z0}}{E_0}, \quad \frac{d\Psi}{dz} = k_p\left(1-\frac{\beta_p}{\beta_z}\right) \approx k_p(1-\beta_g) \approx \frac{k_p}{2\gamma_g^2}, \tag{11}$$

where $m_e c^2\gamma$ is the electron energy, $E_{z0} = E_z(0,z,t)$ the accelerating field at $r = 0$, $E_0 = m_e c\omega_p/e$ the nonrelativistic wave-breaking field, $\Psi = k_p(z-v_p\int_0^z dz/v_z)$ the particle phase with respect to the plasma wave, $\beta_p = v_p/c \approx v_g/c = \beta_g = (1-\omega_p^2/\omega_0^2)^{1/2}$

the phase velocity of the plasma wave and $\gamma_g = (1 - \beta_g^2)^{-1/2} \gg 1$. In Eq. (9), taking into account $C(z) \to -2$ and $S(z) \to 0$ for $z - v_g t \ll -c\tau$, and setting the pulse duration of a drive laser pulse with a Gaussian temporal profile to be the optimum length $k_p c\tau = \sqrt{2}$, the accelerating field is given by

$$\frac{E_{z0}}{E_0} = \frac{1}{2}\sqrt{\frac{\pi}{2}} a_0^{\,2} C_n e^{-(1+4\alpha_d a_0^2 \Psi)/2} \cos \Psi . \tag{12}$$

Integrating the equations of motion, the energy gain and phase of the particle can be obtained as

$$\Psi(z) = \Psi_0 + k_p z / (2\gamma_g^2), \tag{13}$$

$$\gamma(z) = \gamma_0 \pm \gamma_g^2 \sqrt{\frac{\pi}{2}} \frac{a_0^2 C_n}{4\alpha_d^2 a_0^4 + 1} [e^{-(1+4\alpha_d a_0^2 \Psi)/2}(\sin \Psi - 2\alpha_d a_0^2 \cos \Psi) - e^{-(1+4\alpha_d a_0^2 \Psi_0)/2}(\sin \Psi_0 - 2\alpha_d a_0^2 \cos \Psi_0)] , \tag{14}$$

where Ψ_0 is the initial phase of the particle with respect to the wakefield and $m_e c^2 \gamma_0$ is the initial particle energy.

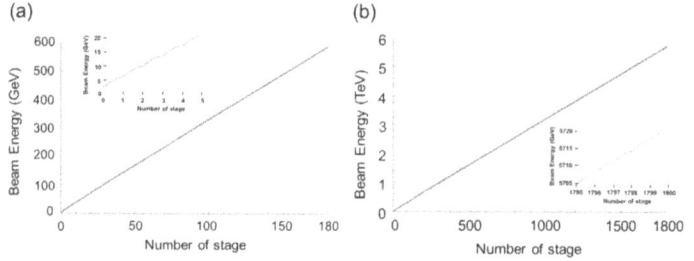

Fig. 1. The evolution of the beam energy (a) for 1.2 TeV LC comprising 180 stages per beam and (b) for 11 TeV LC comprising 1800 stages per beam as a function of the number of capillary LPA stages, where the beam energy reaches 589.3 GeV and 5720 GeV, respectively, from the injection energy of 2.5 GeV. The inset in (a) shows the energy evolution in the initial 5 stages and the inset in (b) that in the final 5 stages, respectively. The parameters of the LPA stage is shown in Table 1.

As an example, for the multistage capillary accelerator with $n_e = 1 \times 10^{17}$ cm^{-3} ($\gamma_g^2 = 1.115 \times 10^4$) driven by a Gaussian laser pulse with $a_0^2 = 1$ and $\tau \approx 79$ fs ($k_p c\tau = \sqrt{2}$), coupled to the EH$_{11}$ mode with the efficiency $C_1 \approx 0.981$, Figure 1 shows the evolution of the particle beam energy $m_e c^2 \gamma$ up to 589.3 GeV with the injection energy of 2.5 GeV over 180 stages (95.4 m) in (a), corresponding the average accelerating gradient of 6.15 GeV/m and up to 5720.4 GeV over 1800 stages (953.7 m) in (b), corresponding the average accelerating gradient of 6.0 GeV/m. This reduction of the accelerating gradient is attributed to the effects of beam loading of $N_b = 1 \times 10^9$ particles in a bunch and radiation reaction force, both of which are induced by transverse beam dynamics of beam particles.

2.2 Beam-driven wakefields and beam loading

In the linear regime, where a plasma density perturbation is $\delta n = n - n_e \ll n_e$, the wakefield $E_{zb}(\xi, r)$ excited by an electron beam with a profile $n_b(\xi, r)$ is described as

$$\left(\frac{\partial^2}{\partial \xi^2} + k_p^2\right)\frac{\delta n}{n_e} = -k_p^2 \frac{n_b}{n_e}, \quad (\nabla_\perp^2 - k_p^2)\frac{E_{zb}}{E_0} = -k_p \frac{\partial}{\partial \xi}\frac{\delta n}{n_e}, \tag{15}$$

where $\xi = z - ct$ is the coordinate in the co-moving frame of a relativistic electron beam with $v_z \simeq c$ and r is the radial, transverse coordinate of an electron beam having a cylindrical symmetry [30]. The Green's function for the beam-driven wakefield excited by a charge bunch with density distribution $\rho_b = \rho_\perp(r)\rho_\parallel(\xi)$ is read as $E_{zb}(\xi, r) = Z(\xi)R(r)$ with $Z(\xi) = -4\pi \int_\xi^\infty d\xi' \rho_\parallel(\xi')\cos k_p(\xi - \xi')$ and $R(r) = (k_p^2/2\pi)$ $\int_0^{2\pi} d\theta \int_0^\infty r' dr' \rho_\perp(r') K_0(k_p |\vec{r} - \vec{r}'|)$, where K_0 is the modified Bessel function of the second kind [31]. For a Gaussian longitudinal density profile $\rho_\parallel(\xi) = q n_b e^{-\xi^2/2\sigma_z^2}$ with the root-mean-square (rms) bunch length σ_z and the particle charge q ($+e$ for a positron bunch and $-e$ for an electron bunch), the longitudinal plasma response is given by

$$Z(\xi) = -(2\pi)^{3/2} q n_b \sigma_z e^{-\frac{k_p^2 \sigma_z^2}{2}}\left(\cos k_p \xi \, \Re\mathrm{erfc}\frac{\xi/\sigma_z + ik_p \sigma_z}{\sqrt{2}} + \sin k_p \xi \, \Im\mathrm{erf}\frac{\xi/\sigma_z + ik_p \sigma_z}{\sqrt{2}}\right). \tag{16}$$

The transverse plasma response for a Gaussian radial density profile $\rho_\perp(r) = e^{-r^2/2\sigma_r^2}$ with the rms bunch radius σ_r is calculated inside the bunch ($r < r'$) as

$$\begin{aligned} R(r) &= \frac{k_p^2}{2\pi}\int_0^{2\pi} d\theta' \int_0^\infty r' dr' e^{-r'^2/2\sigma_r^2} K_0(k_p\sqrt{r^2 + r'^2 - 2rr'\cos(\theta - \theta')}) \\ &= k_p^2 J_0(k_p r)\int_0^\infty r' dr' e^{-r'^2/2\sigma_r^2} K_0(k_p r') = \frac{k_p^2 \sigma_r^2}{2} e^{\frac{k_p^2 \sigma_r^2}{2}} \Gamma(0, \frac{k_p^2 \sigma_r^2}{2}) J_0(k_p r) \end{aligned}, \tag{17}$$

where $\Gamma(\alpha, x) = \int_x^\infty e^{-t} t^{\alpha-1} dt$ is the incomplete Gamma function of the second kind. Here we use the formula $K_0(\sqrt{z^2 + \zeta^2 - 2z\zeta\cos\theta}) = \sum_{n=-\infty}^\infty K_n(z) J_n(\zeta) e^{in\theta}$ $|\zeta| < |z|$ and $\Gamma(\alpha, x) = (2x^{\alpha/2} e^{-x}/\Gamma(1-\alpha))\int_0^\infty e^{-t} t^{-\alpha/2} K_\alpha[2\sqrt{xt}]dt$ for $\mathrm{Re}\,\alpha < 1$. Combining the longitudinal and transverse solutions, the wakefield excited by a bi-Gaussian shaped bunch is

$$\begin{aligned} E_{zb}(r, \xi)/E_0 &= -(q/e)k_p r_e N_b e^{-k_p^2 \sigma_z^2/2} e^{k_p^2 \sigma_r^2/2} \Gamma(0, k_p^2 \sigma_r^2/2) J_0(k_p r) \\ &\times \left(\cos k_p \xi \, \Re\mathrm{erfc}\frac{\xi/\sigma_z + ik_p \sigma_z}{\sqrt{2}} + \sin k_p \xi \, \Im\mathrm{erf}\frac{\xi/\sigma_z + ik_p \sigma_z}{\sqrt{2}}\right), \end{aligned} \tag{18}$$

where $N_b = (2\pi)^{3/2}\sigma_r^2 \sigma_z n_b$ is the total particle number contained in the bunch and $r_e = e^2/m_e c^2$ is the electron classical radius. The net longitudinal electric field on the propagation axis $r = 0$, i.e., the beam loading field, experienced by the particle beam located at $\Psi = \Psi_b$ in the laser co-moving frame, i.e., $k_p \xi \approx \Psi - \Psi_b$, yields $E_{zBL}(\Psi) = E_{z0}(\Psi) + E_{zb}(0, \Psi - \Psi_b)$. At the bunch center $\xi = 0$, the beam particle undergoes the decelerating field,

$$\frac{E_{zb}}{E_0} = -\left(\frac{q}{e}\right)k_p r_e N_b e^{-k_p^2\sigma_z^2/2} e^{k_p^2\sigma_r^2/2}\Gamma\left(0,\frac{k_p^2\sigma_r^2}{2}\right), \tag{19}$$

associated with the energy loss in the phase advance $\Delta\Psi = \Psi - \Psi_0$

$$\Delta\gamma_{BL} = -2\left(\frac{q}{e}\right)\gamma_g^2 k_p r_e N_b e^{-k_p^2\sigma_z^2/2} e^{k_p^2\sigma_r^2/2}\Gamma\left(0,\frac{k_p^2\sigma_r^2}{2}\right)\Delta\Psi, \tag{20}$$

and the rms energy spread due to the beam loading is estimated as

$$\sigma_{\Delta\gamma_{BL}} = \sqrt{\langle\Delta\gamma_{BL}^2\rangle} \approx 2^{1/4}\gamma_g^2 k_p r_e N_b e^{k_p^2\sigma_r^2/2}\Gamma\left(0,\frac{k_p^2\sigma_r^2}{2}\right)S(k_p\sigma_z)\Delta\Psi, \tag{21}$$

where $S(k_p\sigma_z) = \left|(2/\pi)^{1/2} - k_p\sigma_z e^{-k_p^2\sigma_z^2/2}\mathrm{erf}(k_p\sigma_z/\sqrt{2})\right|$ has the minimum $S = 0.35$ at $k_p\sigma_z = 1.26$.

Transverse wakefield excited by the particle bunch is obtained from Eq. (18) according to the Panofsky-Wenzel theorem [32], $\partial E_z/\partial r = \partial(E_r - B_\theta)/\partial\xi$, leading to the beam focusing force

$$F_{rb}(r,\xi)/E_0 = (E_r - B_\theta)/E_0 = (q/e)k_p r_e N_b e^{-k_p^2\sigma_z^2/2} e^{k_p^2\sigma_r^2/2}\Gamma\left(0,k_p^2\sigma_r^2/2\right)J_1(k_p r)$$
$$\times\left(\sin k_p\xi\Re\mathrm{erfc}\frac{\xi/\sigma_z + ik_p\sigma_z}{\sqrt{2}} - \cos k_p\xi\,\Im\mathrm{erf}\frac{\xi/\sigma_z + ik_p\sigma_z}{\sqrt{2}}\right), \tag{22}$$

and the beam focusing strength

$$\frac{\partial F_{rb}}{E_0 k_p\partial r} = \frac{1}{2}\left(\frac{q}{e}\right)k_p r_e N_b e^{-k_p^2\sigma_z^2/2} e^{k_p^2\sigma_r^2/2}\Gamma\left(0,k_p^2\sigma_r^2/2\right)\left[J_0(k_p r) - J_2(k_p r)\right]$$
$$\times\left(\sin k_p\xi\Re\mathrm{erfc}\frac{\xi/\sigma_z + ik_p\sigma_z}{\sqrt{2}} - \cos k_p\xi\,\Im\mathrm{erf}\frac{\xi/\sigma_z + ik_p\sigma_z}{\sqrt{2}}\right). \tag{23}$$

At the bunch center $\xi = 0$, the on-axis focusing strength is

$$\frac{\partial F_{rb}}{E_0 k_p\partial r} = -\frac{1}{\sqrt{\pi}}\left(\frac{q}{e}\right)k_p r_e N_b e^{\frac{k_p^2\sigma_r^2}{2}}\Gamma\left(0,\frac{k_p^2\sigma_r^2}{2}\right)\mathrm{F}\left(\frac{k_p\sigma_z}{\sqrt{2}}\right), \tag{24}$$

where $\mathrm{erf}(ik_p\sigma_z/\sqrt{2}) = (2i/\sqrt{\pi})e^{k_p^2\sigma_z^2/2}\mathrm{F}(k_p\sigma_z/\sqrt{2})$ and $\mathrm{F}(x) = e^{-x^2}\int_0^x e^{z^2}dz$ is the Dawson's integral. A beam particle within either an electron bunch or a positron bunch undergoes a focusing force toward the beam axis [33].

The validity and usefulness of the linear model for beam-driven wakefields have been investigated in Ref. [31], showing the comparison between the wakefield solution derived from the linear fluid theory and that of the particle-in-cell simulation as a function of the normalized beam density $n_b/n_e = \Lambda/(k_p^2\sigma_r^2)$, where $\Lambda = (2/\pi)^{1/2}r_e N_b/\sigma_z$ is the normalized charge per unit bunch length. Since the relativistic electron (positron) bunch keeps a total charge N_b and bunch length σ_z fixed, i.e., $\Lambda = \text{constant}$, during the acceleration process, the beam density n_b / n_e can be changed as $\propto (k_p \sigma_r)^{-2}$ according to the transverse beam dynamics. For the matched beam condition with a constant normalized transverse emittance ε_{n0} and focusing gradient $\bar{K}^2 = \partial F_r / (E_0 k_p \partial r)$ in the plasma wakefield, where $k_p^2 \sigma_r^2 = \gamma^{-1/2} k_p \varepsilon_{n0} / \bar{K}$, the beam density $n_b / n_e \propto \gamma^{1/2}$ can be varied from $n_b / n_e < 1$ to $n_b / n_e > 1$ by a factor of $(\gamma_f / \gamma_0)^{1/2}$ during the acceleration process from the injection energy $m_e c^2 \gamma_0$ to the final energy $m_e c^2 \gamma_f$. This results in the breakdown of the linear fluid theory for the electron bunch with $n_b / n_e \geq 10$ and positron bunch with $n_b / n_e > 1$, associated with appearance of nonlinear effects for very narrow bunches because an electron beam predominantly expels plasma electrons radially outward, while a positron beam sack them in [31]. For the case of interest, $N_b = 1 \times 10^9$ and $k_p \sigma_z = 3$, corresponding to $\Lambda \sim 4 \times 10^{-4} \ll 1$, however, the decelerating wakefield, i.e., the beam loading field, saturates over roughly $n_b / n_e \geq 25$, remaining the sinusoidal wave form for all bunch radii [31], because the blowout radius $k_p r_{bm} \approx 2\sqrt{\Lambda}$, where $\sqrt{\Lambda}$ is the charge neutralizing radius, becomes much larger than the bunch radius $k_p \sigma_r = \sqrt{\Lambda / (n_b / n_e)}$ as $r_{bm} / \sigma_r \approx 2\sqrt{n_b / n_e} \geq 10$ so that any further reduction in the bunch radius can only cause a small change to the beam-driven wakefield [31]. Furthermore, for much higher beam density $n_b / n_e > 100$, although the wakefield deviates from the linear theory, showing strong nonlinear effects such as the wavelength elongation and peak field decrease, the linear prediction of the peak field provides us with the upper limit of the beam-driven wakefield. In such sense, considerations on the beam-driven wakefield and beam loading based on the linear fluid theory are valid and consistent for the broad range of the bunch radius σ_r, correspondingly the normalized beam density n_b / n_e .

Figure 2 shows the beam-driven longitudinal wakefield in (a) and focusing strength in (b) with respect to the amplitude of the accelerating and focusing gradient excited by the laser pulse, respectively. As expected from Eqs. (19) and (24), the beam-driven longitudinal and transverse wakefields are rapidly increased by the bunch radius being converged from an initial value, leading to the equilibrium in about 10 stages.

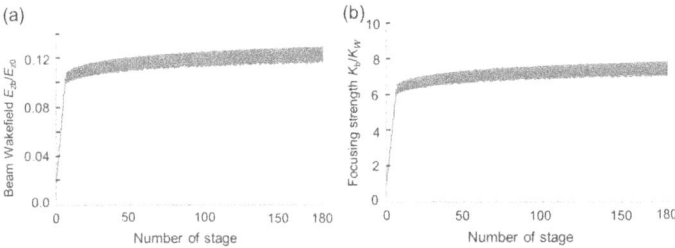

Fig. 2. (a) The evolution of a fraction of the beam-driven decelerating wakefield E_{zb} to the amplitude of the accelerating laser wakefield E_{z0} . (b) The ratio of the beam self-focusing strength K_b to the amplitude of the laser transverse wakefield K_W as a function of the stage number. Both results are obtained from the single particle dynamics simulation for case A.

2.3 Single particle dynamics in the single stage

Consider the motion of electrons (positrons) propagating laser wakefields generated by the single hybrid mode EH_{1n} with the optimum pulse length of $k_p c\tau = \sqrt{2}$ and $\Psi = k_p z/(2\gamma_g^2) \gg k_p c\tau/(2\gamma_g^2) = 1/(\sqrt{2}\gamma_g^2)$. In the wakefield, a relativistic beam particle moving along z axis is exerted by a focusing force at the transverse displacement x and exhibits betatron motion that induces synchrotron (betatron) radiation at high energies. The synchrotron radiation induces the radiation damping of particles and affects the energy spread and transverse emittance via the radiation reaction force. On the contrary to vacuum-based accelerators, an unavoidable drawback for the collider application of plasma-based accelerators would be multiple Coulomb scattering between beam particles and plasma ions, which counteracts the effects of beam focusing and radiation damping on the beam emittance and energy spread during the particle acceleration. The motion of a beam particle traveling along the z-axis is described as

$$\frac{du_x}{cdt} = \frac{F_\perp}{m_e c^2} + \frac{F_x^R}{m_e c^2} + \frac{du_x^S}{cdt}, \quad \frac{du_z}{cdt} = \pm k_p \frac{E_z}{E_0} + \frac{F_z^R}{m_e c^2}, \quad (25)$$

where $\mathbf{u} = \mathbf{p}/m_e c$ is the normalized particle momentum, $\mathbf{F}^R$ the radiation reaction force and $u_x^S \approx \gamma\theta_x$ the transverse kick in momentum projected onto the x-plane due to multiple Coulomb scattering through small deflection angles θ. Here, the focusing force is

$$\frac{F_r}{E_0} = -\sqrt{\frac{\pi}{2}} \frac{a_0^2 u_n C_n}{k_p R_c} J_0\left(\frac{u_n r}{R_c}\right) J_1\left(\frac{u_n r}{R_c}\right) e^{-(1+4\alpha_d a_0^2 \Psi)/2} \sin\Psi. \quad (26)$$

Taking into account $F_\perp \sim \pm(\partial F_r/\partial r)r$ near the z-axis $r \approx 0$, the focusing force is written by $F_\perp/m_e c^2 = \pm(\partial F_r/\partial r)x = -K^2 x$, where $K = (\mp\partial F_r/\partial r)^{1/2}$ is the focusing constant. Near-axis particles experience the accelerating field with the beam loading at $r = 0$, as given by Eqs. (12) and (19),

$$\bar{E}_z \equiv \frac{E_z}{E_0} = \frac{1}{2}\sqrt{\frac{\pi}{2}} a_0^2 C_n e^{-(1+4\alpha_d a_0^2 \Psi)/2} \cos\Psi - \left(\frac{q}{e}\right) N_b k_p r_e e^{-k_p^2 \sigma_z^2/2} e^{k_p^2 \sigma_r^2/2} \Gamma\left(0, \frac{k_p^2 \sigma_r^2}{2}\right), \quad (27)$$

and the focusing strength with beam self-focusing at $r = 0$ from Eqs. (24) and (26)

$$\bar{K}^2 \equiv \mp\frac{\partial F_r}{E_0 k_p \partial r} = \pm\frac{1}{2}\sqrt{\frac{\pi}{2}} a_0^2 C_n e^{-(1+4\alpha_d a_0^2 \Psi)/2} \left(\frac{u_n}{k_p R_c}\right)^2 \sin\Psi + \frac{k_p r_e N}{\sqrt{\pi}} e^{\frac{k_p^2 \sigma_r^2}{2}} \Gamma\left(0, \frac{k_p^2 \sigma_r^2}{2}\right) F\left(\frac{k_p \sigma_z}{\sqrt{2}}\right). \quad (28)$$

Employing the classical expression of the radiation reaction force $\mathbf{F}^R/(m_e c\tau_R) = (d/dt)(\gamma d\mathbf{u}/dt) + \gamma\mathbf{u}[(d\gamma/dt)^2 - (d\mathbf{u}/dt)^2]$ [34], where $\gamma = (1+u^2)$ is the relativistic Lorentz factor of the particle and $\tau_R = 2r_e/3c \approx 6.26\times10^{-24}$ s, and assuming $u_z \gg u_x$ and $dx/dt = cu_x/\gamma \approx cu_x/u_z$, the radiation reaction force is approximately given by [35]

$$\frac{F_x^R}{m_e c^2} \simeq -c\tau_R K^2 u_x (1 + K^2 \gamma x^2), \qquad \frac{F_z^R}{m_e c^2} \simeq -c\tau_R K^4 \gamma^2 x^2. \qquad (29)$$

Since the scale length of the radiation reaction, *i.e.*, $c\tau_R = 2r_e/3 \simeq 1.9$ fm is much smaller than that of the betatron motion, *i.e.*, $\sim \lambda_p \sqrt{\gamma}$, the radiation reaction force is a perturbation for the particle acceleration and betatron motion induced by the accelerating and focusing wakefields.

A beam electron (positron) of the incident momentum $p = \gamma m_e v$, passing a nucleus of charge Ze at impact parameter b in the plasma, suffers an angular deflection $\theta = \Delta p / p \simeq 2e^2 Z / (pbv)$ due to Coulomb scattering [36]. The successive collisions of the relativistic beam particle with $v \sim c$ while traversing the plasma of ion density $n_i = n_e / Z$ results in an increase of the mean square deflection angle at a rate [6, 36]

$$\frac{d\langle \theta^2 \rangle}{cdt} = \frac{8\pi n_i Z^2 r_e^2}{\gamma^2} \ln\left(\frac{b_{max}}{b_{min}}\right) = \frac{2k_p^2 r_e Z}{\gamma^2} \ln\left(\frac{\lambda_D}{R_N}\right), \qquad (30)$$

where $b_{max} = \lambda_D = (T_e / 4\pi n_e e^2)^{1/2}$ is the plasma Debye length at the temperature T_e and $R_N \approx 1.4 A^{1/3}$ fm is the effective Coulomb radius of the nucleus with the mass number A. Here, the logarithm $\ln(b_{max} / b_{min})$ is approximated as $\ln(\lambda_D / R_N) \approx 24.7[1 + 0.047 \log(n_e T_e A^{2/3})]$ for n_e [10^{16} cm^{-3}] and T_e [eV] [37]. The multiple-scattering distribution for the projected angle θ_x is approximately Gaussian for small deflection angles, given by the probability distribution function $P(\theta_x) = 1/(\pi\langle \theta^2 \rangle)^{1/2} \exp(-\theta_x^2 / \langle \theta^2 \rangle)$. Thus, the transverse momentum $u_x^S \approx \gamma\theta_x$ is obtained using the normal distribution with the standard deviation $(\langle \theta^2 \rangle / 2)^{1/2}$ around the mean angle 0 at the successive time step along the particle trajectory.

Since the radiation reaction force and multiple Coulomb scattering are expected to be the perturbation in the motion of the beam particle driven by laser wakefields, the equations of motion is written as [38]

$$\frac{d^2 \bar{x}}{d\bar{t}^2} \pm \frac{\bar{E}_z}{\gamma} \frac{d\bar{x}}{d\bar{t}} + \frac{\bar{K}^2}{\gamma} \bar{x} = 0, \qquad \frac{d\gamma}{d\bar{t}} = \pm\bar{E}_z \qquad (31)$$

where $\bar{x} = k_p x$ and $\bar{t} = \omega_p t$ are the normalized variables of x and t, respectively. If one can assume that $\bar{E}_z$ and $\bar{K}$ are constant along the particle trajectory, introducing a new variable $s = (4\gamma \bar{K}^2 / \bar{E}_z^2)^{1/2}$ to obtain the differential equation

$$s \frac{d^2 \bar{x}}{ds^2} + \frac{d\bar{x}}{ds} + s\bar{x} = 0, \qquad (32)$$

general solutions of which are the Bessel functions of the first kind $J_0(s)$ and the second kind $Y_0(s)$, the solutions of the coupled equations are given by [38]

$$\begin{pmatrix} \bar{x}(s) \\ \beta_x(s) \end{pmatrix} = \mathbf{M}(s|s_0) \begin{pmatrix} \bar{x}_0(s_0) \\ \beta_{x0}(s_0) \end{pmatrix}, \qquad (33)$$

and

$$\gamma = \gamma_0 + \Gamma(\Psi) - \Gamma(\Psi_0),$$ (34)

where

$$\mathbf{M}(s|s_0) = \begin{pmatrix} \dfrac{\pi}{2}s_0\left[J_1(s_0)Y_0(s) - Y_1(s_0)J_0(s)\right] & \dfrac{\pi\gamma_0}{E_{z0}}\left[J_0(s)Y_0(s_0) - Y_0(s)J_0(s_0)\right] \\[3mm] -\dfrac{\pi E_z s_0 s}{4\gamma}\left[J_1(s)Y_1(s_0) - Y_1(s)J_1(s_0)\right] & \dfrac{\pi E_z \gamma_0}{2E_{z0}\gamma}s\left[J_1(s)Y_0(s_0) - Y_1(s)J_0(s_0)\right] \end{pmatrix},$$ (35)

$$\Gamma(\Psi) = \pm\sqrt{\dfrac{\pi}{2}}a_0^2 C_n \gamma_g^2 \dfrac{e^{-(1+4\alpha_d a_0^2\Psi)/2}}{1+4\alpha_d^2 a_0^4}\left(\sin\Psi - 2\alpha_d a_0^2\cos\Psi\right),$$ (36)

$\beta_x = \beta_g(d\bar{x}/d\bar{t})$ and subscripts "0" denote the initial values. When the electron stays in the focusing region of the wakefield, i.e., $\partial F_r/\partial r > 0$, the electron exhibits betatron oscillation at the frequency given by $\omega_\beta = ds/dt = \omega_p \bar{K}/\gamma^{1/2}$. Contrarily, when the electron moves to the defocusing region where $\partial F_r/\partial r < 0$ and s becomes imaginary, the amplitude of the electron trajectory increases monotonically as a result of the Bessel functions being transformed to the modified Bessel functions, leading to ejection of the particle from the wakefield [38]. The particle orbit and energy are obtained from the coupled equations describing the single particle motion in the segmented phase, where $\bar{E}_z$ and $\bar{K}$ are assumed to be constant over the phase advance $\Delta\Psi$. Provided the initial values of $\bar{x}_0$ and β_{x0} are specified from the energy γ_0, relative energy spread $\Delta\gamma/\gamma_0$ and normalized emittance ε_{n0} of the injected beam, $\gamma(s)$, $\bar{x}(s)$ and $\beta_x(s)$ are first calculated from the coupled equations using $s(\Psi)$, where $\Psi = \Psi_0 + \Delta\Psi$ is the phase at the next step. Then, the effects of the radiation reaction and multiple Coulomb scattering are corrected as follows:

$$\beta_x(s) = \beta_x^B(s) + \Delta\beta_x^R(s_0) + \Delta\beta_x^S(s_0),\quad \gamma(\Psi) = \gamma^A(\Psi) + \Delta\gamma^R(\Psi_0),$$ (37)

where $\beta_x^B(s)$ and $\gamma^A(\Psi)$ are the solutions obtained from Eqs. (33) and (34), respectively, $\Delta\beta_x^R(s_0)$ and $\Delta\gamma^R(\Psi_0)$ correction terms for the effect of the radiation reaction force, given by

$$\Delta\beta_x^R = -2C_R\gamma_g\beta_{x0}\bar{K}_0^2(1+\gamma_0\bar{K}_0^2\bar{x}_0^2)\Delta\Psi,\quad \Delta\gamma^R(\Psi_0) = -2C_R\gamma_g\gamma_0^2\bar{K}_0^2\bar{x}_0^2\Delta\Psi,$$ (38)

with $C_R = k_p c\tau_R\gamma_g = (2/3)k_p r_e\gamma_g = 11.8\times10^{-9}[\mu m]/\lambda_L$ and $\bar{K}_0^2 = \bar{K}^2(\Psi_0)$, and $\Delta\beta_x^S(s_0) = \theta_x$ a projected angle due to multiple Coulomb scattering, the standard deviation of which is obtained from Eq. (30) for $\lambda_L = 1\,\mu m$ and $\beta_g \simeq 1$ as

$$\sigma_{\theta_x} = \sqrt{\dfrac{\langle\theta^2\rangle}{2}} \approx \dfrac{2.66\times10^{-4}}{\gamma_0}\left[\gamma_g\Delta\Psi\ln\left(\dfrac{\lambda_D}{R_N}\right)\right]^{1/2}.$$ (39)

The radiated power of the electron in the classical limit is given by $P_{rad} = (2e^2\gamma^2/3c)[(d\mathbf{u}/dt)^2 - (d\gamma/dt)^2] = (2e^2\gamma^2/3m_e^2c^3)[|\mathbf{F}_{ext}|^2 - |\mathbf{F}_{ext}\cdot\mathbf{u}/\gamma|^2]$ [35], where $\mathbf{F}_{ext}$ is the external force and $m_e cd\gamma/dt = \mathbf{F}_{ext}\cdot\mathbf{u}/\gamma$ is used. For a relativistic electron with $u_x^2 \ll \gamma^2$ and $u_z \sim \gamma$, taking into account $\mathbf{F}_{ext} = F_\perp \mathbf{e}_x + F_\parallel \mathbf{e}_z$ with $F_\perp = -m_e c^2 K^2 x$ and $F_\parallel = -eE_z$, the radiated power can be written as $P_{rad} = 2e^2\gamma^2 F_\perp^2/(3m^2c^3) = mc^4\tau_R\gamma^2 K^4 x^2$, which means the radiative damping rate $v_R = P_{rad}/(\gamma m_e c^2) = \tau_R c^2 \gamma K^4 x^2$. Thus, a total radiation energy loss along the particle orbit is estimated as

$$\Delta\gamma_{rad} = \frac{1}{m_e c^2}\int_0^t dt P_{rad}(t) = \sum \left|\Delta\gamma^R(\Psi_0)\right|. \qquad (40)$$

Numerical calculations of the single particle dynamics can be carried out throughout the segments in phase Ψ for a set of test particles under the initial conditions, and then the underlying beam parameters can be obtained as an ensemble average over test particles; for instance, the mean energy is calculated as $\langle\gamma\rangle = \sum_i \gamma_i/N_p$, where γ_i is the energy of the i-th particle and N_p the number of test particles, and the energy spread is defined as $\sigma_\gamma = (\langle\gamma^2\rangle - \langle\gamma\rangle^2)^{1/2}$. The normalized transverse emittance is obtained from

$$\varepsilon_n = \left[\left\langle(x-\langle x\rangle)^2\right\rangle\left\langle(u_x-\langle u_x\rangle)^2\right\rangle - \left\langle(x-\langle x\rangle)(u_x-\langle u_x\rangle)\right\rangle^2\right]^{1/2}, \qquad (41)$$

where $u_x = \gamma\beta_x$ is the dimensionless momentum.

2.4 Numerical results on single particle dynamics

The particle orbit and energy can be numerically tracked by using the solutions of the single particle motion, Eqs. (33) and (34), associated with the perturbation arising from the effects of the radiation reaction and multiple Coulomb scattering, as given by Eqs. (38) and (39), respectively. The particle tracking simulation has been carried out by using an ensemble of 10^4 test particles, for which the initial values at the injection and the deflection angles due to multiple Coulomb scattering at each segment in a phase step $\Delta\Psi_{stage}/9$, where $\Delta\Psi_{stage}$ is the phase advance per stage, are obtained from the random number generator for the normal distribution, assuming that the particle beam with rms bunch length $\sigma_z = 50\,\mu m$ ($k_p\sigma_z = 3$) containing 10^9 electrons or positrons (0.16 nC) is injected into the capillary accelerator operated at the plasma density of $n_e = 1\times10^{17}$ cm^{-3} from the external injector at an energy $m_e c^2\gamma_0 = 2.5$ GeV with the relative energy spread $\delta\gamma/\gamma_0 = 0.1$ and the initial normalized transverse emittance ε_{n0} in the condition initially matched to laser wakefields, namely, the initial bunch radius $\bar{x}_0 = k_p\sigma_0 = \sqrt{k_p\varepsilon_{n0}}/(\gamma_0\bar{K}^2)^{1/4}$ and momentum $\gamma_0\beta_0 = (\gamma_0\bar{K}^2)^{1/4}\sqrt{k_p\varepsilon_{n0}}$ with the focusing strength $\bar{K}^2$, given by Eq. (28). The simulation results for case A comprising 180 stages with $\varepsilon_{n0} = 10\,\mu m$, the initial phase $\Psi_i = 0°$ and final phase $\Psi_f = 81°$, case B comprising 100 stages with $\varepsilon_{n0} = 10\,\mu m$, $\Psi_i = -50°$ and $\Psi_f = 42.75°$, and case C comprising 100 stages with $\varepsilon_{n0} = 0.1\,\mu\mathrm{m}$, $\Psi_i = -51°$ and $\Psi_f = 40°$ are shown in Fig. 3 with regard to the normalized phase space in the $k_p x - \gamma\beta_x$ ($\bar{x} - u_x$) coordinates at the final energies 589 GeV (case A), 539 GeV (case B) and 522 GeV (case C), respectively, the evolution of the rms bunch radius and relative energy spread over the whole stages. In this simulation, the multiple Coulomb scattering has been considered for a helium plasma with $A = 4$, $Z = 2$ and $T_e = 100\,\mathrm{eV}$.

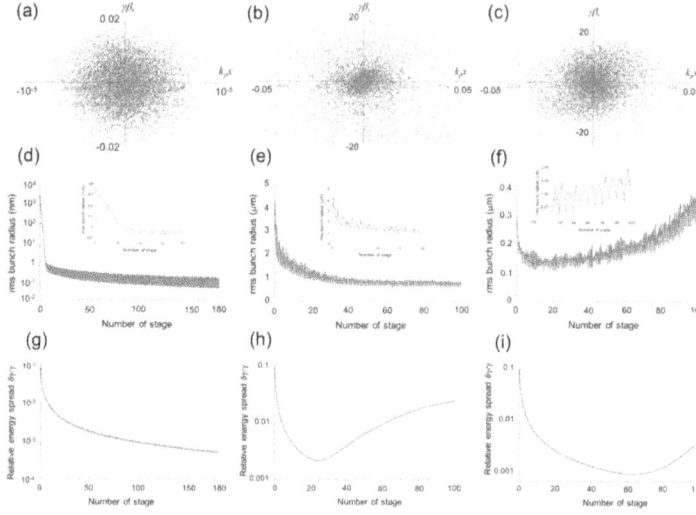

Fig. 3. The results of single particle dynamics simulation on normalized phase space at the final stage (a, b, c), rms bunch radius (d, e, f) and relative energy spread (g, h, i) as a function of the stage number for case A (180 stages), B (100 stages) and C (100 stages), respectively, corresponding to the normalized emittance converging, conserving and growing cases. The insets show the evolution of the bunch radius in the initial 20 stages for case A and B, and in the final 11 stages for case C.

3. Beam Dynamics in the Multistage Capillary Accelerator

3.1 Evolution of betatron amplitude

A gas-filled capillary waveguide made of metallic or dielectric materials can make it possible to comprise a seamless staging without the coupling section, where a fresh laser pulse and accelerated particle beam from the previous stage are injected so as to minimize coupling loss in both laser and particle beams and the emittance growth of particle beams due to the mismatch between the injected beam and plasma channel. For dephasing limited laser wakefield accelerators, the total linac length will be minimized by choosing the coupling distance to be equal to a half of the dephasing length. A side coupling of laser pulse through a curved capillary waveguide diminishes the beam matching section consisting of a vacuum drift space and focusing magnet beamline [7]. Furthermore, the proposed scheme comprising seamless capillary waveguides can provide us with suppression of synchrotron radiation from high-energy electron (positron) beams generated by betatron oscillation in plasma focusing channels and delivery of remarkably small normalized emittance from the linac to the beam collision region in electron-positron colliders.

For $s \gg 1$, the asymptotic form of betatron motion Eqs. (33) and (35) yields

$$\bar{x}(s) \sim \left[\bar{x}_0^2 + \left(\frac{2\gamma_0\beta_{x0}}{s_0\bar{E}_{z0}}\right)^2\right]^{1/2} \sqrt{\frac{s_0}{s}} \cos(s - s_0 + \delta_0), \qquad (42)$$

$$\beta_x(s) \sim \frac{\bar{E}_z}{2\gamma}\left[\bar{x}_0^2 + \left(\frac{2\gamma_0\beta_{x0}}{s_0\bar{E}_{z0}}\right)^2\right]^{1/2} \sqrt{ss_0} \sin(s - s_0 - \delta_0), \qquad (43)$$

where $\tan\delta_0 = 2\gamma_0\beta_{x0}/(s_0\bar{x}_0\bar{E}_{z0})$. The variation of the betatron amplitude with respect to the initial amplitude in the k-th stage is given by

$$\left|\frac{\bar{x}(\Psi)}{\bar{x}(\Psi_{ki})}\right| \sim \sqrt{\frac{s_{ki}}{s}} = (\frac{\gamma_{ki}}{\gamma})^{1/4}\left|\frac{\bar{E}_z(\Psi)\bar{K}(\Psi_{ki})}{\bar{E}_z(\Psi_{ki})\bar{K}(\Psi)}\right|^{1/2} \sim (\frac{\gamma_{ki}}{\gamma})^{1/4}\left|\frac{\bar{E}_z(\Psi)}{\bar{E}_z(\Psi_{ki})}\right|^{1/2}, \qquad (44)$$

where $\Psi_{ki} \leq \Psi \leq \Psi_{(k+1)i}$ ($k = 1, 2, \cdots$) is the particle phase Ψ with respect to the plasma wave, Ψ_{ki} the initial phase and γ_{ki} is corresponding to the initial energy of the particle in the k-th stage, respectively, assuming an approximately constant focusing strength $\bar{K}(\Psi) \sim \bar{K}(\Psi_{ki})$ over the stage. As expected, the betatron amplitude is simply proportional to $(\gamma_{ki}/\gamma)^{1/4}$ for the constant accelerating field $\bar{E}_z(\Psi)$ during the stage. In the capillary accelerator system comprising the periodic accelerating structure, $i.e.$, $\bar{E}_z(\Psi_{ki} + \Delta\Psi) = \bar{E}_z(\Psi_{li} + \Delta\Psi)$ for a phase advance $\Delta\Psi$ in the k-th and l-th stages, the ratio of the accelerating field amplitude is given by

$$\left|\bar{E}_z(\Psi)/\bar{E}_z(\Psi_{ki})\right| = e^{-2\alpha_d a_0^2(\Psi - \Psi_{ki})}\left|\cos\Psi/\cos\Psi_{ki}\right|. \qquad (45)$$

In the accelerator system consisting of N_s stages, the final betatron amplitude at the N_s-th stage yields

$$\left|x(\Psi_f)\right| \sim \left|x_0\right|\left(\frac{\gamma_0}{\gamma_f}\right)^{1/4} R^{N_s/2} \exp[-\alpha_d a_0^2(\Psi_f - \Psi_0)], \qquad (46)$$

where Ψ_f, $m_e c^2\gamma_f$ are the final phase and energy of the particle at the N-th stage, respectively, $\Psi_i = \Psi_{1i}$, $\Psi_f = \Psi_{1f}$ are the initial and final phase in the single stage, respectively, and $R = \left|\cos\Psi_f/\cos\Psi_i\right|$ is the ratio of the amplitude between the final and initial phases.

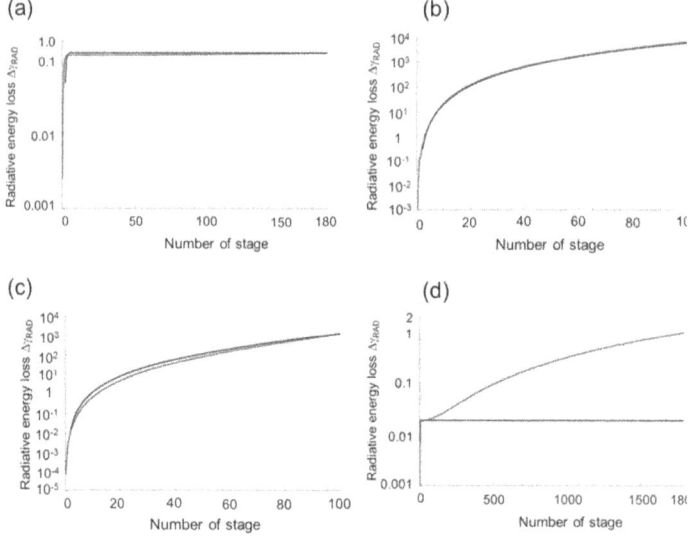

Fig. 4. The evolution of radiative energy loss $m_e c^2 \Delta \gamma_{\mathrm{RAD}}$ (red solid line) as a function of the stage number (a) for case A, (b) for case B and (c) for case C up to the final beam energy of 589 GeV, 539 GeV and 522 GeV, respectively, and (d) for case A up to the final beam energy of 5720 GeV. Blue solid lines indicate the analytical predictions given by Eq. (50) for $\Sigma = 2.0$, $\bar{K}^2 = 0.52$ in (a), $\Sigma \sim 0$, $\bar{K}^2 = 0.2$ in (b), $\Sigma = -0.018$, $\bar{K}^2 = 0.46$ in (c) and $\Sigma = 2.0$, $\bar{K}^2 = 0.62$ in (d), respectively.

3.2 Betatron radiation loss

Radiation energy due to betatron motion of the single particle moving through N_s stages during the time t_f is estimated from

$$\Delta E_{\mathrm{RAD}} = \int_0^{t_f} P_{\mathrm{RAD}}(t)\,dt = 2m_e c^3 \tau_R K^4 \gamma_g^2 k_p^{-1} \sum_{k=1}^{N_s} \int_{\Psi_{ki}}^{\Psi_{kf}} \gamma^2 x^2 d\Psi . \tag{47}$$

Here we assume that the focusing strength K is constant over the accelerator system. Substituting Eq. (34) for $\gamma(\Psi)$ and Eq. (46) for $x(\Psi)$ to Eq. (47), the radiation energy is obtained from

$$\Delta \gamma_{\mathrm{RAD}} = \frac{\Delta E_{\mathrm{RAD}}}{m_e c^2} \approx \frac{2}{5}\left(\frac{C_R}{\gamma_g}\right)\frac{\bar{K}^4 \bar{x}_0^2 \gamma_0^{1/2}}{|\bar{E}_z(\Psi_f)|} \sum_{k=1}^{N_s} R^k e^{-2k\alpha_d a_0^2 \Delta\Psi} \gamma_k^{5/2}\left[1-\left(\frac{\gamma_{k-1}}{\gamma_k}\right)^{5/2}\right], \tag{48}$$

where $\Delta\Psi = \Psi_{kf} - \Psi_{ki}$ is a phase advance in the single stage. Considering the energy of k-th stage $m_e c^2 \gamma_k$ is approximately given by

$$\gamma_k \sim k\sqrt{\frac{\pi}{2}}a_0^2 \gamma_g^2 C_n e^{-(1+4\alpha_d a_0^2 \Delta\Psi)/2} \sin\Delta\Psi , \tag{49}$$

the radiation energy over $N_s \gg 1$ is approximately calculated as

$$\Delta\gamma_{\mathrm{RAD}} \sim (2\pi)^{3/4} C_R a_0^3 C_n^{3/2} \gamma_g^4 e^{-3(1+4\alpha_d a_0^2 \Delta\Psi)/4} \frac{\overline{K}^4 \overline{x}_0^2 \gamma_0^{1/2}}{\cos\Psi_f} \left(\frac{\sin\Delta\Psi}{|\Sigma|}\right)^{5/2} \left[\Theta(N_s|\Sigma|) - \Theta(|\Sigma|)\right], \quad (50)$$

where $\Sigma \equiv 2\alpha_d a_0^2 \Delta\Psi - \ln R$ and $\Theta(z)$ is a function, given by $\Theta(z) = (3\sqrt{\pi}/4)\mathrm{erf}(\sqrt{z}) - e^{-z}\sqrt{z}(z-3/2)$ with the error function $\mathrm{erf}(\sqrt{z})$ for $\Sigma > 0$, $\Theta(z) = e^{z}[(3/2)\mathrm{F}(\sqrt{z}) + \sqrt{z}(z-3/2)]$ with the Dawson's integral $\mathrm{F}(\sqrt{z})$ for $\Sigma < 0$ and $\Theta(z) = (2/5)z^{5/2}$ for $\Sigma = 0$, i.e., $[\Theta(N_s|\Sigma|) - \Theta(|\Sigma|)]/|\Sigma|^{5/2} = N_s^{5/2} - 1$. It is noted that the initial and final phases at the single stage must be chosen so as to be $R < \exp(2\alpha_d a_0^2 \Delta\Psi)$.

Figure 4 shows the evolution of radiative energy loss over the whole stages for case A, case B and case C, corresponding to $\Sigma = 2.0 > 0$, $\Sigma = 0.049 \sim 0$ and $\Sigma = -0.018 < 0$, respectively, in comparison with the analytical estimates obtained from Eq. (50) with the dimensionless focusing strength $\overline{K}^2 = 0.52$ (case A), $\overline{K}^2 = 0.2$ (case B) and $\overline{K}^2 = 0.46$ (case C), respectively.

3.3 *Evolution of the normalized transverse emittance*

Consider the evolution of the normalized transverse emittance for the particle beam acceleration in the multistage capillary accelerator. According to Appendix A, taking into account the transverse emittance of the particle beam with initial energy spread that dominates decoherence, the normalized emittance for $t \gg t_{\mathrm{dec}}$, where $t_{\mathrm{dec}} \simeq \pi\langle\gamma\rangle/(\omega_\beta\sigma_\gamma)$ is the decoherence time [35], is given by

$$\overline{\varepsilon}_{nx} = \frac{1}{2}\left|\frac{\overline{E}_z}{\overline{E}_{z0}}\right|\overline{K}_0\sqrt{\langle\gamma_0\rangle}\langle\overline{x}_{m0}^2\rangle = \frac{1}{2}\left|\frac{\overline{E}_z}{\overline{E}_{z0}}\right|\overline{K}_0\sqrt{\langle\gamma_0\rangle}\left(\langle\overline{x}_0^2\rangle + \frac{\langle u_{x0}^2\rangle}{\overline{K}_0^2\langle\gamma_0\rangle}\right), \quad (51)$$

where $\overline{\varepsilon}_{nx} = k_p\varepsilon_{nx}$ is the dimensionless normalized emittance. If the beam is initially matched to the laser wakefield focusing channel, i.e., $\overline{x}_0 = 2\gamma_0\beta_{x0}/(s_0\overline{E}_{z0}) = u_{x0}/(\overline{K}_0\sqrt{\gamma_0})$, such that in the absence of radiation the beam radial envelope undergoes no betatron oscillation, the normalized emittance can be expressed as

$$\overline{\varepsilon}_{nx} = \left|\frac{\overline{E}_z}{\overline{E}_{z0}}\right|\overline{K}_0\sqrt{\langle\gamma_0\rangle}\langle x_0^2\rangle = \left|\frac{\overline{E}_z}{\overline{E}_{z0}}\right|\overline{\varepsilon}_{nx0}. \quad (52)$$

This indicates that in the absence of radiation and multiple Coulomb scattering, the transverse normalized emittance of an initially matched beam is conserved in laser wakefield acceleration when the amplitude of accelerating field has no variation, i.e., $|\overline{E}_z| = |\overline{E}_{z0}|$. Note that decreasing the accelerating field at the final phase results in a decrease of the normalized emittance of the injected beam matched to the laser wakefield at the initial phase in the single stage. For the multistage laser wakefield acceleration without a vacuum drift space in the coupling section, properly choosing the initial and final phases enables continuous reduction of the normalized emittance in the absence of synchrotron radiation and multiple Coulomb scattering with plasma ions. In Eq. (52), since the initial values of the displacement $\langle \overline{x} \rangle$ and normalized momentum $\langle u_x \rangle$ at the next stage are expressed as

$$\langle x_1^2 \rangle \simeq \frac{1}{2} \frac{\overline{E}_{z1} \overline{K}_0}{\overline{E}_{z0} \overline{K}_1} \sqrt{\frac{\langle \gamma_0 \rangle}{\langle \gamma_1 \rangle}} \langle x_m^2 \rangle_0 , \quad \langle u_{x1}^2 \rangle \simeq \frac{1}{2} \frac{\overline{E}_{z1}}{\overline{E}_{z0}} \overline{K}_0 \overline{K}_1 \sqrt{\langle \gamma_0 \rangle \langle \gamma_1 \rangle} \langle x_m^2 \rangle_0 , \tag{53}$$

the initial amplitude of betatron oscillation at the next stage is

$$\langle \overline{x}_m^2 \rangle_1 = \langle \overline{x}_1^2 \rangle + \frac{\langle u_{x1}^2 \rangle}{\overline{K}_1 \langle \gamma_1 \rangle} = \frac{\overline{E}_{z1}}{\overline{E}_{z0}} \frac{\overline{K}_0}{\overline{K}_1} \sqrt{\frac{\langle \gamma_0 \rangle}{\langle \gamma_1 \rangle}} \langle x_m^2 \rangle_0 . \tag{54}$$

Thus, the emittance at the k-th stage is calculated as

$$\overline{\varepsilon}_{nx}^k = \frac{1}{2} \left| \frac{\overline{E}_z(\Psi_f)}{\overline{E}_z(\Psi_i)} \right| \overline{K}(\Psi_i) \sqrt{\langle \gamma_{k-1} \rangle} \langle \overline{x}_m^2 \rangle_{k-1} = \frac{1}{2} \left| \frac{\overline{E}_z(\Psi_f)}{\overline{E}_z(\Psi_i)} \right|^k \left| \frac{\overline{K}(\Psi_i)}{\overline{K}(\Psi_f)} \right|^k \overline{K}(\Psi_f) \sqrt{\langle \gamma \rangle_0} \langle \overline{x}_m^2 \rangle_0 . \tag{55}$$

Assuming $\overline{K}(\Psi_f) \approx \overline{K}(\Psi_i) = \overline{K}_0$, the dimensionless normalized emittance at the k-th stage yields

$$\overline{\varepsilon}_{nx}^k \simeq \frac{1}{2} \left| \frac{\overline{E}_z(\Psi_f)}{\overline{E}_z(\Psi_i)} \right|^k \overline{K}_0 \sqrt{\langle \gamma \rangle_0} \langle \overline{x}_m^2 \rangle_0 = \frac{1}{2} R^k e^{-2\alpha_d a_0^2 k \Delta\Psi} \overline{K}_0 \sqrt{\langle \gamma \rangle_0} \langle \overline{x}_m^2 \rangle_0 = R^k e^{-2\alpha_d a_0^2 k \Delta\Psi} \overline{\varepsilon}_{n0} , \tag{56}$$

where $\left| \overline{E}_z(\Psi_f) / \overline{E}_z(\Psi_i) \right| = R e^{-2\alpha_d a_0^2 \Delta\Psi}$ is the ratio of the accelerating field amplitude at the final phase Ψ_f to that at the initial phase Ψ_i with $R = \left| \cos\Psi_f / \cos\Psi_i \right|$ and $\Delta\Psi = \Psi_f - \Psi_i$. Setting $\overline{K}_0 \sqrt{\langle \gamma_0 \rangle} \langle x_m^2 \rangle = 2\overline{\varepsilon}_{n0}$, the normalized emittance increases or decreases monotonically, depending on $R > e^{2\alpha_d a_0^2 \Delta\Psi}$ or $R < e^{2\alpha_d a_0^2 \Delta\Psi}$ as the particles moves along the stage in absence of radiation and multiple Coulomb scattering.

For an application of laser-plasma accelerators to electron-positron colliders, it is of most importance to achieve the smallest possible normalized emittance at the final stage of the accelerator system, overwhelming the emittance growth due to multiple Coulomb scattering off plasma ions, which is increased in proportion to the square root of the beam energy. We consider the effect of multiple Coulomb scattering on the emittance growth and evaluate an achievable normalized emittance at the end of the accelerator system comprising N_s stages. Using the growth rate of the mean square deflection angle Eq. (30) due to multiple Coulomb scattering, the growth of the normalized transverse emittance is estimated as

$$\frac{d\varepsilon_n^{SCAT}}{dz} = \frac{1}{2} \frac{\gamma}{k_\beta} \frac{d\langle \theta^2 \rangle}{dz} = \frac{k_p^2 r_e Z}{K\sqrt{\gamma}} \ln\left(\frac{\lambda_D}{R_N}\right) = \frac{k_p r_e Z}{\overline{K}\sqrt{\gamma}} \ln\left(\frac{\lambda_D}{R_N}\right) , \tag{57}$$

where $k_\beta = K / \sqrt{\gamma}$ is the wave number of betatron oscillation [6, 36]. In the single stage, transverse normalized emittance of the particles undergoing the wakefield evolves the growth in the same manner as the injected particle beam without multiple Coulomb scattering as

$$\bar{\varepsilon}_{nx}^1 = \left| \frac{\bar{E}_z(\Psi_f)}{\bar{E}_z(\Psi_i)} \right| \bar{\varepsilon}_{n0} + \frac{k_p r_e Z \ln(\lambda_D / R_N)}{\bar{K}_i \bar{E}_z(\Psi_i)} (\sqrt{\langle \gamma_{1f} \rangle} - \sqrt{\langle \gamma_{1i} \rangle}). \tag{58}$$

At the N_s-th stage, the normalized emittance can be obtained from

$$\bar{\varepsilon}_{nx}^{N_s} = \left| \frac{\bar{E}_z(\Psi_f)}{\bar{E}_z(\Psi_i)} \right|^{N_s} \bar{\varepsilon}_{n0} + \frac{k_p r_e Z \ln(\lambda_D / R_N)}{\bar{K}_i \bar{E}_z(\Psi_i)} \sum_{k=1}^{N} \left| \frac{\bar{E}_z(\Psi_f)}{\bar{E}_z(\Psi_i)} \right|^{N_s - k} \sqrt{\gamma_k} \left(1 - \sqrt{\frac{\gamma_{k-1}}{\gamma_k}} \right). \tag{59}$$

Assuming that the beam energy at the k-th stage is approximately given by Eq. (49), Eq. (59) can be calculated as

$$\bar{\varepsilon}_{nx}^{N_s} = e^{-N_s \Sigma} \bar{\varepsilon}_n^0 + 2(\frac{2}{\pi})^{1/4} \frac{C_S e^{(1+4\alpha a_0^2 \Delta\Psi)/4} Z \ln(\lambda_D / R_N)}{a_0 C_n^{1/2} \bar{K}_i \cos\Psi_f} \left(\frac{\sin \Delta\Psi}{\Sigma} \right)^{1/2} [F(\sqrt{N_s \Sigma}) - e^{-(N_d - 1)\Sigma} F(\sqrt{\Sigma})], \tag{60}$$

where $C_S = k_p r_e \gamma_g = 17.7 \times 10^{-9} [\mu m] / \lambda_L$, $\Sigma = 2\alpha_d a_0^2 \Delta\Psi - \ln R > 0$, i.e., $R < e^{2\alpha_d a_0^2 \Delta\Psi}$ and $F(x)$ is the Dawson's integral. For $R > e^{2\alpha_d a_0^2 \Delta\Psi}$, setting $\Pi = -\Sigma$, the growth of normalized emittance is expressed by

$$\bar{\varepsilon}_{nx}^{N_s} = e^{N\Pi} \bar{\varepsilon}_n^0 + (2\pi)^{1/4} \frac{C_S e^{(1+4\alpha a_0^2 \Delta\Psi)/4} Z \ln(\lambda_D / R_N)}{a_0 C_n^{1/2} \bar{K}_i \cos\Psi_f} \left(\frac{\sin \Delta\Psi}{\Pi} \right)^{1/2} e^{N\Pi} [\mathrm{erf}(\sqrt{N_s \Pi}) - \mathrm{erf}(\sqrt{\Pi})]. \tag{61}$$

For $R = e^{2\alpha_d a_0^2 \Delta\Psi}$, i.e., $|\bar{E}_z(\Psi_f) / \bar{E}_z(\Psi_i)| = 1$, the normalized emittance at the N_s-th stage is simply calculated as

$$\bar{\varepsilon}_{nx}^N \sim \bar{\varepsilon}_n^0 + 2(\frac{2}{\pi})^{1/4} \frac{C_S Z e^{(1+4\alpha_d a_0^2 \Delta\Psi)/4} \ln(\lambda_D / R_N)}{a_0 C_n^{1/2} \bar{K}_i \cos\Psi_i} (N_s \sin \Delta\Psi)^{1/2}, \tag{62}$$

Figure 5 shows the evolution of the normalized emittance over the whole stages for case A, case B and case C, corresponding to $\Sigma = 2.0 > 0$, $\Sigma = 0.049 \sim 0$ and $\Sigma = -0.018 < 0$, respectively, in comparison with the analytical estimates obtained from Eq. (60), (62) and (61) with the initial dimensionless focusing strength $\bar{K}_i = 8$ (case A), $\bar{K}_i = 0.1$ (case B) and $\bar{K}_i = 0.1$ and the growth rate per stage $\Pi = -1.8\Sigma$ (case C), respectively. As expected, the normalized emittance in the multistage accelerator operated with the constant accelerating field is conserved to the initial normalized emittance, then limited by an increasing growth due to multiple Coulomb scattering. For $R > e^{2\alpha_d a_0^2 \Delta\Psi}$, the initial emittance of the injected beam dominates an exponential growth of the normalized emittance, while for $R < e^{2\alpha_d a_0^2 \Delta\Psi}$, an exponential decrease of the initial normalized emittance is followed by a slow decrease of the normalized emittance originated from multiple Coulomb scattering.

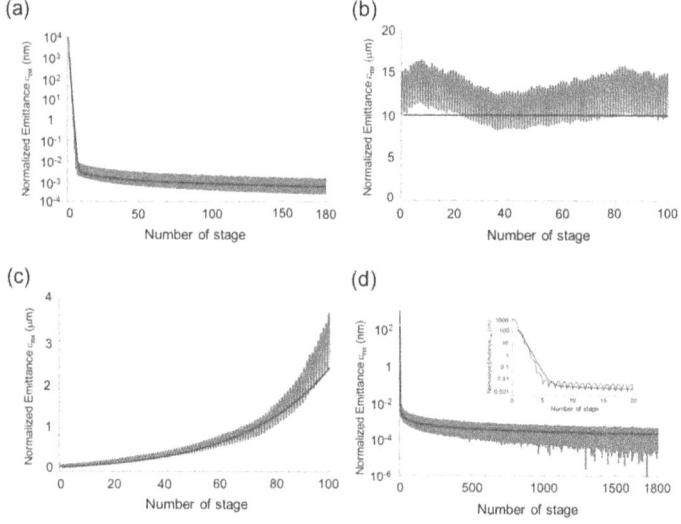

Fig. 5. The evolution of normalized emittance ε_{nx} as a function of the stage number (a) for case A, (b) for case B and (c) for case C up to the final beam energy of 589 GeV, 539 GeV and 522 GeV, respectively, and (d) for case A with $\varepsilon_{n0} = 1\,\mu m$ up to the final beam energy of 5720 GeV. Blue solid lines indicate the analytical predictions given by Eq. (60) with $\Sigma = 2.0$, $\bar{K}_i = 8$ in (a), by Eq. (62) with $\bar{K}_i = 0.1$ in (b), and by Eq. (61) with $\Pi = 0.032$, $\bar{K}_i = 0.1$ in (c) and by Eq. (60) with $\Sigma = 2.0$, $\bar{K}_i = 8$ in (d), respectively. The inset in (d) shows the simulation result (red solid curve) and its analytical prediction (blue solid curve) in the initial 20 stages.

4. Consideration on Luminosity and Beam-beam Interaction

Electron and positron beams being reached to the final energies in the multistage capillary plasma accelerator are extracted at a phase corresponding to the minimum transverse normalized emittance, followed by propagating a drift space in vacuum and a final focusing system to the beam-beam collisions at the interaction point. Outside plasma in a vacuum drift space, the particle beam changes the spatial and temporal dimensions of the bunch proportional to the propagation distance due to the finite emittance and energy spread of the accelerated bunch. The evolution of the bunch envelope in vacuum without the external focusing force is expressed by the equation $d^2\sigma_b / dz^2 = \varepsilon_n^2 /(\gamma^2\sigma_b^3)$, where σ_b is the rms bunch envelope radius and z is the propagation coordinate. Taking into account the solution given by $\sigma_b^2 = \sigma_0^2[1 + (z - z_0)^2 / Z_b^2]$ [38], where σ_0 is an initial radius and $Z_b = \sigma_0^2\gamma / \varepsilon_n$ is the characteristic distance of the bunch size growth, the bunch radius after propagation of the distance L_{col} between the plasma accelerator and interaction point is estimated to be

$$\sigma_{b*} = \left[\sigma_{bf}^2 + \left(\frac{L_{col}\varepsilon_{nf}}{\gamma\sigma_{bf}}\right)^2\right]^{1/2}, \tag{63}$$

where σ_{b*} is the rms bunch radius at the interaction point and σ_{bf}, ε_{nf} the rms bunch radius and normalized emittance at the exit of the plasma accelerator, respectively. A finite energy spread and divergence of beam particles in the bunch result in bunch lengthening during the propagation distance L_{col} to the extent of $\Delta\sigma_z \sim (\delta\gamma/\gamma)(L_{col}/\gamma^2) + (\sigma_{b*}^2/2L_{col})$, where $\delta\gamma/\gamma$ is the relative energy spread. Thus, after propagation to the interaction point the particle bunch is lengthened as

$$\sigma_{z*} = \sigma_{zf} + \left(\frac{\delta\gamma}{\gamma}\right)_f \frac{L_{col}}{\gamma^2} + \frac{\sigma_{b*}^2}{2L_{col}}, \tag{64}$$

where σ_{z*} and σ_{zf} are the rms bunch length at the interaction point and at the exit of plasma accelerator, respectively, and $(\delta\gamma/\gamma)_f$ is the relative energy spread at the exit of plasma accelerator.

In collisions between high energy electron and positron bunches from the accelerators, the beam particles emit synchrotron radiation due to the interaction, so-called beamstrahlung, with the electromagnetic fields generated by the counterpropagating beam. The beamstrahlung effect leads to substantial beam energy loss and degradation on energy resolution for the high-energy experiments in electron-positron linear colliders [25]. Intensive research on beamstrahlung radiation has been explored [24-27], being of relevance to the design of e^+e^- linear colliders in the TeV c.m. energies, for which two major effects should be taken into account: namely, the disruption effect bending particle trajectories by the oncoming beam generated-electromagnetic fields and beamstrahlung effect yielding radiation loss of the particle energies induced by bending their trajectories due to the disruption. Since beamstrahlung radiation is generated from the interaction with intense fields in a relatively short bunch length, the photon emission has the quantum mechanical nature with a small number of photons emitted per particle and a significant fraction of the initial particle energy, exhibiting a spectrum quite different from classical radiation [25]. The quantum mechanical nature of beamstrahlung can be characterized by the Lorentz-invariant parameter $\Upsilon \simeq 2\gamma E/F_c$ in units of the Schwinger critical field $F_c = m_e^2 c^3/e\hbar$ ($\simeq 4.4 \times 10^{13}$ G $\simeq 1.3 \times 10^{16}$ V/cm) [25], expressed by using electromagnetic fields in the laboratory frame. Here classical radiation corresponds to $\Upsilon \ll 1$, for which the classical synchrotron critical photon energy is given by $\hbar\omega_c = 3\gamma m_e c^2 \Upsilon/2$, while $\Upsilon \gg 1$ corresponds to extreme quantum radiation in which the peak position of the synchrotron power spectrum approaches the incident particle energy. In the case of disruption neglected, where two colliding beams are not deformed during the collision, the radiative energy loss due to beamstrahlung for a Gaussian beam can be estimated in terms of the beamstrahlung parameter [25]

$$\Upsilon_* = \frac{5N_b r_e^2 \gamma}{6\alpha(1+R_A)\sigma_{y*}\sigma_{z*}} = \frac{5C_s^2 \gamma}{6\alpha\gamma_g^2(1+R_A)k_p\sigma_{z*}}\left(\frac{N_b}{k_p\sigma_{y*}}\right), \tag{65}$$

where $\alpha = e^2/\hbar c$ ($\simeq 1/137.036$) is the fine-structure constant and $R_A = \sigma_{x*}/\sigma_{y*}$ is the beam aspect ratio at the interaction point. According to the beamstrahlung simulations, the average number of emitted photon per electron and the average fractional energy loss are

$$n_\gamma \approx 2.54 B_\gamma \Upsilon_* \left(1 + \Upsilon_*^{2/3}\right)^{-1/3}, \qquad \delta_b \approx 1.24 B_\gamma \Upsilon_*^2 \left[1 + (3\Upsilon_* / 2)^{2/3}\right]^{-2}, \qquad (66)$$

where $B_\gamma = \alpha^2 \sigma_{z*} / (r_e \gamma) = \alpha^2 \gamma_g k_p \sigma_{z*} / (C_s \gamma)$ [25]. Using these parameters, the average c.m. energy loss and rms c.m. energy spread can be calculated as

$$\frac{\Delta W}{W_0} \approx \left(0.44 + 0.01 \log_{10} \Upsilon_*\right) \delta_b \left(1 + \frac{\delta_b}{n_\gamma}\right), \qquad (67)$$

$$\frac{\sigma_W}{W_0} \approx \left[0.21 + \frac{0.01}{2}(4 - \log_{10} \Upsilon_*)(3 - \log_{10} \Upsilon_*)\right] \delta_b \left(1 + \frac{14 + 4\log_{10} \Upsilon_*}{n_\gamma}\right)^{1/2}, \qquad (68)$$

where $W_0 = 2\gamma m_e c^2$ is the c.m. energy [25]. In the quantum beamstrahlung regime, the collider design must consider the c.m. energy loss and spread such that their requirements can be reached as well as that of the luminosity. The geometric luminosity is given by

$$\mathcal{L}_0 = \frac{f_c N_b^2}{4\pi \sigma_{x*} \sigma_{y*}} = \frac{10^8 \pi f_c N_b^2}{\gamma_g^2 R_A (k_p \sigma_{y*})^2} [\text{cm}^{-2}\text{s}^{-1}], \qquad (69)$$

where f_c is the collision frequency. Complying with the luminosity guideline for the future $e^+ e^-$ linear colliders, approximately given as $\mathcal{L}_0 \approx (W_0[\text{TeV}])^2 \times 10^{34} [\text{cm}^{-2}\text{s}^{-1}]$ [6, 7], the transverse particle line density is constrained as

$$\frac{N_b}{k_p \sigma_{y*}} = 1.022 \times 10^7 \gamma_g \gamma_* \sqrt{\frac{R_A}{\pi f_c}}, \qquad (70)$$

where $\gamma_* m_e c^2 = W_0 / 2$ is the nominal beam energy required. For $\Upsilon_* \gg 1$, the c.m. energy loss $\Delta W / W_0 = (\gamma - \gamma_*) / \gamma_* \approx 0.33 B_\gamma \Upsilon_*^{2/3}$ and beamstrahlung parameter Eq. (67) provide the equation on the accelerator beam energy required $m_e c^2 \gamma$:

$$\frac{\gamma_*}{\gamma} = 1 - 0.0688 \left(\frac{\gamma_g \gamma_*}{f_c} \frac{R_A k_p \sigma_{z*}}{(1 + R_A)^2}\right)^{1/3} \left(\frac{\gamma_*}{\gamma}\right)^{1/3}. \qquad (71)$$

It is noted that the beam energy reduction due to beamstrahlung scales as $\propto (k_p \sigma_{z*})^{1/3} R_A^{-1/3}$ for $R_A \gg 1$, while the particle density required for a given luminosity scales as $\propto R_A^{1/2}$.

It is pointed out that an appreciable disruption effect turns out to the luminosity enhancement as a result of significant increase of the transverse particle density $N_b / (k_p \sigma_{y*})$ through the pinch effect arising from the attraction of the oppositely charged beams [26, 27]. For Gaussian beams, the disruption parameters for the horizontal and vertical directions are [25, 27]

$$D_x = \frac{D_y}{R_A}, \quad D_y = \frac{2r_e\sigma_{z*}N_b}{\gamma(1+R_A)\sigma_{y*}^2} = \frac{2C_s k_p\sigma_{z*}N_b}{\gamma_g\gamma(1+R_A)(k_p\sigma_{y*})^2}, \tag{72}$$

which is defined as the ratio of the bunch length to the focal length of a thin lens. For round beams with $R_A = 1$, *i.e.*, $D \equiv D_x = D_y$, the luminosity enhancement factor being defined as the ratio of the effective luminosity $\mathcal{L}$ induced by the disruption to the geometric luminosity in the absence of disruption $\mathcal{L}_0$ has been found from the simulations on the beam-beam interaction of electron-positron beams:

$$H_D \equiv \frac{\mathcal{L}}{\mathcal{L}_0} = 1 + D^{1/4}\left(\frac{D^3}{1+D^3}\right)\left[\ln\left(\sqrt{D}+1\right) + 2\ln\left(\frac{0.8}{A}\right)\right], \tag{73}$$

where $A = \varepsilon_n\sigma_{z*}/(\gamma\sigma_{y*}^2) = \varepsilon_n D/(r_e N_b)$ is the inherent divergence of the incoming beam [27]. Here, the maximum disruption angle $\theta_{\text{disrupt}}^{\max}$, which is the important information for the linear collider design to avoid the debris from the beam-beam collision, has been estimated at $A = 0$ as $\theta_{\text{disrupt}}^{\max} \simeq 0.87 D\sigma_{y*}/\sigma_{z*}$ for $D \ll 1$ and $\theta_{\text{disrupt}}^{\max} \simeq 1.84\sqrt{D}\sigma_{y*}/\sigma_{z*}$ for $D \gg 1$, respectively [27].

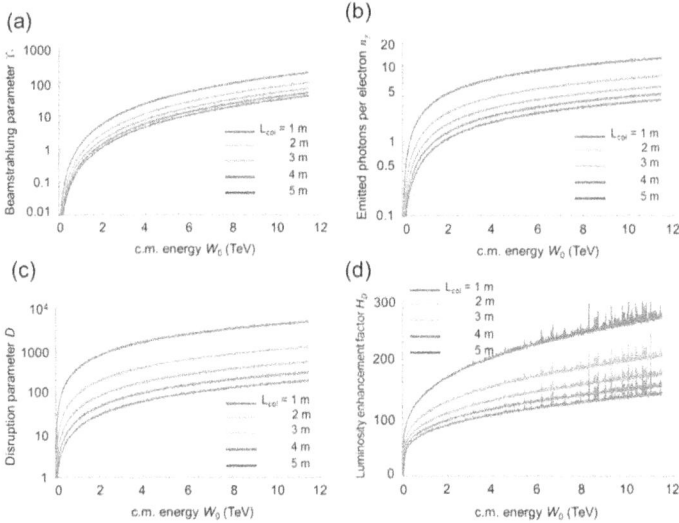

Fig. 6. Beam-beam interaction parameters at various distances of the interaction point L_{tol} from the exit of LPA as a function of c.m. energy W_0: (a) beamstrahlung parameter Υ_*, (b) number of emitted photons per electron (positron) n_γ, (c) disruption parameter D and (d) luminosity enhancement factor H_D.

These parameters on electron-positron beam-beam interactions can be evaluated from using the outputs of numerical calculations on energy $m_e c^2\gamma_f$, relative energy spread $(\delta\gamma/\gamma)_f$, bunch radius σ_{bf}, transverse normalized emittance ε_{nf} at the N_s-th final stage, taking into account the evolution of bunch radius σ_{y*} and σ_{z*} in the vacuum

36

propagation distance L_{col}, as obtained from Eqs. (65) and (66). Figures 6 and 7 show important parameters on the beam-beam interaction and collider performance at various distances of the interaction point L_{col} from the exit of the plasma linac as a function of c.m. energy W_0.

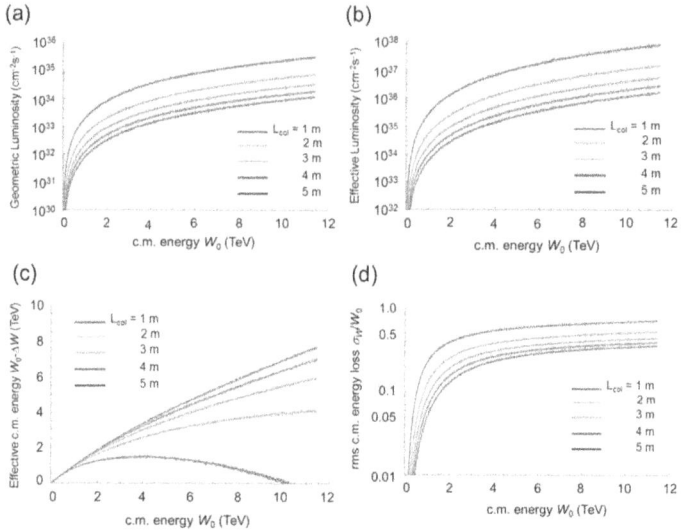

Fig. 7. Collider parameters on (a) geometric luminosity $\mathcal{L}_0$, (b) effective luminosity $\mathcal{L}$, (c) effective c.m. energy $W_0 - \Delta W$ and (d) rms c.m. energy spread σ_W / W_0 as a function of c.m. energy W_0 for various distances of the interaction point from the exit of LPA.

5. Coherent Amplification Network Laser

An application such as LPA requires a high peak and average power laser system in order to be a feasible driver for a LC. The Chirped Pulse Amplification (CPA) of short pulse lasers [39] allows for high peak power systems of sufficient energy and short pulse duration to achieve $a_0^2 = 1$ but the repetition rate of these systems is still quite low. The highest energy systems of modern laser facilities are typically on the order of 1 shot/min or at most a few Hz. This is primarily due to the thermal management required of their amplification media. The ability to cool along the length of a fiber makes an improved heat removal possible within a fiber-based amplification system. So high repetition rates are made possible due to the better optimization of thermal management within the laser amplification chain. In addition there is the benefit of fulfilling another requirement for a practical use within a LC design when compared to crystal or glass based laser systems: greater wall-plug efficiency. There are two contributions to this efficiency: the diode laser lasers pumping the fiber amplifier are far more efficient than the systems employed in traditional high power laser; and the effective transfer of energy from the pump photons into the laser pulse within the fiber amplifier. Despite these benefits, the primary problem for a fiber-based CPA is the low energy within the individual pulses that must be maintained in order to prevent distortion or destruction of the laser within the long interaction length of the fiber. Therefore the high repetition rate desired for a LC is possible with fiber-based lasers providing up to MHz pulse trains but these while being high average power laser systems do not provide adequate peak power required of individual pulses.

Just as the coherence of laser light has revolutionized so many areas with their application, here it provides the solution to the high peak and average challenge through coherent beam combining (CBC). The small energy of a single fiber can be overcome by combining several such amplifiers in parallel. This laser architecture concept is known as the coherent amplification network (CAN) and has been introduced for the application to accelerators in Refs. [40, 41] with its design parameters further explored in Ref. [42].

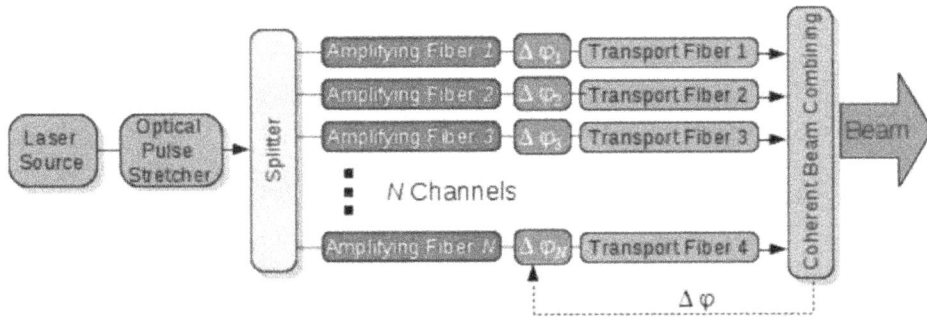

Fig. 8 Schematic of the coherent amplification network (CAN) laser architecture. A femtosecond seed laser pulse is repeatedly split and amplified through this network composed of amplifying and transport fibers. The resulting pulses are then recombined coherently. (Figure from Ref. 42.)

As depicted in Fig. 8, a short pulse seed laser source is chirped within an optical pulse stretcher before being split into a parallel array of fiber-based amplifiers. A single high average and peak power laser source results through the coherent combination of these phased-locked channels. The CBC requires an active phase-locking of the separate fiber channels in order to maintain control of the delay and relative phase between the channels. A kHz interferometric sampling of the beams at the point of recombination is adequate to maintain the required coherence between the channels. While there are several possible design geometries for CBC, one that is being demonstrated for short pulsed femtosecond lasers through the development of a 61-fiber prototype based on flexible Ytterbium-doped amplifying fibers within the XCAN project [43]. From the initial laser source the laser system remains within fiber until the CBC stage which minimizes instabilities within the system. A precision mount for the fibers permits the beams to exit each channel with minimal deviation in separation, pointing, and depth of field. A microlens array recollimates the individual beams exiting the fibers before a single lens then combines the beams at its focus where a pinhole selects the central peak. After the beam combination step is sent to a typical grating compressor so that the amplified combined beam can be compressed back to its transform-limited pulse duration. In the case of the 61-fiber prototype, the design parameters have the goal to achieve a pulse energy of 3 mJ, with 300 fs pulse duration at a repetition rate of 200 kHz and a laser wavelength of 1054 nm after the grating compressor. Along the way to this goal, intermediate systems consisting of 7 and 19 fibers have already been reported [44, 45]. The CAN laser is envisioned to benefit from increased scaling in the number of parallel channels, both in cost per unit channel and in the reduction in noise due to small errors in the combination channels. Ultimately, the development of CAN lasers of 10^6 fiber-channels is foreseeable for major applications such as a LC facility.

The bandwidth limitations of a fiber-based amplifier places the minimum pulse duration near 300 fs. In order to reach the design requirements closer to a 80 fs pulse duration, a post-compression stage can be implemented such as a Thin Film Compressor (TFC) proposed in Ref. [46] or an unguided gas-cell based compressor [47]. Both methods work by increasing the intrinsic bandwidth of the pulse through a nonlinear interaction within either a thin transparent film or a dilute gas. The TFC for this level of compression relies on the high intensity of the pulse and makes a single pass of the unfocused beam through the nonlinear medium during its transport to the final interaction point. Ideally the beam to be compressed should be a flat-top in order to use the simplest film geometry. The resultant pulse is then brought to the optimized pulse duration by reflecting from dispersion-controlled mirrors, also referred to as chirped mirrors. The second option of an unguided gas-cell relies on focusing of the beam through the gas and passes through the nonlinear medium multiple times. The slightly modified pulse is recompressed following each short interaction using similar dispersion-compensating mirrors. This process is repeated until the desired compression factor is achieved. A post-compression factor of 4, from a pulse of 300 fs to 75 fs, is achievable for this type of CAN laser system.

The development of an efficient and scalable CAN laser system gives access to an economical and flexible tool capable to meet the requirements to drive the multiple stages of acceleration within a LC facility.

6. Discussions and Conclusions

In this work, we have investigated the dynamics of electron and positron beams in a laser-plasma-based linear collider with the TeV c.m. energies, comprising seamlessly consecutive multi-stages of gas-filled capillary accelerators. Here the numerical simulation model for particle tracking has been first derived by integrating the basic coupled equations of a single particle motion in laser longitudinal and transverse wakefields, using the analytic solutions that are exactly valid in a segmented region with a constant wakefield and then corrected in order to account for the radiation reaction force and deflection due to multiple Coulomb scattering exerted on the single particle motion as the perturbation. In this model, the collective effects of the bunched beam dynamics arising from beam-driven wakefields in plasma have been taken into account at each step of the particle orbit, leading to the beam loading energy loss and self-focusing.

The depletion of the laser pulse energy due to the generation of wakefields is considered as the amplitude attenuation of the laser vector potential with an intact shape in the pump depletion length L_{pd} for the quasi-linear laser wakefield, while the amplitude damping of the electromagnetic hybrid mode EH_{11} during propagation in the capillary can be neglected for the condition of the mode attenuation length $L_1^i = 1/k_1^i \gg L_{pd} \geq L_{stage} = \Delta \Psi_{stage} / k_p$, where L_{stage} is the stage length with phase advance $\Delta \Psi_{stage}$. Furthermore, the simple analytic formulae presented can account for the evolution of radiative energy loss and normalized emittance over numerous stages, i.e., $N_s \gg 1$. Note that an important finding is the criteria given by the amplitude ratio of the accelerating field between the initial and final phase in the stage, i.e., $\left| \cos \Psi_i / \cos \Psi_f \right| > \exp(-2\alpha_d a_0^2 \Delta \Psi)$, under which the normalized emittance converges on the equilibrium with the emittance growth due to multiple Coulomb scattering within 10 stages for the present case. After the equilibrium, since the bunch radius reduces adiabatically as $\gamma^{-1/4}$ and the self-focusing strength arising from the beam-driven wakefield increases concurrently when the particle beam energy increases, the normalized emittance continues to decrease until the end of the plasma linac. As a consequence, the radiative energy loss due to betatron radiation can be suppressed to an extremely small fraction. This scenario allows us to transport both beams into the interaction point through no extra focusing devices, which often induce the degradation of beam qualities prior to their interactions. In this scheme, the vacuum drift region from the end of the plasma linac to the interaction point may be used for control of the transverse beam size that strongly affect the luminosity and c.m. energy through the beamstrahlung radiation and disruption.

A typical design example of the LPA stage using a gas-filled capillary operated in the electromagnetic hybrid mode EH_{11} is shown in Table 1. An embodiment of the LPA stage may be envisioned by exploiting a tens kW-level high average power laser such as a coherent amplification network of fiber lasers [41]. This scheme restricts the laser intensity such that the plasma response is within the quasi-linear regime, i.e., $a_0^2 \leq 1$ for two reasons; the one is avoidance of the nonlinear plasma response such as in the bubble regime, where symmetric wakefields for the electron and positron beams cannot be obtained for the application to electron-positron colliders [6, 7] and the degradation of the beam quality due to the self-injection from the background plasma electrons. The other is an inherent demand that the laser intensity guided in a capillary should be lower enough than the threshold of material damage on the capillary wall. The maximum intensity that can be guided through a capillary tube without wall ionization can be estimated by taking into account the normalized maximum flux for EH_{1m} mode at the capillary wall, $F_w^m \simeq [(\varepsilon_r + 1)/2\sqrt{\varepsilon_r - 1}]u_m^2 J_1^2(u_m)/(k_0 R_c)^2$, defined as the ratio of the radial component of the Poynting vector at $r = R_c$ to the longitudinal component of the on-axis Poynting vector [18]. For the single mode EH_{11}, the normalized wall flux amounts to $C_1 F_w^1 \sim 8.0 \times 10^{-6}$ for a glass capillary with the relative permittivity $\varepsilon_r \simeq 2.25$. The energy fluence traversing the capillary wall yields $\mathcal{F}_{wall} \sim C_1 F_w^1 I_L \tau_L \sim 0.9$ J/cm^2 for the peak intensity $I_L = 1.37 \times 10^{18}$ W/cm^2 ($a_0^2 = 1$) and the pulse duration $\tau_L = 79$ fs. The experimental study of laser-induced breakdown in fused silica (SiO_2) [48] suggests that the fluence breakdown threshold is scaled to be $\mathcal{F}_{th} \sim 50$ J/cm^2 for $\tau_L = 79$ fs .

Table 1. Parameters of the laser-plasma accelerator stage

Plasma density n_e [cm^{-3}]	1×10^{17}
Plasma wavelength λ_p [μm]	106
Capillary radius R_c [μm]	84
Capillary stage length [m]	0.53
Laser wavelength λ [μm]	1
Laser spot radius r_0 [μm]	54
Laser pulse duration τ [fs]	79
Normalized vector potential a_0	1
Electromagnetic hybrid mode	EH$_{11}$
Coupling efficiency C_1	0.98
Bunch initial and final phase (Ψ_i, Ψ_f)	$(0, 0.45\pi)$
Average accelerating gradient [GeV/m]	6
Laser peak power P_L [TW]	63
Laser pulse energy U_L [J]	5
Repetition frequency f_c [kHz]	10
Laser average power per stage [kW]	50
Laser depletion η_{pd} [%]	15

Table 2. Parameters for 1 – 10 TeV novel laser-plasma linear colliders.

Center-of-mass energy	W_0 [TeV]	1.2	3.5	5.8	11
Beam energy	[GeV]	589	1741	2884	5720
Particle per bunch	N_b [10^9]	1	1	1	1
Collision frequency	f_c [kHz]	10	10	10	10
Total beam power	[MW]	1.89	5.58	9.24	18.3
Geometric luminosity	$\mathcal{L}_0$ [10^{34} cm^{-2}s^{-1}]	0.017	0.15	0.28	1.16
Effective luminosity	$\mathcal{L}$ [10^{34} cm^{-2}s^{-1}]	1.41	16.6	32.9	168
Effective c.m. energy	W_0-ΔW [TeV]	1.14	2.91	4.55	7.68
c.m. energy spread	σ_W/W_0 [%]	8.1	24.2	28.5	32.8
Bunch length	σ_z [μm]	50	50	50	50
Beam radius at IP	σ_{b*} [nm]	22	7.3	5.3	2.6
Beam aspect ratio	R_A	1	1	1	1
Normalized emittance at IP	ε_{nf} [pm]	0.27	0.12	0.068	0.022
Distance to IP	L_{col} [m]	4	4	5	5
Beamstrahlung parameter	Υ_*	0.48	4.2	9.5	38.5
Beamstrahlung photons	n_γ	0.85	1.9	2.3	3.6
Disruption parameter	D_*	27	83	94	196
Luminosity enhancement	H_D	84	110	117	145
Number of stages per beam	N_s	180	540	900	1800
Linac length per beam	[m]	95.4	286	477	954
Power requirement for lasers	[MW]	18	54	90	180

Table 2 summarizes key parameters on the performance of electron-positron linear collider, assuming the round beam option, *i.e.*, the beam aspect ratio $R_A = 1$, for simplicity. A sort of engineering discussions on the total power requirement, referred to as a wall plug power, and the accelerator efficiencies will be reserved for the future work till the advance of relevant technologies such as CAN lasers [40-47] and seamless long staging of LPAs [49, 50] as well as the electron/positron external injectors matched to the LPA.

In conclusion, a new scheme of TeV electron-positron linear collider comprising properly phased multi-stage LPAs using a gas-filled capillary can provide an alternative approach in collider applications to the scheme proposed by Schroeder et al. [9] using hollow plasma channels relying on a technically hypothetical plasma density structure. The novel scheme presented resorts two major mechanisms pertaining to laser wakefield acceleration, that is, dephasing and strong focusing force as well as very high-gradient accelerating field, which are largely reduced in a hollow plasma channel having an analogy to vacuum-based accelerating structures. More intriguingly in physics point of view, it is noted that the normalized emittance at the end of LPA stages in multi-TeV energies may reach the quantum mechanical limit $\varepsilon_{n\min} = \lambda_e / 2 \simeq 0.2$ pm determined from the uncertainty principle [51], where $\lambda_e = \hbar / m_e c$ is the electron Compton wavelength. In this point, the quantum mechanical consideration on the beam dynamics of the radiative process would be investigated in the future work.

Acknowledgments

K. Nakajima was supported by Center for Relativistic Laser Science, the Institute for Basic Science, Republic of Korea, under IBS-R012-D1.

Appendix A: Emittance expression in terms of the ensemble averaged quantities

The definition of transverse normalized emittance given by Eq. (41) is expressed as

$$\varepsilon_{nx}^2 = \langle \delta x^2 \rangle \langle \delta u_x^2 \rangle - \langle \delta x \delta u_x \rangle^2, \tag{A.1}$$

where $\delta x = x - \langle x \rangle$ and $\delta u_x = u_x - \langle u_x \rangle$ are the deviation from the mean transverse displacement $\langle x \rangle$ and normalized momentum $\langle u_x \rangle = \langle \gamma \beta_x \rangle$, respectively. The particle orbits undergoing the betatron motion is written for $s \gg 1$ from Eq. (42)

$$\bar{x} = \bar{x}_m \sqrt{\frac{s_0}{s}} \cos \varphi \qquad u_x = \gamma \beta_x = \frac{1}{2} \bar{x}_m \bar{E}_z \sqrt{s s_0} \sin \varphi, \tag{A.2}$$

where $\varphi = \bar{\omega}_\beta t$ is the betatron phase and $\bar{x}_m = [\bar{x}_0^2 + (2\gamma_0 \beta_{x0} / s_0 \bar{E}_{z0})^2]^{1/2}$. Thus, the ensemble averaged quantities can be obtained as

$$\langle \delta \bar{x}^2 \rangle = \left\langle \frac{s_0}{s} \right\rangle \left[\frac{1}{2} \langle \bar{x}_m^2 \rangle (1 + \langle \cos 2\varphi \rangle) - \langle \bar{x}_m \rangle^2 \langle \cos \varphi \rangle^2 \right], \tag{A.3}$$

$$\langle \delta u_x^{\,2} \rangle = \frac{\bar{E}_z^2}{4} \langle ss_0 \rangle [\frac{1}{2} \langle \bar{x}_m^{\,2} \rangle (1 - \langle \cos 2\varphi \rangle) - \langle \bar{x}_m \rangle^2 \langle \sin \varphi \rangle^2], \tag{A.4}$$

$$\langle \delta \bar{x} \delta u_x \rangle^2 = \frac{\bar{E}_z^2}{16} \langle s_0^2 \rangle \langle \bar{x}_m^{\,2} \rangle^2 \langle \sin 2\varphi \rangle^2$$

$$- \frac{\bar{E}_z^{\,2}}{4} \langle s_0 \rangle \langle (\frac{s_0}{s})^{1/2} \rangle \langle (ss_0)^{1/2} \rangle \langle \bar{x}_m^{\,2} \rangle \langle \bar{x}_m \rangle^2 \langle \sin \varphi \rangle \langle \cos \varphi \rangle \langle \sin 2\varphi \rangle, \tag{A.5}$$

$$+ \frac{\bar{E}_z^{\,2}}{4} \langle (\frac{s_0}{s})^{1/2} \rangle^2 \langle (ss_0)^{1/2} \rangle^2 \langle \bar{x}_m \rangle^4 \langle \cos \varphi \rangle^2 \langle \sin \varphi \rangle^2$$

Assuming that the energy distribution about the mean energy $\langle \gamma \rangle$, i.e., the $\delta \gamma = \gamma - \langle \gamma \rangle$ distribution, is Gaussian with width σ_γ, the energy is approximated about its mean value to the first order in $\delta \gamma / \langle \gamma \rangle$, i.e., $\gamma = \langle \gamma \rangle + \delta \gamma$, $\omega_\beta \simeq \omega_{\beta 0} (1 - \delta \gamma / 2\langle \gamma \rangle)$, and $\varphi \simeq \omega_{\beta 0} \bar{t} (1 - \delta \gamma / 2\langle \gamma \rangle)$. The ensemble averaged quantities can be calculated as averages over the distribution of energy deviations [35]:

$$\langle \cos \varphi \rangle \simeq \frac{1}{\sqrt{2\pi}\sigma_\gamma} \int_{-\infty}^{\infty} d\,\delta\gamma \exp(-\frac{\delta\gamma^2}{2\sigma_\gamma^2}) \cos(\varphi_0 + \delta\varphi) = e^{-v_\varepsilon^2 t^2} \cos \varphi_0, \tag{A.6}$$

$$\langle \sin \varphi \rangle \simeq \frac{1}{\sqrt{2\pi}\sigma_\gamma} \int_{-\infty}^{\infty} d\,\delta\gamma \exp(-\frac{\delta\gamma^2}{2\sigma_\gamma^2}) \sin(\varphi_0 + \delta\varphi) = e^{-v_\varepsilon^2 t^2} \sin \varphi_0, \tag{A.7}$$

$$\langle \cos 2\varphi \rangle \simeq \frac{1}{\sqrt{2\pi}\sigma_\gamma} \int_{-\infty}^{\infty} d\,\delta\gamma \exp(-\frac{\delta\gamma^2}{2\sigma_\gamma^2}) \cos(2\varphi_0 + 2\delta\varphi) = e^{-4v_\varepsilon^2 t^2} \cos 2\varphi_0, \tag{A.8}$$

$$\langle \sin 2\varphi \rangle \simeq \frac{1}{\sqrt{2\pi}\sigma_\gamma} \int_{-\infty}^{\infty} d\,\delta\gamma \exp(-\frac{\delta\gamma^2}{2\sigma_\gamma^2}) \sin(2\varphi_0 + 2\delta\varphi) = e^{-4v_\varepsilon^2 t^2} \sin 2\varphi_0, \tag{A.9}$$

$$\left\langle \frac{s_0}{s} \right\rangle = \frac{1}{\sqrt{2\pi}\sigma_\gamma} \frac{\bar{K}_0 \bar{E}_z}{\bar{K}\bar{E}_{z0}} \int_{-\infty}^{\infty} d\,\delta\gamma \exp(-\frac{\delta\gamma^2}{2\sigma_\gamma^2}) \frac{(\langle \gamma_0 \rangle + \delta\gamma)^{1/2}}{(\langle \gamma \rangle + \delta\gamma)^{1/2}} = \frac{\bar{K}_0 \bar{E}_z}{\bar{K}\bar{E}_{z0}} \sqrt{\frac{\langle \gamma_0 \rangle}{\langle \gamma \rangle}}, \tag{A.10}$$

$$\langle ss_0 \rangle = \frac{4}{\sqrt{2\pi}\sigma_\gamma} \frac{\bar{K}_0 \bar{K}}{\bar{E}_{z0}\bar{E}_z} \int_{-\infty}^{\infty} d\,\delta\gamma \exp(-\frac{\delta\gamma^2}{2\sigma_\gamma^2}) (\langle \gamma_0 \rangle + \delta\gamma)^{1/2} (\langle \gamma \rangle + \delta\gamma)^{1/2} = \frac{4\bar{K}_0 \bar{K}}{\bar{E}_{z0}\bar{E}_z} \sqrt{\langle \gamma_0 \rangle \langle \gamma \rangle}, \tag{A.11}$$

where $\delta\varphi = -\varphi_0 \delta\gamma / (2\langle \gamma \rangle)$ and $v_\varepsilon = \omega_{\beta 0} \sigma_\gamma / (\sqrt{8} \langle \gamma \rangle)$ is the frequency corresponding to decoherence time $t_{\rm dec} \simeq \pi \langle \gamma \rangle / (\omega_\beta \sigma_\gamma)$ [35], defined as the time when the phase difference between the low energy part of the beam and the high energy part is $\Delta\varphi \simeq \omega_\beta \int dt \sigma_\gamma / \langle \gamma \rangle = \pi$.

References

1. T. Tajima and J. M. Dawson, Laser electron accelerator, *Phys. Rev. Lett.* **43**, 267 (1979).
2. K. Nakajima, Laser-driven plasma electron acceleration and radiation, in *Review of Accelerator Science and Technology*, Vol. 9, eds. A. W. Chao, W. Chou, (World Scientific, Singapore, 2016), pp. 19-61.
3. T. Tajima, K. Nakajima, G. Mourou, Laser acceleration, *Rivista del Nuovo Cimento*, **40**, 33 (2017).
4. F. Grüner, S. Becker, Schramm, T. Eichner, M. Fuchs, R. Weingartner, D. Habs, J. Meyer-ter-Vehn, M. Geissler, M. Ferrario, L. Serafini, B. Vander Geer, H. Backe, W. Lauth, S. Reiche, Design considerations for table-top, laser-based VUV and X-ray free electron lasers, *Appl. Phys. B* **86**, 431 (2007).
5. K. Nakajima, Compact X-ray sources-Towards a table-top free-electron laser, *Nature Phys.* **4**, 92 (2008).
6. C. B. Schroeder, E. Esarey, C. G. R. Geddes, C. Benedetti, W. P. Leemans, Physics considerations for laser-plasma linear colliders, *Phys. Rev. ST Accel. Beams* **13**, 101301 (2010).
7. K. Nakajima, A. Deng, X. Zhang, B. Shen, J. Liu, R. Li, Z. Xu, T. Ostermayr, S. Petrovics, C. Klier, K. Iqbal, H. Ruhl, T. Tajima, Operating plasma density issues on large-scale laser-plasma accelerators toward high-energy frontier, *Phys. Rev. ST Accel. Beams* **14**, 091301 (2011).
8. C. B. Schroeder, E. Esarey, W. P. Leemans, Beamstrahlung considerations in laser-plasma-accelerator-based linear colliders, *Phys. Rev. ST Accel. Beams* **15**, 051301 (2012).
9. C. B. Schroeder, C. Benedetti, E. Esarey, W. P. Leemans, Laser-plasma-based linear collider using hollow plasma channels, *Nucl. Instrum. Methods* A **829**, 113 (2016).
10. K. Nakajima, H. Y. Lu, X. Y. Zhao, B. F. Shen, R. X. Li, Z. Z. Xu, 100-GeV large scale laser plasma electron acceleration by a multi-PW laser, *Chin. Opt. Lett.* **11**, 013501 (2013).
11. J. S. Liu, C. Q. Xia, W. T. Wang, H. Y. Lu, C. Wang, A. H. Deng, W. T. Li, H. Zhang, X. Y. Liang, Y. X. Leng, X. M. Lu, C. Wang, J. Z. Wang, K. Nakajima, R. X. Li, Z. Z. Xu, All-optical cascaded laser wakefield accelerator using ionization-induced injection, *Phys. Rev. Lett.* **107**, 035001 (2011).
12. H. T. Kim, K. H. Pae, H. J. Cha, I. J. Kim, T. J. Yu, J. H. Sung, S. K. Lee, T. M. Jeong, J. Lee, Enhancement of electron energy to the multi-GeV regime by a dual-stage laser-wakefield accelerator pumped by petawatt laser pulses, *Phys. Rev. Lett.* **111**, 165002 (2013).
13. H. T. Kim, V. B. Pathak, K. H. Pae, A. Lifschitz, F. Sylla, J. H. Shin, C. Hojbota, S. K. Lee, J. H. Sung, H. W. Lee, E. Guillaume, C. Thaury, K. Nakajima, J. Vieira,

L. O. Silva, V. Malka, C. H. Nam, Stable multi-GeV electron accelerator driven by waveform-controlled PW laser pulses, *Sci. Rep.* **7**, 10203 (2017).

14. W.P. Leemans, A. J. Gonsalves, H.-S. Mao, K. Nakamura, C. Benedetti, C. B. Schroeder, Cs. T´oth, J. Daniels, D. E. Mittelberger, S. S. Bulanov, J.-L. Vay, C. G. R. Geddes, E. Esarey, Multi-GeV electron beams from capillary-discharge-guided subpetawatt laser pulses in the self-trapping regime, *Phys. Rev. Lett.* **113**, 245002 (2014).

15. H. Lu, M. Liu, W. Wang, C.Wang, J. Liu, A. Deng, J. Xu, C. Xia, W. Li, H. Zhang, X. Lu, C. Wang, J. Wang, X. Liang, Y. Leng, B. Shen, K. Nakajima, R. Li and Z. Xu, Laser wakefield acceleration of electron beams beyond 1 GeV from an ablative capillary discharge waveguide, Appl. Phys. Lett. 99, 091502 (2011).

16. A. J. Gonsalves, K. Nakamura, J. Daniels, C. Benedetti, C. Pieronek, T. C. H. de Raadt, S. Steinke, J. H. Bin, S. S. Bulanov, J. van Tilborg, C. G. R. Geddes, C. B. Schroeder, Cs. Tóth, E. Esarey, K. Swanson, L. Fan-Chiang, G. Bagdasarov, N. Bobrova, V. Gasilov, G. Korn, P. Sasorov and W. P. Leemans, Petawatt laser guiding and electron beam acceleration to 8 GeV in a laser-heated capillary discharge waveguide, Phys. Rev. Lett. 122, 084801 (2019).

17. J. Osterhoff, A. Popp, Z. Major, B. Marx, T. P. Rowlands-Rees, M. Fuchs, M. Geissler, R. Horlein, B. Hidding, S. Becker, E. A. Peralta, U. Schramm, F. Gruner, D. Habs, F. Krausz, S. M. Hooker, S. Karsch, Generation of stable Low-divergence electron beams by laser-wakefield acceleration in a steady-state-flow gas cell, Phys. Rev. Lett. 101, 085002 (2008).

18. B. Cros, C. Courtois, G. Matthieussent, A. Di Bernardo, D. Batani, N. Andreev, S. Kuznetsov, Eigenmodes for capillary tubes with dielectric walls and ultraintense laser pulse guiding, Phys. Rev. E 65, 026405 (2002).

19. K. Nakajima, Laser electron acceleration beyond 100 GeV, Eur. Phys. J. Special Topics 223, 999 (2014).

20. T. Behnke, J. E. Brau, B. Foster, J. Fuster, M. Harrison, J. M. Paterson, M. Peskin, M. Stanitzki, N. Walker and H. Yamamoto (eds.), The International Linear Collider Accelerator, in *The International Linear Collider Technical Design Report, Vol. 1, Executive Summary* (ILC international linear collider, 2013) pp. 9-26.

21. R. Budriünas, T. Stanislauskas, J. Adamonis, A. Aleknavičius, G. Veitas, D. Gadonas, S. Balickas, A. Michailovas and A. Varanavičius, 53 W average power CEP-stabilized OPCPA system delivering 5.5 TW few cycle pulses at 1 kHz repetition rate, Opt. Express 25, 5797 (2017).

22. D. Schulte, Application of advanced accelerator concepts for colliders, in Review of Accelerator Science and Technology, Vol. 9, eds. A. W. Chao and W. Chou, (World Scientific, Singapore, 2016), pp. 209-233.

23. S. Cheshkov, T. Tajima, W. Horton and K. Yokoya, Particle dynamics in multistage wakefield collider, Phys. Rev. ST Accel. Beams 3, 071301 (2000).

24. K. Yokoya, Quantum correction to beamstrahlung due to the finite number of photons, Nucl. Instrum. Methods A 251, 1 (1986).

25. R. J. Noble, Beamstrahlung from colliding elecfron-positron beams with negligible disruption, Nucl. Instrum. Methods A 256, 427 (1987).

26. R. Hollebeek, Disruption limits for linear colliders, Nucl. Instrum. Methods 184, 333 (1981).

27. P. Chen and K. Yokoya, Disruption effects from the interaction of round e+e- beams, Phys. Rev. D 38, 987 (1988).

28. A. Curcio, M. Petrarca, D. Giulietti and M. Ferrario, Numerical and analytical models to study the laser-driven plasma perturbation in a dielectric gas-filled capillary waveguide, Opt. Lett. 41, 4233 (2016).

29. B. A. Shadwick, C. B. Schroeder and E. Esarey, Nonlinear laser energy depletion in laser-plasma accelerators, Phys. Plasmas 16, 056704 (2009).

30. E. Esarey, C. B. Schroeder, W. P. Leemans, Physics of laser-driven plasma-based electron accelerators, Rev. Mod. Phys. 81, 1229 (2009).

31. W. Lu, C. Huang, M. M. Zhou, W. B. Mori, T. Katsouleas, Limits of linear plasma wakefield theory for electron or positron beams, Phys. Plasmas 12, 063101 (2005).

32. W. K. H. Panofsky, W. A. Wenzel, Some considerations concerning the transverse deflection of charged particles in radio-frequency fields, Rev. Sci. Instrum. 27, 967 (1956).

33. P. Chen, J. J. Su, T. Katsouleas, S. Wilks, J. M. Dawson, Plasma focusing for high-energy beams, IEEE Trans. Plasma Science PS-15, 218 (1987).

34. J. D. Jackson, Classical Electrodynamics, 3rd edn. (John Wiley & Sons, New York, 1999).

35. P. Michel, C. B. Schroeder, B. A. Shadwick, E. Esarey, W. P. Leemans, Radiative damping and electron beam dynamics in plasma-based accelerators, Phys. Rev. E 74, 026501 (2006).

36. N. Kirby, M. Berry, I. Blumenfeld, M. J. Hogan, R. Ischebeck, R. Siemann, Emittance growth from multiple coulomb scattering in a plasma wakefield accelerator, in Proc. of Particle Accelerator Conference 2007, (Albuquerque, New Mexico, USA, 2007), pp. 3097-3099.

37. A. Deng, K. Nakajima, J. Liu, B. Shen, X. Zhang, Y. Yu, W. Li, R. Li, Z. Xu, Electron beam dynamics and self-cooling up to PeV level due to betatron radiation in plasma-based accelerators, Phys. Rev. ST Accel. Beams 15, 081303 (2012).

38. A. G. Khachatryan, A. Irman, F. A. van Goor, K.-J. Boller, Femtosecond electron-bunch dynamics in laser wakefields and vacuum, Phys. Rev. ST Accel. Beams 10, 121301 (2007).

39. D. Strickland and G. Mourou, Compression of amplified chirped optical pulses, Opt. Commun., 56, 219 (1985).

40. G. Mourou, B. Brocklesby, T. Tajima, and J. Limpert, The future is fibre accelerators, Nat. Photonics, 7, 258 (2013).

41. J. Wheeler, G. Mourou, T. Tajima, Laser technology for advanced acceleration: accelerating beyond TeV, in Review of Accelerator Science and Technology, Vol. 9, eds. A. W. Chao, W. Chou, (World Scientific, Singapore, 2016), pp. 151-163.

42. R. Soulard, M. N. Quinn, and G. Mourou, Design and properties of a coherent amplifying network laser, Appl. Opt., 54, 4640 (2015).

43. L. Daniault, S. Bellanger, J. Le Dortz, J. Bourderionnet, Lallier, C. Larat, M. Antier-Murgey, J. C. Chanteloup, A. Brignon, C. Simon-Boisson, and G. Mourou, XCAN — A coherent amplification network of femtosecond fiber chirped-pulse amplifiers, Eur. Phys. J. Spec. Top., 224, 2609 (2015).

44. J. Le Dortz, A. Heilmann, M. Antier, J. Bourderionnet, C. Larat, I. Fsaifes, L. Daniault, S. Bellanger, C. Simon Boisson, J.-C. Chanteloup, E. Lallier, and A. Brignon, Highly scalable femtosecond coherent beam combining demonstrated with 19 fibers, Opt. Lett., 42, 1887 (2017).

45. A. Heilmann, J. Le Dortz, L. Daniault, I. Fsaifes, S. Bellanger, J. Bourderionnet, C. Larat, E. Lallier, M. Antier, E. Durand, C. Simon-Boisson, A. Brignon, and J.-C. Chanteloup, Coherent beam combining of seven fiber chirped-pulse amplifiers using an interferometric phase measurement, Opt. Express, 26, 31542 (2018).

46. G. Mourou, S. Mironov, E. Khazanov, and A. Sergeev, Single cycle thin film compressor opening the door to Zeptosecond-Exawatt physics, Eur. Phys. J. Spec. Top., 223, 1181 (2014).

47. L. Lavenu, M. Natile, F. Guichard, Y. Zaouter, X. Delen, M. Hanna, E. Mottay, and P. Georges, Nonlinear pulse compression based on a gas-filled multipass cell, Opt. Lett., 43, 2252 (2018).

48. D. Du, X. Liu, G. Korn, J. Squier, G. Mourou, Laser-induced breakdown by impact ionization in Si02 with pulse widths from 7 ns to 150 fs, Appl. Phys. Lett. 64, 3071 (1994).

49. J. Luo, M. Chen, W. Y. Wu, S. M. Weng, Z. M. Sheng, C. B. Schroeder, D. A. Jaroszynski, E. Esarey, W. P. Leemans, W. B. Mori, and J. Zhang, Multistage coupling of laser-wakefield accelerators with curved plasma channels, Phys. Rev. Lett. 120, 154801 (2018).

50. K. Nakajima, Seamless multistage laser-plasma acceleration toward future high-energy colliders, Light: Science & Applications 7, 21 (2018).

51. Z. Huang, P. Chen, R. D. Ruth, Radiation reaction in a continuous focusing channel, *Phys. Rev. Lett.* **74**, 1759 (1995).

Plasma Based Acceleration in Crystals and Nanostructures: Advantages and Limitations[*]

V. Lebedev

Fermi National Accelerator Laboratory, MS 312,
Batavia, IL, 60510, USA
val@fnal.gov

The paper presents a short overview of main physics phenomena which determine how the beam can be accelerated in plasma, crystal or a nanotube.

Keywords: Accelerators; crystals; carbon nanotubes; nanostructures.

1. Traditional Linacs versus Plasma-Based Linacs

Traditional linacs are based on the electromagnetic field excited inside cavities with well conducting walls. In difference to a plasma, which for small amplitudes responds at the plasma frequency only, cavities respond at a large number of modes oscillating at different frequencies. An intense bunch coming through these cavities gets energy from the fundamental accelerating mode and, consequently, reduces it amplitude. However, it also excites other modes which typically have larger frequencies and therefore are called the high order modes (HOMs). An excitation of these HOMs limits the efficiency of energy transfer for the case of single bunch operation. To mitigate these problem as well as to obtain reasonably small momentum spread in the accelerated bunch and high efficiency of energy transfer, the multiple bunches are used in accelerated. ILC[1] represents a typical example of such high efficiency acceleration where the energy transfer from a cavity field to a single bunch is about ~0.07%.

Although there are no HOMs in a plasma in a linear regime, the high efficiency acceleration in a plasma is also problematic due to beam interaction with plasma electrons located at the beam path.[2] Therefore, only acceleration in the strong bubble regime, where electrons are removed from the axis, suits such a choice. In this case the plasma response has low quality factor. Consequently, that limits the operation to the single bunch mode.

Theoretically, the energy transfer efficiency close to 100% is potentially achievable in the bubble regime. That requires accurate profiling of charge density along the bunch. However, even if this can be achieved, then the acceleration to high energy is limited by the beam break-up instability.[3] An increase of efficiency makes the bunch more unstable and limits the achievable energy gain. In the strong bubble regime the beam transverse stability is supported by very strong transverse focusing coming from the plasma itself, or

---

[*]Fermi National Accelerator Laboratory is operated by Fermi Research Alliance, LLC under Contract No. DE-AC02-07CH11359 with the United States Department of Energy.

in other words from the ions located at the beam path. The motion of these ions is not excited by plasma wave due to their large masses. Unfortunately, this focusing does not work for positrons. Therefore, an acceleration of bright positron bunches is presently infeasible.

2. Properties of Acceleration in a Plasma

In this section we shortly consider relationships connecting parameters of the beam and plasma in the strong bubble regime of acceleration. We will use the Lu equation[4] for description of bubble properties.[3] In the strong bubble regime, the plasma electrons are pushed out from the beam path so that there is a cavity formed around the bunch (or laser pulse) exciting the plasma. While electrons are absent in the cavity the motion of ions is still weakly affected. All the electrons are pushed out of the cavity and located in the thin layer surrounding it. The cavity has almost spherical shape and the electric field at its axis can be approximated by the following equation[3]:

$$E(\xi) \approx E_0 R_b k_p \frac{0.394\xi/\xi_b}{\sqrt[3]{1-(\xi/\xi_b)^2}} \ , \tag{1}$$

where R_b is the cavity radius, $\xi_b \approx 0.847 R_b$ is its half-length, the coordinate ξ is directed along the axis and is referenced to the bubble center, $k_p = \sqrt{4\pi n_e e^2/mc^2}$, is the plasma wave number, e and m are the electron charge and mass, n_e is the electron density, r_e is the classical radius of electron, and $E_0 = \sqrt{4\pi n_e e^2/r_e}$. Figure 1 shows dependence of dimensionless electric field, $R_E(\xi)=E(\xi)/(E_0 R_b k_p)$, on the coordinate from the bubble center.

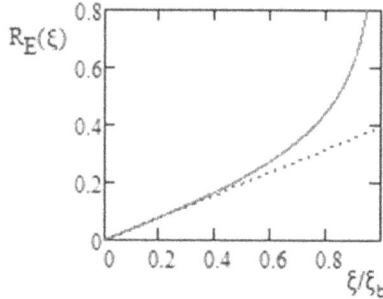

Fig. 1. Dependence of the dimensionless electric field on distance from the bubble center.

The power transferred to the bubble is uniquely determined by the parameter $R_b k_p$:

$$P = \frac{mc^3}{64 r_e}\left(k_p R_b\right)^4 \ . \tag{2}$$

Assuming that 100% of this energy is transferred to the particles accelerated by this wave, one can obtain an upper boundary of the maximum number of accelerated particles.

The beam focusing in the plasma is supported by the uncompensated space charge of plasma ions located at the beam path. The focusing strength is orders of magnitude stronger than what can be achieved with conventional magnets. The corresponding equilibrium beta-function is:

$$\beta_f = \frac{1}{k_p}\sqrt{2\gamma\frac{M}{m}} \ , \tag{3}$$

where M is the mass of accelerated particles, and γ is their relativistic parameter.

We also introduce the normalized acceptance of the bubble as following:

$$\varepsilon_n = \gamma\frac{R_b^2}{\beta_f} = \sqrt{\frac{m\gamma}{2M}}\frac{(k_p R_b)^2}{k_p} \ . \tag{4}$$

Scattering off the plasma ions results in the transverse emittance growth of the accelerated beam. The corresponding emittance increase is equal to:

$$\delta\varepsilon_n \approx \frac{Z^2 r_e L_c}{Z_v E_{acc}/E_0}\sqrt{\frac{M}{2m}}\left(\sqrt{\gamma_{fin}} - \sqrt{\gamma_{in}}\right) \ . \tag{5}$$

Here Z_v is the number of free electrons per atom, L_c is the Coulomb logarithm, and γ_{in} and γ_{fin} are the relativistic factors at the beginning and the end of the plasma-based linac.

All the above equations were written so that to express them through dimensionless parameters. The parameter $R_b k_p$ characterizes the strength of the perturbation. In further estimates we put $R_b k_p = 2$. In this case choosing a coordinate of accelerated bunch not to close to the bubble end we can estimate the accelerating field $E/E_0 = 1$ (see Fig. 1). Assuming these two parameters we obtain other relevant parameters. Their values are presented in Table 1 for the cases of typical plasma acceleration and for the acceleration in a solid silicon.

Table 1: Plasma and beam parameters for the cases of typical plasma acceleration and for an acceleration in solid silicon.

	Hydrogen plasma	Solid silicon
Nuclear density	10^{17}	$5 \cdot 10^{22}$
Z/A	1/1	14/28
Number of valence electrons	0	4
Electron density, n_e	10^{17}	$2 \cdot 10^{23}$
Basic wavelength, $2\pi k_p^{-1}$	100 μm	750 Å
Acceptance, mm mrad	$47\sqrt{\gamma m/M}$	$0.034\sqrt{\gamma m/M}$
"Maximum" field, E_0	300 MeV/cm	430 GeV/cm
Relative emittance growth, $(d\varepsilon_n/\varepsilon_n)$	$\sim 10^{-7}$	~ 0.003
Energy stored in plasma, mJ/GeV	60	0.04
Maximum number of accelerated particles[*]	$1.8 \cdot 10^8$	$1.3 \cdot 10^5$

* 100% energy efficiency is implied

Comparing the two columns one can see that there are the following challenges for acceleration in a solid medium.

First, to excite a plasma wave one needs to have the rms beam sizes of the driving beam less or about k_p^{-1} or <120 Å. Note that the longitudinal size may be larger if the envelope self-modulation instability of the drive bunch is used (like in the AWAKE experiment[5]), but in that case the excitation efficiency is strongly suppressed and the modulation depth is reduced making impossible an acceleration of a bright bunch.

Second, if the full amplitude of electric field is excited the field is so large that it will induce the impact ionization of inner shell of the atoms. It is not a problem in a hydrogen plasma implied in the first column.

Third, due to much larger accelerating gradient the maximum number of accelerated particles is quite small $\leq 10^5$.

Fourth, an acceleration of muons is greatly complicated by small acceptances of the plasma channel. This is correct for both the longitudinal and transverse planes. In particular, $\varepsilon_n \sim 34$ nm for 10 GeV muons and grows as $\sqrt{\gamma}$.

First proposals for multistage acceleration in plasma-based accelerators are under investigation now and face many challenges. Multi-stage acceleration in crystals does not seem feasible. There are two leading reasons. The first one is the focusing chromaticity at transition between consecutive accelerating stages; and the second one is stability of relative transverse and longitudinal positions of the driving and accelerating bunches. Therefore, continuous one-stage acceleration is critical and crystals and carbon nanostructures might have great potential in that regard.

The power efficiency is another important characteristic to be considered in the technology choice for an accelerator. Initial considerations of plasma acceleration in the bubble regime looked very promising. Computer simulations showed that in some conditions the power efficiency can achieve ~90% if accurate profiling of particle distribution along the bunch can be realized. However, more detailed consideration,[3] which accounts for the relationship between the longitudinal and transverse wakes in the plasma bubble, shows that an increase of power efficiency reduces the transverse beam stability. High power efficiency implies large charge in the accelerated bunch which, in its turn, generates strong transverse head-tail effects in the course of bunch acceleration. We will characterize the transverse wake strength by parameter η_t equal to the wake deflecting force at the bunch end for uniformly displaced bunch to the focusing force coming from plasma. Then the relationship between the transverse and longitudinal wakes binds η_t and the power efficiency of energy transfer from plasma to the beam, η_p, so that[3]:

$$\eta_t \approx \frac{\eta_P^2}{4(1-\eta_p)} . \tag{6}$$

An analysis of beam stability in a beam acceleration to very high energy requires $\eta_t \leq 0.01$. That yields $\eta_p \approx 18\%$.

3. Channeling in Crystals and Nanotubes

Particle channeling in a crystal greatly reduces the multiple scattering in a medium. It focuses positive particles to the channel center with gradient ~$4 \cdot 10^{17}$ V/cm^2 for a silicon crystal. An excitation of the plasma wave in this crystal in the strong bubble regime expels all "free" electron from the axis and introduces a charge density perturbation with the transverse size about two orders of magnitude larger than the channel width. For a silicon crystal the defocusing coming from the plasma wave is about half of the focusing coming from channeling. Figure 2 demonstrates an electric field between two crystal planes in silicon (left) and its change when plasma wave in a strong bubble regime is excited (right, four planes are shown). As one can see from the right pane only the central channel may be used for particle acceleration because the defocusing field of the plasma wave exceeds the focusing field of the channeling in other channels. The acceptance of the channel is about 4 orders of magnitude smaller compared to the acceptance of a plasma wave in the plasma of the same electron density. Here the particle charge is changed from positive to negative. Typical acceptance of single channel is about 5 pm for 10 GeV muons. We need to stress that channeling of negative particles implies particle focusing around the crystal plane/axis. That greatly amplifies multiple scattering. Also, an acceleration to high energy requires focusing in both transverse planes, and, consequently, the axial channeling must be used.

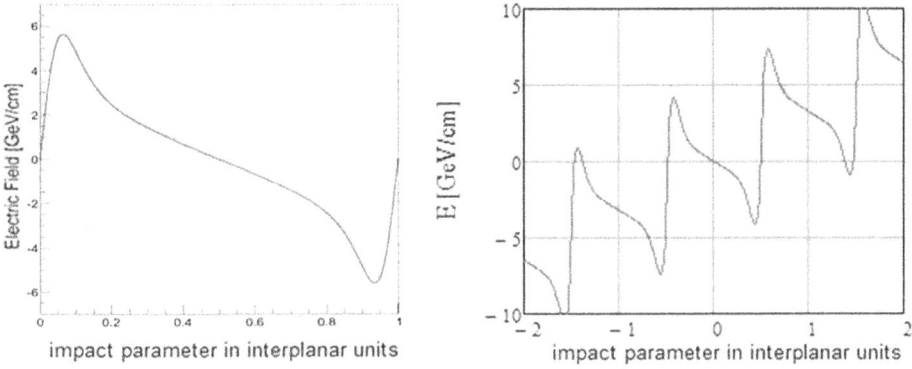

Fig. 2. Dependence of electric field on distance between two crystal planes in silicon taken from Ref. 6 (left) and its change when a plasma wave in a strong bubble regime is excited (right).

Although channeling removes scattering on the nuclei, the multiple scattering on electrons is still present. For silicon a reduction in scattering is about 2 orders of magnitude but the channel acceptance is 4 orders of magnitude smaller than in the plasma bubble operation in the absence of channeling. Note that both the normalized emittance growth due to scattering on the electrons,

$$\delta\varepsilon_n \approx \frac{r_e L_c}{(E_{acc}/E_0)}\sqrt{\frac{\gamma M}{2m}} \, , \tag{7}$$

and the normalized acceptance of channeling

$$\varepsilon_n \approx k_p a_0^2 \sqrt{\frac{m\gamma}{2M}} \tag{8}$$

grow as $\sqrt{\gamma}$. That implies that the scattering in the channel does not limit the maximum energy if

$$\frac{r_e L_c}{\left(E_{acc}/E_0\right)}\frac{M}{m} \le k_p a_0^2 \ . \tag{9}$$

To fulfil this requirement in the acceleration of μ+, the required electric field has to be $E_{acc}/E_0 > 10$; i.e. the acceleration at the maximum possible rate is desirable. It is more forgiving for positrons which have smaller mass.

The hollow beam channel[7] was proposed for electron-positron plasma accelerators to overcome scattering on the ions and to prevent the pinching of ions by the electric field of bright electron bunch. However, this method usage is limited to an acceleration of small intensity beams. Obtaining the transverse stability in the presence of large transverse impedance of the plasma channel requires considerable momentum spread in the bunch and very strong focusing which can be supported by plasma focusing only. External focusing, supplied by conventional magnets, will be 3–4 orders of magnitude weaker.

An acceleration in carbon nonotubes is somewhat similar to the hollow channel. It also suits well for acceleration of relatively small intensity bunch. The average density in a bunch of nanotubes is one to two orders of magnitude smaller than in a crystal. Although that decreases the maximum achievable electric field, it still stays at a very high value. What is more important, usage of nanotubes should greatly reduce the multiple scattering which only happens when a particle approaches the tube wall to be reflected from it. Similar to channeling in crystals, only positively charged particles are reflected from the nanotube walls. Although the reflections from walls represent very non-linear focusing, the beam transverse emittance is conserved in the acceleration of small intensity beam: the transverse size and transverse momentum are not changed. However, with the intensity increasing, absence of the focusing in the center of the channel may lead to the emittance growth due to the beam breakup instability.

4. Conclusions

Plasma-based acceleration is a very interesting subject. A beam acceleration in a solid material represents a tremendous step in technology of plasma acceleration. Presently it looks as an extremely challenging avenue for construction of high luminosity collider. However, acceleration of small intensity bunch to extremely high energy looks as an interesting possibility. Although the basic principles and limitations for such accelerator are similar to the already demonstrated plasma-based accelerators the experimental demonstration of such a system, if possible, will require considerable time and effort.

References

1. T. Behnke *et al.* (eds.), *The International Linear Collider. Technical Design Report. Volume 1: Executive Summary*, FERMILAB-TM-2554, ISBN 978-3-935702-74-4, http://lss.fnal.gov/archive/test-tm/2000/fermilab-tm-2554.pdf.

2. V. Lebedev, A. Burov and S. Nagaitsev, Luminosity limitations in linear colliders, based on plasma acceleration, *Rev. Accel. Sci. Tech.* **9**, 187–207 (2016), doi:10.1142/S1793626816300097.

3. V. Lebedev, A. Burov and S. Nagaitsev, Efficiency versus instability in plasma accelerators, *Phys. Rev. Accel. Beams* **20**, 121301 (2017), arXiv:1701.01498.

4. W. Lu, C. Huang, M. Zhou, W. B. Mori, and T. Katsouleas, Nonlinear theory for relativistic plasma wakefields in the blowout regime, *Phys. Rev. Lett.* **96**, 165002 (2006).

5. E. Adli *et al.*, Acceleration of electrons in the plasma wakefield of a proton bunch, *Nature* **561**, 363–367 (2018) doi:10.1038/s41586-018-0485-4.

6. E. Bagli, V. Guidi and V. A. Maisheev, *Phys. Rev. E* **81**, 026708 (2010).

7. C. B. Schroeder, C. Benedetti, E. Esarey and W. P. Leemans, Beam loading in a laser-plasma accelerator using a near-hollow plasma channel, *Phys. Plasmas* **20**, 123115 (2013); https://doi.org/10.1063/1.4849456.

Carbon Nanotube Accelerator — Path Toward TeV/m Acceleration: Theory, Experiment, and Challenges[*]

Young-Min Shin

Department of Physics, Northern Illinois University (NIU),

DeKalb, IL 60115, USA

alcolpeter@gmail.com

Aspirations of modern high energy particle physics call for compact and cost efficient lepton and hadron colliders with energy reach and luminosity significantly beyond the modern HEP facilities. Strong interplanar fields in crystals of the order of 10–100 V/Å can effectively guide and collimate high energy particles. Besides continuous focusing crystals plasma, if properly excited, can be used for particle acceleration with exceptionally high gradients $O(TeV/m)$. However, the angstrom-scale size of channels in crystals might be too small to accept and accelerate significant number of particles. Carbon-based nano-structures such as carbon-nanotubes (CNTs) and graphenes have a large degree of dimensional flexibility and thermo-mechanical strength and thus could be more suitable for channeling acceleration of high intensity beams. Nano-channels of the synthetic crystals can accept a few orders of magnitude larger phase-space volume of channeled particles with much higher thermal tolerance than natural crystals.

This paper presents conceptual foundations of the CNT acceleration, including underlying theory, practical outline and technical challenges of the proof-of-principle experiment. Also, an analytic description of the plasmon-assisted laser acceleration is detailed with practical acceleration parameters, in particular with specifications of a typical tabletop femtosecond laser system. The maximally achievable acceleration gradients and energy gains within dephasing lengths and CNT lengths are discussed with respect to laser-incident angles and the CNT-filling ratios.

Keywords: Carbon nanotubes; CNT; channeling; crystals; TeV/m; acceleration.

1. Introduction

1.1. *Background*

Presently, international high energy community actively discusses options for a post-Large Hadron Collider (LHC) machine, such as the International Linear Collider (ILC), Compact Linear Collider (CLIC), or Future Circular Collider (FCC). These machines can sufficiently expand the center-of-mass (CM) energy reach and luminosity for the next generation particle physics experiments.[1,2,3] The next energy frontier super-colliders are envisioned to achieve ~ 100 TeV in CM energy and $O(10^{35}$ cm^{-2}s$^{-1})$ luminosity. To be feasible, such CM energy and luminosity should be achieved within a realistic facility footprint and within reasonable AC wall plug power envelope, with the maximum possible efficiency of the power conversion to the beams. Ideally, the future machine should be about 10 kilometer or shorter, and operate with beam power of less than a few tens of

[*]This work was supported in part by the DOE Contract No. DE-AC02-07CH11359 to the Fermi Research Alliance LLC.

MWs[4]. How can such colliders reach the energies of interest, namely, 100–1000 TeV? Currently available acceleration technologies cannot be extended to meet the challenge. To obtain the energies of interest within the given footprint, one has to develop new methods of ultra-fast acceleration with tight phase-space control of high-power beams for future lepton and hadron colliders and it can only be possible with multidisciplinary and multilateral approaches.

1.2. Crystal acceleration —TeV/m gradient

Electromagnetic wakefield waves in an ionized plasma media, excited by short relativistic bunches of charged particles or by short high power laser pulses, have been of great interest due to the promise to offer extremely high acceleration gradients of G (max. gradient) $= m_e c \omega_p / e \approx 96 \times n_0^{1/2}$ [V/m], where $\omega_p = (4\pi n_p e^2 / m_e)^{1/2}$ is the electron plasma frequency and n_p is the ambient plasma density of [cm^{-3}], m_e and e are the electron mass and charge, respectively, and c is the speed of light in vacuum. However, a practically obtainable plasma density (n_p) in ionized gas is about $\sim 10^{18}$ cm^{-3}, which in principle corresponds to wakefields up to ~ 100 GV/m,[5,6,7] and it is realistically very difficult to create a stable gas plasma with a charge density beyond that. Metallic crystals offer naturally existing dense plasma media which is completely full with a large number of conduction electrons available for the wakefield interactions. The density of charge carriers (conduction electrons) in solids $n_0 = \sim 10^{20} - 10^{23}$ cm^{-3} is significantly higher than what was considered above in gaseous plasma, and correspondingly the wakefield strength of conduction electrons in solids, if excited, can possibly reach 10 TV/m in principle.

Figure 1 shows our simulation graphs of a beam-driven wakefield acceleration in a homogeneous plasma model with solid-level charge densities, 10^{25} m^{-3} and 1.6×10^{22} cm^{-3}, which might be in the lower and upper limits of electron density of solid-state media (acceleration gradient and energy gain versus beam charge density).[8] The densities are considered for the assessment as the corresponding plasma wavelengths, 10 μm and 0.264 μm, are also in the spectral range of available beam-modulating sources such as IR/UV lasers or magnetic undulators (inverse FELs)[9] for strong beam-plasma coupling in the beam-driven acceleration. As shown in the figure, with the beam-driven acceleration the acceleration gradient ranges from 0.1 TeV/m to 10 TeV/m with the solid plasma densities at the linear beam-plasma coupling condition ($n_b \sim n_p$). The gradient corresponds to 10–30 MeV of energy gain with a channel length of 10 plasma wavelengths. Also, it appears that the beam is strongly focused and collimated by the large transverse fields generated from the oscillating plasma over the distance (Figs. 1(c) and (d)). Apparently, this effective plasma simulation model shows TeV/m range gradient with solid state level plasma density. However, in spite of the exceptionally large amplitude of accelerating and focusing fields, in reality if charged particles are injected into either an amorphous solid or a random orientation of a single phase crystal, they will encounter a large stopping power from nuclear and electron scatterings. Also, irregular head-on collisions at atomic sites completely spread the particles all over, accompanied with phase-space volume expansion. Particles will thus be quickly repelled from a driving field due to fast pitch-angle diffusion of large scattering rates/angles from non-uniformly distributed atoms.

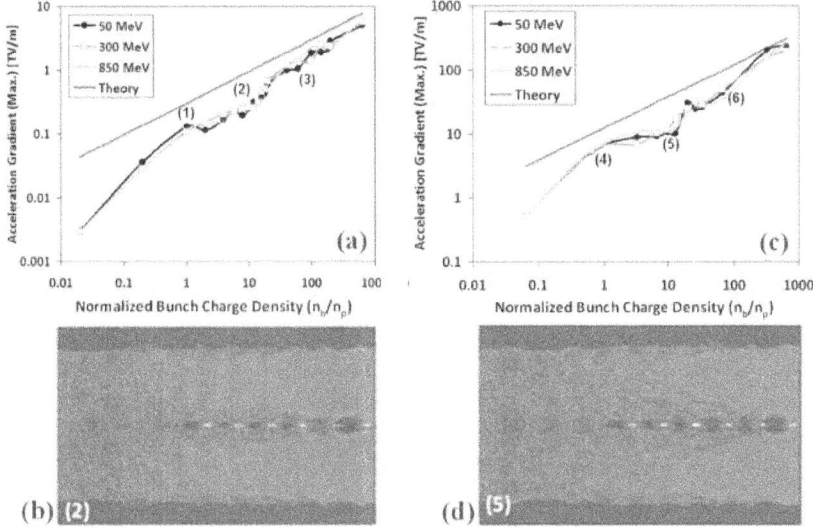

Fig. 1. Acceleration gradient versus normalized charge density graphs (top: (a) and (c)) and beam/plasma charge distributions (bottom: (b) and (d)) of multi-bunched beam with (a) and (b) $n_p = 10^{19}$ cm^{-3} and (c) and (d) $n_p = 1.6 \times 10^{22}$ cm^{-3}.

The channels between atomic lattices or lattice planes aligned in a crystal orientation of natural crystals like silicon or germanium are sparse spaces with relatively low electron densities.[10] Charged particles, injected into a crystal orientation of a mono-crystalline (homogeneous and isotropic) target material, undergo much lower nucleus and electron scatterings. The channeled particles can be accelerated in two different ways, depending upon how they gain energies from driving sources (lasers or particle-beams) through plasma wakefields of conduction electrons, or diffracted EM fields confined in the crystal channels. Plasma wakefields in crystals, if excited by a strong driving source, either a short particle-bunch or a high power laser, can accelerate particles along the space in the lattice channel. For the channeling acceleration with confined x-ray diffraction, the atomic channel can hold $> 10^{13}$ V/cm transverse and 10^9 V/cm longitudinal fields of diffracted traveling EM-waves from an x-ray laser coupled to a crystal at the Bragg diffraction angle ($\lambda/2a = \sin\theta_B$), where a is the lattice constant and θ_B is the diffraction angle). However, to accelerate channeled particles with high gradients by confined x-ray fields, the acceleration requires coherent hard x-rays ($\hbar\omega \approx 40$ keV) of power density $\geq 3 \times 10^{19}$ W/cm^2 to overcome radiation losses of channeled muons,[11] which exceed those conceivable today. The x-ray pumping method thus fits for heavy particles, e.g. muons and protons, which have relatively smaller radiation losses. For electrons (and positrons), the beam-driven acceleration is more favorably applicable to channeling acceleration as the energy losses of a drive beam can be transformed into the acceleration energy of a witness beam.[12] The maximum particle energies, $E_{max} \approx (M_b/M_p)^2(\Lambda G)^{1/2}\{G/(z^3\times100$ [GV/cm])$\}^{1/2}10^5$ [TeV], can reach 0.3 TeV for electrons/positrons, 10^4 TeV for muons, and 10^6 TeV for protons (M_b is the mass of the channeled particle, M_p is the proton rest mass, Λ is the de-channeling length per unit energy, and z is the charge of the channeled particle).[13] Here, channeling radiation of betatron oscillations between atomic planes is the major source of energy dissipation. The de-channeling length (Λ) is also a critical factor, in particular in the low energy regime, since it scales as $E^{1/2}$.[14] The idea of accelerating charged particles in solids along major crystallographic directions was suggested by several scientists such as Pisen Chen, Robert Noble, Richard Carrigan, and Toshiki Tajima in the 1980's and 1990's[15,16,17,18] for the possible advantage that periodically aligned electrostatic potentials in crystal lattices are capable of providing a channeling effect[19,20,21,22,23,24,25] in combination with low emittance determined by an Ångström-scale aperture of the atomic "tubes". The basic concepts of atomic accelerator with short pulse driving sources like high power lasers or ultra-short bunches have been considered theoretically. However, the idea has never been demonstrated by experiment or simulation due to the extremely tight interaction condition of the Angstrom-size atomic channels in natural crystals and the complexity of electron dynamics in solid-plasma.

2. Channeling Acceleration in Carbon Nanotubes

Carbon nanotubes (CNTs) are a synthetic nanostructure, which is a roll of a graphene sheet, and its tube diameter can be easily increased up to sub-micron by optimizing fabrication processes (chemical vapor deposition, CVD). For channeling applications of high-power beams, carbon nanostructures have various advantages over crystals.[26] Particles are normally de-channeled when the transverse forces are larger than the maximal electric field acting on channeled particles from crystal atoms, which is described by the critical angle (ε_{cn} (normalized rms acceptance) = $(1/2)\gamma a\theta_c$, where θ_c is the critical angle). The dechanneling rate is significantly reduced and the beam acceptance is dramatically increased by the large size of the channels, e.g., a 100 nm wide CNT channel has larger acceptance than a silicon channel by three orders of magnitude. Some previous studies[27] on the radiative interaction in a continuous focusing channel present an efficient method to damp the transverse emittance of the beam without diluting the longitudinal phase space significantly. CNTs can efficiently cool channeled particles similar to natural crystals. The strong focusing in a CNT channel results in a small beta function, so that it is quite feasible to create a beam of small transverse emittance.

If the channel size is increased from angstroms to nanometers (Figs. 2(a) and (b)), the maximally reachable acceleration gradient would be lowered from ~ 100 TeV/m to ~ 1 TeV/m due to the decrease of effective plasma charge density. However, the nanotube channels still provide sufficiently large transverse and longitudinal fields in the range of TV/m. For the crystal channels in angstrom scale, the lattice dissociation time of atomic structures ($\Delta t \approx \sqrt{(m_i/m_e)}(2\pi/\omega_p)$, where m_i and m_e are the masses of ion (carbon) and electron respectively[28]) is in the range of sub-100 fs with 1 TV/m fields, corresponding to 10^{19} W/cm^3. For beam driven acceleration, a bunch length with a sufficient charge density would need to be in the range of the plasma wavelength to properly excite plasma wakefields, and channeled particle acceleration with the wakefields must occur before the ions in the lattices move beyond the restoring threshold and the atomic structure is fully destroyed. It is extremely difficult to compress a particle bunch within a time scale of femto-seconds since the bunch charge required for plasma wave excitation and the beam power corresponding to the time scale will exceed the damage threshold of the crystal. The disassociation time is, however, noticeably extended, to the order of pico-seconds, by increasing the channel size to nanometers because the effective plasma density and corresponding plasma frequency are decreased by a few orders of magnitude, as shown in Fig. 2(a). The constraint on the required bunch length is thus significantly mitigated (Fig. 2(c)) and the level of power required for an external driving source could be lowered by a few orders of magnitude, although the acceleration gradient will be lowered accordingly.[29]

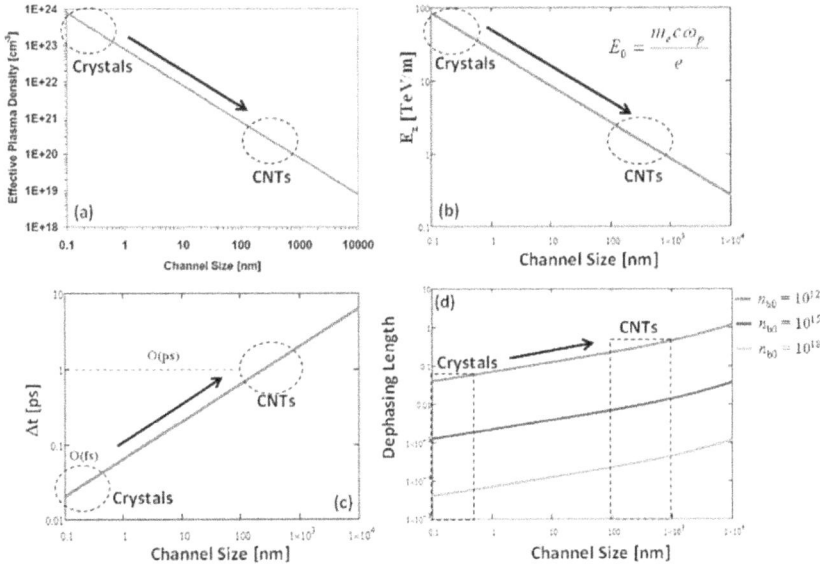

Fig. 2. (a) Effective plasma density, (b) acceleration gradient, (c) dissociation time scale, and (d) dephasing length versus channel size (carbon-based).

Furthermore, dephasing length[30] is appreciably increased with the larger channel, which enables channeled particles to gain a sufficient amount of energy (Fig. 2(d)). The atomic channels in natural crystals, even if they provide extremely high potential gradients, are limited to angstroms and are unchangeable due to the fixed lattice constants. The fixed atomic spaces make the channeling acceleration parameters impractically demanding, but CNTs could relax the constraints to more realistic regimes. The carbon structures comprised entirely of covalent bonds ($sp2$) are extremely stable and thermally and mechanically stronger than crystals, steel, or even diamonds (*sp3* bond). It is known that thermal conductivity of CNTs is about 20 times higher than that of natural crystals (e.g. silicon) and the melting point of a freestanding single-tip CNT is 3,000–4,000 Kelvin. Carbon based channels thus have significantly improved physical tolerance against intensive thermal and mechanical impacts from high power beams.

3. Theory

3.1. *Beam-driven acceleration*

Figure 3 shows time-tagged snapshots of a two-beam accelerating system with a ~$10\lambda_p$ hollow plasma channel that is modeled with $n_p \sim 10^{25}$ m^{-3}. The simulation condition also includes the drive-witness coupling distance of ~ $1.6\lambda_p$, $\sigma = \sim 0.1\lambda_p$. and a linear regime bunch charge density of $n_b \sim n_p$. For this simulation, the plasma channel is designed with a tunnel of $r = 0.1\lambda_p$. Just like a uniformly filled one, the drive bunch generates tailing wakes in the hollow channel due to the repulsive space charge force. The plasma waves travel along the hollow channel with the density modulation in the same velocity with the witness beam. Note that the bunch shape of the drive beam in the tunnel remains relatively longer than in the cylindrical plasma column. The energy versus distance plots in Fig. 3 (bottom) shows that sinusoidal energy modulation apparently occurs in the plasma channel perturbed by two bunches. Here, the relative position of the drive beam corresponds to the first maximum energy loss, while that of the witness beam does to the first maximum energy gain. The traveling wakes around the tunnel continuously transform acceleration energy from the drive beam to the witness one.

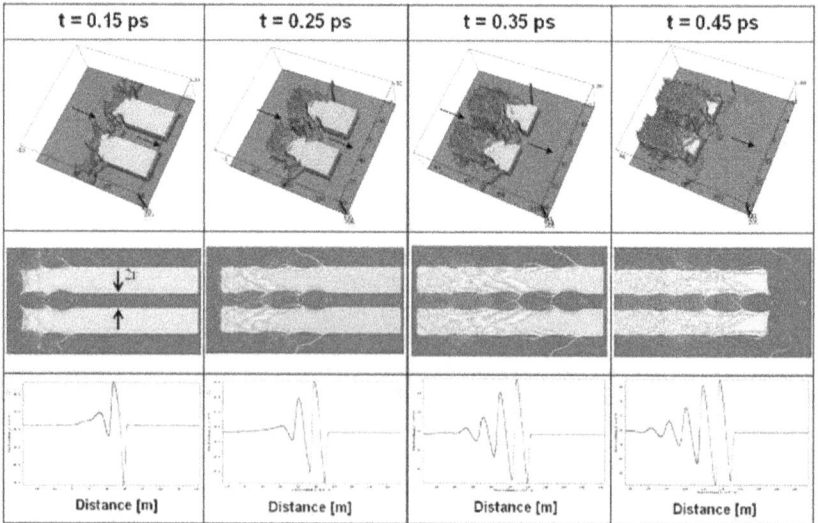

Fig. 3. Time-tagged charge distribution of a hollow plasma acceleration ($n_p = n_b$) with a drive and witness beam (top) 3D charge distribution (middle) 2D distribution (bottom) spatial energy distribution.

The energy gain and acceleration gradient are fairly limited by the radius and length of the tunnel with respect to plasma wavelength and bunch charge density. Bunch parameters of the beam-driven acceleration system have thus been analyzed with various tunnel radii, as shown in Fig. 4. For the analysis, the bunch charge density was swept from 1 to 300, normalized by plasma density, n_p, for five different tunnel radii from 0.2 to $0.6\lambda_p$ and relativistic beam energy 20MeV. While in the linear regime $n_b = \sim 1\text{--}10n_p$, the maximum acceleration gradient drops off with an increase of the tunnel radius from 0.2 to 0.6, it increases in the blowout regime, $n_b = \sim 10\text{--}100n_p$. The maximum acceleration gradient is increased from ~ 0.82 TeV/m of $r = 0.2\ \lambda_p$ to ~ 1.02 TeV/m of $r = 0.6\ \lambda_p$ with $n_b = 100n_p$, corresponding to $\sim 20\%$ improvement. The energy transformer ratio follows a similar tendency with the acceleration gradient curve in the linear and blowout regimes. In the linear regime ($n_b/n_p \sim 1\text{--}10$), scattering is negligibly small, which does not perturb particle distribution of the bunch within the hollow channel. The repulsive space charge force between the bunch and the plasma is increased in the inversely proportional to their spacing. The channeled bunch thus undergoes the higher acceleration gradient as the channel gets smaller, as shown in Fig. 4. However, in the blowout regime ($n_b/n_p \sim 10\text{--}$ 100), the repulsive space charge force from an excessive amount of the bunch charge density against the plasma channel is strong as to heavily perturb the bunch and to scatter electrons out of the bunch. The strength of space charge force is decreased with the channel radius, so the electrons in the bunch is less scattered with an increase of the channel radius. The gradient is thus lowered with an increase of the channel radius accordingly. The similar tendency also appears on the transformer ratio, as shown in Fig. 4(b). The energy is more efficiently converted from the drive beam to the witness one with the larger channel in the blowout regime, although the transformer ratio does not similarly follow the tendency of accelerating gradient with the channel size in the linear regime. The un-similarity between the acceleration gradient and transformer ratio with respect to the channel size in the linear regime might be attributed to the insufficient channel length, $10\lambda_p$. The plasma oscillation from the small bunch charge density is not strong enough to properly convert the beam energies from deceleration to acceleration. The result implies that a hollow channel thus has a higher gradient than a homogeneously filled plasma column in the blowout regime, and the plasma wakefield acceleration gradient is effectively increased by enlarging the channel size. This result opens the possibility of controlling beam parameters of plasma accelerators for higher gradient and large energy conversion efficiency.

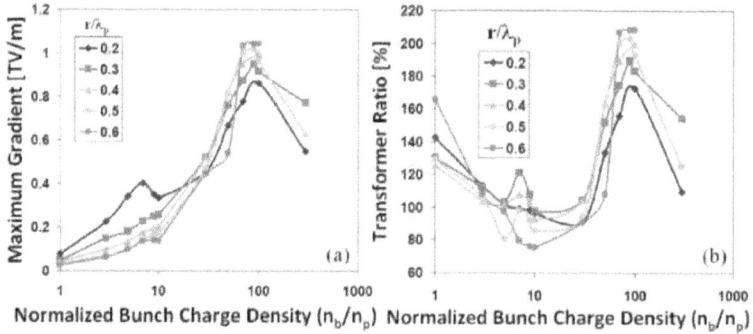

Fig. 4. (a) maximum acceleration gradient and (b) transformer ratio versus bunch charge distribution normalized by bunch charge density with various tunnel radii ($r = 0.2–0.6\lambda_p$).

3.2. Laser-driven acceleration

Figure 5 depicts the concept of the plasmon-driven acceleration in a laser-pumped nanotube. In the substrate target, particles channeled in the nanotube are repeatedly accelerated and focused by the confined fields of the laser-excited plasmon along the nanotubes embedded in the nano-holes under the phase-velocity matching condition. The energy gain of accelerated particles, if any, is limited by the dephasing length. Continuous phase velocity matching between particles and quantized waves can be extended by tapering the longitudinal plasma density in a target. In a CNT-target, the longitudinal plasma density profile can be controlled by selectively adjusting the tube dimensions. When an intense short pulse laser illuminates the near-critical density plasma, the inductive acceleration field moves with a speed v_g, which is less than v_p, depending on the plasma density: $v_g = c\sqrt{1 - \omega_p^2/\omega^2}$, where c is the speed of light, ω_p is the electron plasma frequency, and ω is the laser frequency. The accelerating ions have a progressively higher speed along the targets, so the inductively accelerated ions are kept accelerating for a long time inside the near-critical density plasma target. The distance between the two adjacent targets are also adjusted accordingly. The acceleration mechanism of the laser-excited sub-λ plasmon is conceptually depicted in Fig. 5, illustrating accelerating particles in a laser-pumped CNT channel. The laser irradiating a target modulates the electron gas along a tube wall and quickly induces a plasma oscillation a one-dimensional Fermi-liquid, Luttinger-liquid. The photo-excited density fluctuation induces electromagnetic fields in the CNT and the oscillating evanescent fields penetrate in the tube within the attenuation length.[31,32] The charged particles channeling through a tube are accelerated by the induced plasmons at their phase-matching condition. The energy gain remains until the particles begin to outrun the plasma wave and are dephased from the wave. A proper target thickness would therefore be mostly determined by the dephasing length if there is no additional phase-matching mechanism implemented in the single target acceleration.

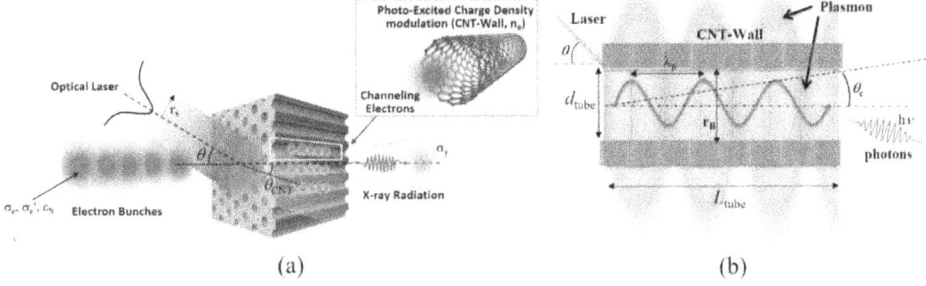

Fig. 5. Conceptual drawings of (a) optically pumped nanotube (b) x-ray radiation of photo-excited CNT.

The effective density of an electron plasma over CNTs is mostly controlled by tube diameter, number of walls, and spacing between the tubes in a unit area ($\sim\lambda_L^2$). The lattice constant of carbon-bonding in a honey-comb unit cell on a CNT wall is $a_0 \sim 1.4$ Angstrom and wall-to-wall spacing is normally 3.4 Angstrom. A local tube wall density is about 8×10^{22} cm^{-3} and the typical diameter of a CNT ranges up to a few hundred nanometers, which is usually related to the tube length. Given that a single CNT is sized from a few tens of nanometers up to 1 μm in diameter, which would be effectively the same as a few hundred square nanometers to a few square microns of unit area on a target, the effective electron plasma density averaged over a volume of CNT ranges from 1×10^{21}–6×10^{23} e/cm^3. The density corresponds to 10^{20}–10^{22} e/cm^3 over a CNT-embedded unit area ($\sim \lambda_L^2$), as depicted in Fig. 5.

In the given condition, a CNT channel can be described by a homogenized model with effective dielectric parameters. Let us consider an array of parallel nanotubes of areal density N_c, with axes parallel to z (Fig. 5). The separation between the nanotubes is $d = N_c^{-1/2}$, the free electron density inside a nanotube is n_e, and the tube radius is r_c. A nanotube can be treated as a superposition of overlapping cylinders of free electrons and immobile ions. The dispersion/absorption relation of the periodic array is given by

$$\kappa = k_r + ik_i , \tag{1}$$

where

$$k_r(\omega) = \frac{\omega}{c}\sqrt{\frac{\left(\varepsilon_L-\dfrac{\omega'^2_p}{\omega^2-\omega_p^2/2}\right)\left(\varepsilon_L-\dfrac{\omega'^2_p}{\omega^2-\omega_p^2}\right)}{\varepsilon_L-\left(\omega'^2_p/\omega^2-\omega_p^2\right)\left(\cos^2\theta+\dfrac{\omega^2}{\omega^2-\omega_p^2/2}\sin^2\theta\right)}} \tag{2}$$

and

$$k_i(\omega) = \frac{\omega^3 v\omega'^2_p}{2c^2\left(\omega^2-\omega_p^2\right)^2 k_r}\left(\frac{\left(\varepsilon_L-\dfrac{\omega'^2_p}{\omega^2-\omega_p^2/2}+\left(\varepsilon_L-\dfrac{\omega'^2_p}{\omega^2-\omega_p^2}\right)\dfrac{\left(\omega^2-\omega_p^2\right)^2}{\left(\omega^2-\omega_p^2/2\right)^2}\right)}{\varepsilon_L-\dfrac{\omega'^2_p}{\omega^2-\omega_p^2}\left(\cos^2\theta+\dfrac{\omega^2-\omega_p^2}{\omega^2-\omega_p^2/2}\sin^2\theta\right)}\right.$$
$$\left. -\frac{\left(\varepsilon_L-\dfrac{\omega'^2_p}{\omega^2-\omega_p^2/2}\right)\left(\varepsilon_L-\dfrac{\omega'^2_p}{\omega^2-\omega_p^2}\right)+\left(\cos^2\theta+\dfrac{\left(\omega^2-\omega_p^2\right)^2}{\left(\omega^2-\omega_p^2/2\right)^2}\sin^2\theta\right)}{\left(\varepsilon_L-\dfrac{\omega'^2_p}{\omega^2-\omega_p^2}\left(\cos^2\theta+\dfrac{\omega^2-\omega_p^2}{\omega^2-\omega_p^2/2}\sin^2\theta\right)\right)^2}\right) \tag{3}$$

Here, $\omega_p = \sqrt{\dfrac{n_e e^2}{\varepsilon_0 m}}$, $n_e = Z n_0$, and $\omega'^2_p = s\omega^2_p$ (n_0 is the ion density of a single CNT and $s = \pi r_r^2 / d^2$ is the aerial CNT-filling ratio). For $\varepsilon_L = 5.5$ (for graphite), $n_e = 10^{21}$ cm^{-3}, $r_c = 50$ nm, $d = 150$ nm, $\pi r_c^2 N_c \sim 0.35$, and $v/\omega_p \sim 0.001$, the dispersion/absorption relations of a CNT-confined plasmon with respect to a p-polarized laser is plotted in Fig. 2.

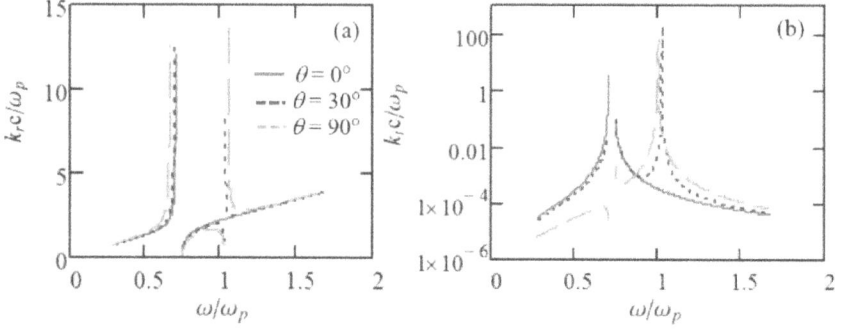

Fig. 6. Normalized (a) dispersion (k_r) and (b) absorption (k_i) graphs of the laser-excited SPP in a CNT array with respect to laser-incident angles ($\theta_{in} = 0°$ (red), 30° (blue), and 90° (green)).

It is apparent that confined modes are excited at the harmonic excitation conditions with the light line (laser photon). At the resonance condition, the laser-light is thus coupled into the sub-wavelength CNT and the electronic density on the wall is modulated with the laser wavelength ($\lambda_L = \lambda_p > r_{cnt}$, where r_{cnt} is the radius of CNT). The laser-excited plasmonic wave moves along the tube within the absorption length and subsequently forms a standing plasma wave (plasma oscillation) when the photon-plasmon energy transfer reaches equilibrium. With a sufficiently narrow energy spread, the particles channeled in the tube, if simultaneously injected into the CNTs during the excitation, can be accelerated by the quantized fields confined in the sub-λ CNT at the phase-velocity matching condition.

In general, with $\lambda_L = 1.056$ μm and 5 mJ of pulse energy, $P_L = 125$ GW of laser power (pulse duration: $\tau = 40$ fs) would be a maximally affordable laser power from a tabletop-scale femtosecond laser system. The minimum laser spot size on a target is determined by the damage threshold of the target material. An anodized aluminum oxide (AAO) membrane can be a good target material as it is a naturally formed capillary substrate with periodic nano-holes. Straight CNTs can be vertically grown along the holes with a high dimensional aspect ratio up to 1 : 1000, e.g. 100 nm in diameter and 100 μm in length. The CNT-embedded AAO substrate (AAO-CNT) would efficiently transport the charged particles through a nano-channel in the range of a micrometer in length. Channeling through AAO-CNTs was already demonstrated with an H+ ion beam by Zhu.[33] The substrate material, aluminum oxide (Al$_2$O$_3$), is known to have the highest ablation threshold, ranging from 2.5–3 × 10^{13} W/cm^2. A laser beam illuminated on an AAO-CNT target can be focused down to a spot size of $r_L = 350$ μm while still avoiding target damage. The laser spot size varies in distance with respect to the Rayleigh length which is a function of the laser wavelength, so that

$$r_s(z) = r_L \sqrt{1 + \left(\frac{z}{z_R}\right)^2} \qquad (4)$$

where $z_R = \frac{\kappa r_L^2}{2}$ is the Rayleigh length and $\kappa = \frac{2\pi}{\lambda_L}$ is the wave number of the driving laser. As shown in Fig. 7(a), the laser beam size remains within 450 μm over 0.7 m, so that the laser intensity will remain relatively constant over the distance. Let us consider the interaction of a laser with nanotubes. The laser beam has a Gaussian field distribution along the propagating direction such as

$$E_L = A_0 e^{-r^2/4r_s^2} e^{-i(\omega t - kz)} , \qquad (5)$$

where $A_0 = \sqrt{\frac{2I_L}{\varepsilon_0 c}} = 15$ [GV/m] with $I_L = 2.75 \times 10^{13}$ W/cm², as plotted in Fig. 7(b).

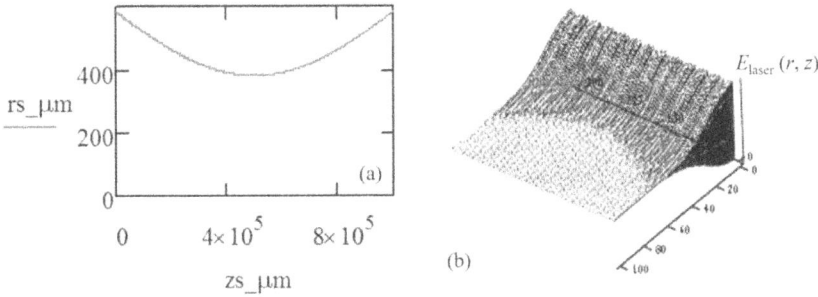

Fig. 7. (a) Laser beam size versus propagating distance and (b) electric field plot of laser with a Gaussian distribution.

As illustrated in Fig. 5, if the laser beam is coupled into the CNT array with an incident angle, θ, then the electric field of a p-polarized laser beam is

$$\vec{E}_L = \hat{x} E_x + \hat{z} E_z \qquad (6)$$

where $E_x = A_x e(x, z)$ and $E_z = A_z e(x, z)$ ($A_x = A_0 \sin\theta$ and $A_x = A_0 \cos\theta$). At a laser-coupling condition, the electron plasma density of a nanotube modulated by the laser-excited LLP is defined as

$$n_e = Z n_0 \left(1 + a_0 e^{-\frac{r^2}{2r_s^2}}\right)^{-2} e^{i(\kappa z - \omega_{laser} t)} \qquad (7)$$

where

$$a_0 = \frac{e A_0 \cos\theta \cdot \sqrt{s}}{m(\omega_{laser}^2 - \omega_p^2/2)r_c} . \qquad (8)$$

Also, the transmitted wave has the electric field components

$$E_x = \frac{Z n_0 e}{\varepsilon_0} a_0 \xi e^{-\frac{r^2}{2r_s^2}} e^{i(\kappa z - \omega_{laser} t)} \cos\theta_{CNT} , \quad E_z = \frac{Z n_0 e}{\varepsilon_0} a_0 \xi e^{-\frac{r^2}{2r_s^2}} e^{i(\kappa z - \omega_{laser} t)} \sin\theta_{CNT} , \qquad (9)$$

where ξ is the distance from the tube axis and θ_{CNT} is the refraction angle of the transmitted wave in the substrate, defined as

$$\theta_{CNT} = \sin^{-1}\left(\frac{\sin\theta}{n_{CNT}}\right). \qquad (10)$$

Here, $n_{CNT} = \frac{k_r}{\omega_{laser}/c}$, where $k_r = k_r(\omega = \omega_{laser})$. The electric fields averaged over the tube radius, r_c, therefore, are

$$\langle E_x \rangle = \frac{1}{r_c} \int_0^{r_c} E_x(\xi) d\xi = \frac{Zn_0 e}{2\varepsilon_0} a_0 r_c e^{-\frac{r^2}{2r_s^2}} e^{i(\kappa z - \omega_{laser} t)} \cos \theta_{CNT} \tag{11}$$

and

$$\langle E_z \rangle = \frac{1}{r_c} \int_0^{r_c} E_z(\xi) d\xi = \frac{Zn_0 e}{2\varepsilon_0} a_0 r_c e^{-\frac{r^2}{2r_s^2}} e^{i(\kappa z - \omega_{laser} t)} \sin \theta_{CNT}. \tag{12}$$

Note that the averaged fields have no dependence upon the tube radius. The transverse field, $\langle E_x \rangle$, and longitudinal one, $\langle E_z \rangle$, act to focus and accelerate the ions channeling through a nanotube at the phase-velocity matching condition respectively. The energy gain of accelerated ions along the CNT is given by integrating $\langle E_z \rangle$ over the acceleration length as follows,

$$W_z = \left(\frac{Zn_0 e}{2\varepsilon_0} \right) \left(a_0 r_c e^{-\frac{r^2}{2r_s^2}} \right) \left(\frac{1 - e^{-k_i z}}{k_i} \right) \sin \theta_{CNT}. \tag{13}$$

Figure 8 shows axial distributions of the normalized plasma density and electric fields with $\theta = 50°$, $n_0 = 3.2 \times 10^{20}$ cm^{-3} ($Z = 6$, n_e ($r = 0$, $z = 0$, and $t = 0$) $= 2 \times 10^{21}$ cm^{-3}) and CNT aerial filling ratio of 4.9%. Figure 5 shows angle-dependent optical response parameters ($\kappa = k_r + ik_i$, n_r, and θ_{CNT}) of a CNT-implanted substrate illuminated by a laser of $\lambda_{laser} = 1.054$ μm and $P_{laser} = 125$ GW.

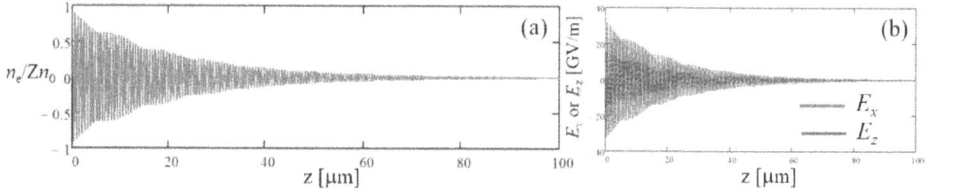

Fig. 8. (a) Normalized electron plasma density and (b) electric field amplitudes (E_x: red, E_z: blue) versus distance (z) graphs.

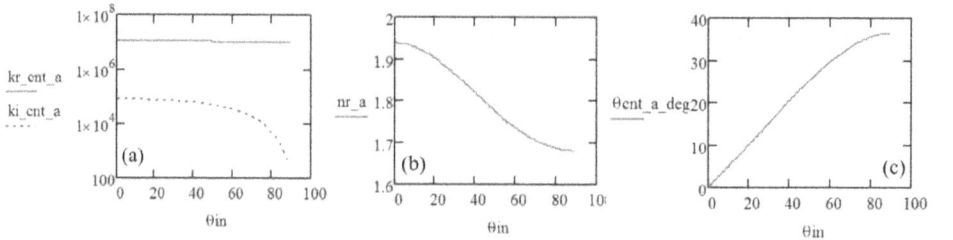

Fig. 9. (a) Dispersion (k_r: red) and absorption (k_i: blue) and (b) index of refraction (n_r) (c) refraction angle (θ_{CNT}) versus incident angle (θ_{in}) graphs.

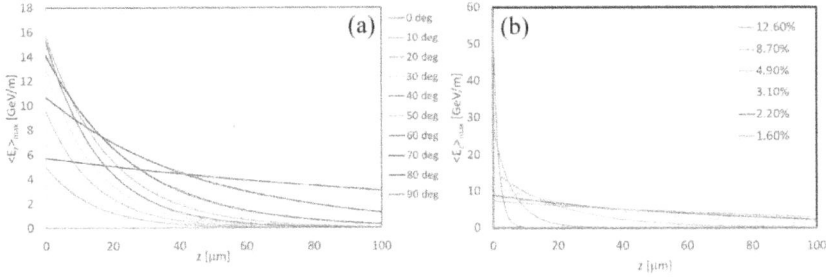

Fig. 10. Acceleration field ($\langle E_z \rangle_{max}$) versus distance ($z$) graphs with respect to (a) incident angle (θ_{in}) ($s = 4.9\%$) and (b) aerial CNT-filling ratio (with $\theta_{in} = 50°$).

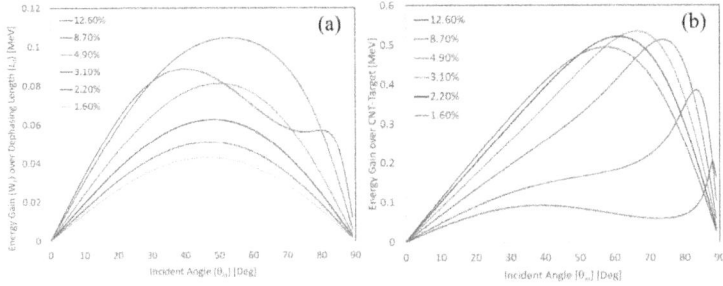

Fig. 11. Energy gain (W_z) versus incident angle (θ_{in}) over (a) dephasing length and (b) CNT-implanted target ($z = 100\ \mu m$) with respect to aerial CNT-filling ratio (s).

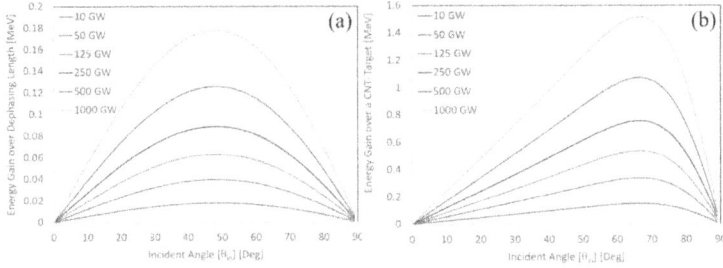

Fig. 12. Energy gain (W_z) versus incident angle (θ_{in}) over (a) dephasing length and (b) CNT-implanted target ($z = 100\ \mu m$) with respect to laser power (P_{laser}).

Figure 10 shows the maximum electric field ($\langle E_z \rangle_{max}$) over a tube distance under two various laser-plasmon coupling conditions with parametric scans of incident angles (with $s = 4.9\%$) and aerial CNT-filling ratios (with $\theta_{in} = 50°$). The peak field reaches 60 GeV/m with the large filling ratio, while steeply attenuating with distance. However, the field modestly attenuates with a smaller filling ratio, although the peak field drops to 5–6 GeV/m (over $\theta_{in} = 50$–$60°$). Figure 11 shows energy gain versus laser incident angle for channeling particles through a nanotube in a unit area on a substrate with respect to the aerial CNT-filling ratio (s). The ions phase-matched with the confined field are accelerated by the longitudinal field until they overrun the plasma wave.

The Bremsstrahlung radiation loss length is

$$\lambda_R^{-1} = 4Z_{eff}^2 r_p^2 r_e^{-3} \alpha^7 \varphi^{-3} K_0 \left(\frac{2\pi r}{b} \right) \ln \left(\frac{2\gamma a m_e}{m_p} \right), \qquad (14)$$

where $r_p = e^2/m_p c^2$ is the classical radius of a particle with mass m_p (r_e for electron), $\phi = b/a_B$ (b: transverse lattice constant), a is the fine-structure constant, b is the lattice constant, a_B is the Bohr radius, and K_0 is the modified Bessel function of the second kind. For electrons, the energy loss via a CNT wall ($b \sim 2\text{--}3$ Å) is about 7–15 GeV/m, which is larger than the acceleration gradients. Therefore, the electrons moving along the carbon-layers of the CNT walls would not gain the energy for overwhelmingly large radiation losses. However, CNTs are a hollow channel, which is a fairly empty space inside of the tube. The electrons moving along the inner space of the tube undergo a significantly lower Bremsstrahlung radiation loss since the tube radius (r_{CNT}) is usually a few orders of magnitude larger than a crystal lattice space (b). Therefore, the acceleration in a CNT would not be much affected by the Bremsstrahlung radiation and it is rather limited by a dephasing length.

According to laser-plasma acceleration theory, the dephasing length along the unit volume across the substrate is given by

$$L_d = \frac{\lambda_p^3}{\lambda_{laser}^2}, \tag{15}$$

where $\lambda_p = \dfrac{2\pi c}{\omega_p}$ and $\lambda_{laser} = \dfrac{2\pi c}{\omega_{laser}}$. The energy gain is fairly limited by the dephasing length: the particles are too rapidly kicked up and pushed away from the plasma wave by the large electric field before being fully synchronized with the plasmon. The particles could be continuously accelerated if they remain synchronized with the plasmonic wave. The continuous phase-matching condition is normally established by tapering the plasma density. In such a designed acceleration, the only limiting factor of energy gain would be the target thickness. A typical range of maximum AAO-CNT target thickness is 50–100 μm. Figures 11 and 12 show energy gains over the dephasing length and target thickness (or tube length) varying with incident laser angles (θ_{in}) with respect to aerial CNT-filling ratio (s) and laser power, respectively. In Fig. 11, the maximum energy gain over the 100 μm thick target reaches 0.5 MeV with $s = 3.1$ % and $\theta_{in} = 60\text{--}70°$. Obviously, the energy gain increases proportionally to the square root of the laser power (Fig. 12) in the linear regime. However, within a realistic scale of a femtosecond laser system, the highest laser power is limited within 100–125 GW and the most obtainable energy gain over a 100 μm thick target implanted with a CNT array ($r_c = 50$ nm and $s = 3.1\%$) will be 0.5–0.6 MeV with that laser system (125 GW, $\lambda_{laser} = 1.054$ μm, $r_{laser} = 380$ μm), corresponding to a 5–6 GeV/m gradient. In addition, another limiting factor of the energy gain could be the coherent Betatron radiations occurring when a longitudinal motion of channeling electrons is perturbed by the plasmon-transverse field component (E_x). A fractional ratio of the transverse field component depends on a laser-incident angle — the radiation loss will be lowered as the laser-injecting angle is increased. In this condition, the electrons would be rather accelerated by the longitudinal field component (E_z) than undulated by the transverse field component (E_x). The electron-acceleration or -undulation regime can thus be selectively chosen by adjusting the laser-incident angle.

X-ray channeling acceleration was well described in Ref. 11, including a theoretical model of beam-wave synchronization conditions. Looking back to the acceleration system, Figure 13 shows that an x-ray incident to a crystal with a Bragg angle undergoes anomalous optical transmission (Bormann effect)[34, 35] inducing x-ray diffraction among inter-planar lattice spaces, obtained from $E_v = hc/2a/\sin\theta_B$ (h is the Planck constant) in terms of a lattice constant, a, and Bragg angles, θ_B. Channeling particles coupled in the same orientation can be accelerated by the guided traveling waves at the m-th diffraction order where they are synchronized with a periodicity of a superlattice with a Brillouin wave number $k_s = 2\pi/s$ where s is the periodic length, i.e. $\omega/(k_z + k_s) = c$, where ω and k_z are the light frequency and longitudinal wave number. As channeled particles are confined to the rows of atoms by electric fields of the order 1–10 V/Å, crystals can be used in a collider not only for acceleration sections but also for bending sections. In the channeling acceleration process, the Bormann effect must occur to hold strong inter-planar fields in the crystallographic planes and the transmission becomes stronger with a thicker crystal. In our simulation, the crystal plate is modeled with a lattice constant of 3.1 Å and a Bragg angle of 2.866°, designed from a coupling condition of the 10th diffraction order. The incident plane wave slant to the crystal face forms tilt wave-fronts with respect to crystal orientation. It creates the longitudinal and transverse electric field components in the lattice spaces. The field plot depicts that transverse fields of the diffracted waves are a few orders of magnitude larger than longitudinal fields. Figure 13 is the normalized transmission of two crystal plate structures depicting 4 and 10 lattice layers. In the transmission spectrum, resonant diffraction peaks appear at the wavelengths corresponding to the Bragg angles. The simulation shows the Bormann effect occurring in the crystal for the 10-layered plate has a much stronger transmission than the 4-layered one. It clearly shows that the required photon energy for the transmission can be significantly lowered with a larger lattice constant and a larger angle. Increasing the size of the unit cell from 3 to 15 Å drops the x-ray energy down to an order of 0.1 ~ 10 keV, depending upon the Bragg angle. Enlarging a unit cell to the nano-scale can take the energy level of the driving photon source down to the ultraviolet or even the visible light spectrum. With the analytic calculation on radiation losses and x-ray powers, we extensively examined the channeling acceleration condition mainly with respect to their unit cell sizes and lattice constants.

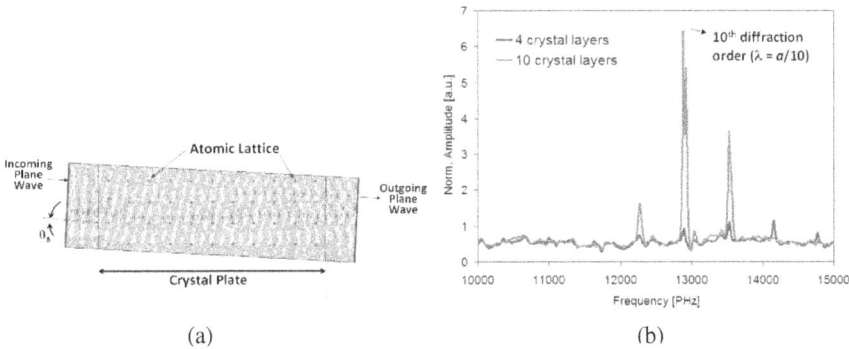

(a) (b)

Fig. 13. (a) 2D-contour plot of diffracted electric field distribution in the Bormann effect at θ_B ~ 2.866°, obtained from EM-wave simulation (b) simulated transmission spectra of crystals with 4- and 10-lattice layers.

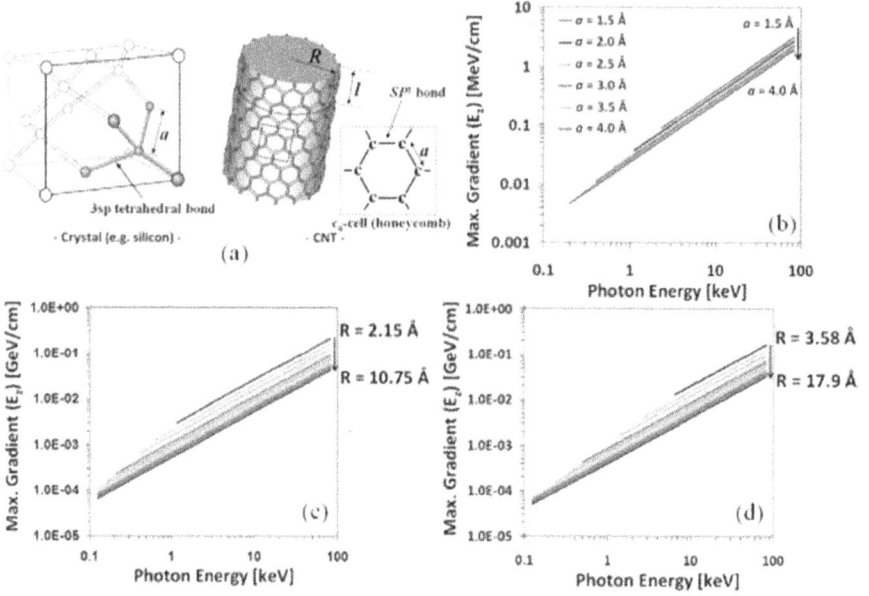

Fig. 14. (a) Conceptual drawings of a crystal (silicon) and a CNT with dimensional parameters. Maximum gradients versus photon energy graphs of (b) crystals and CNTs of (c) $a = 1.5$ Å and (d) $a = 2.5$ Å.

Figure 14 (b) is the maximum acceleration gradient (eE_z) versus photon energy (wavelength) with respect to the lattice constant (a) that can be provided by a crystal (below the damage threshold). The lower cutoffs of the highest gradient curves are restricted by the radiation loss. The lower cutoffs are calculated with the condition that the atomic number, Z, is 30, the effective charge, Z_{eff}, is 10, the superlattice constant ratio (s/a) is 20, the transverse mode number (N) is 2, and number of atoms in a unit cell is 1.5. Also, we assume that 1 TeV of particle energy is lost over the radiation loss length (total energy loss (E_l) = 1 TeV) and channeling muons (μ-) of $\beta \cong 0.995$ and $\gamma \cong 10$ is distributed in the transverse atomic space of $0 \sim a/2$. With this condition, the acceleration gradients of crystals with $a = 1.5 \sim 4.0$ Å are calculated with a photon energy of $0.2 \sim 41.34$ keV ($0.3 \sim 62.2$ Å). Figure 3 shows the result that a smaller unit cell leads to a higher field gradient. Note that the smaller lattice structure increases the cutoff of the acceleration gradient as particles lose more energy by radiation at wavelengths comparable to the structure dimensions. Figures 14(c) and (d) show the maximum acceleration gradient (upper limit) versus the photon energy for two CNT models where $a = 1.5$ Å and 2.5 Å (Z = 6). The unit cell volume of a single-wall tube is defined as $V_{\text{cnt}} = \pi R^2 l$, where R and l is the radius and length of the tube unit cell respectively. These are determined by the number of c_6-cells (honeycombs) in the circumference. In general, expressions $R_{\text{armchair}} = \frac{n_{cnt}(3a)}{2\pi}$ and $l_{\text{armchair}} = \sqrt{3}a$ describe the armchair and expressions $R_{\text{zigzag}} = \frac{n_{cnt}(\sqrt{3}a)}{2\pi}$ and $l_{\text{zigzag}} = 3a$ describe the zigzag, where n_{cnt} is the number of the c_6-cells along the tube circumference. The expressions correspond to unit-cell volumes $V_{\text{armchair}} = \frac{9\sqrt{3}}{4\pi} n_{cnt}^2 a^2$ and $V_{\text{zigzag}} = \frac{9}{4\pi} n_{cnt}^2 a^2$.

Therefore, the armchairs are about 73% larger than the zigzags with the same number of c_6-cells. The plasma density of a single-wall nanotube (SWNT) is given by $n_{cnt} = N_{cnt}/V_{cnt}$, where N_{cnt} is the number of the ions and their inner-shell electrons in a tube unit-cell. The graphs are plotted with the 3 to 15 c_6-cells along the circumference ($R = 2.15 \sim 10.75$ Å with $a = 1.5$ Å and $R = 3.58 \sim 17.9$ Å with $a = 2.5$ Å). The upper thresholds of accelerating gradients with $a = 1.5$ Å and $a = 2.5$ Å are reduced to $\sim 15\%$ for the crystal (silicon, Fig. 14(b)). This corresponds to an acceleration gradient of $E_{acc} = \sim 150$ MeV/cm ($a = 1.5$ Å) and $= \sim 70$ MeV/cm ($a = 2.5$ Å). The reduction of the acceleration gradient is mainly owing to the smaller atomic number, Z, and larger unit-cell size of CNTs. It should be noted that the cutoff acceleration gradient significantly drops with smaller radiation losses for muons owing to the larger beam tunnel and lower plasma density.

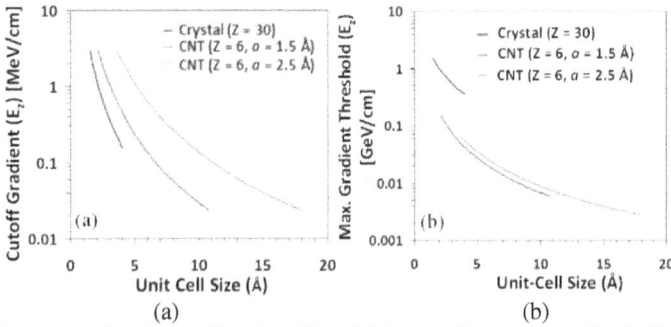

Fig. 15. (a) Minimum acceleration gradient (cutoff) and (b) upper limit (damage threshold) of acceleration gradient versus unit-cell size graphs of a crystal (Z = 30) and CNTs (Z = 6), $a = 1.5$ Å and $a = 2.5$ Å.

Figure 15 depicts the summarized graphs of the (a) cutoff gradients and (b) the upper gradient threshold of the crystal and the CNTs with respect to the tube unit-cell sizes. As shown in Fig. 15(a), CNTs and crystals have similar levels of energy loss within the same range of the unit cell dimension, but the cutoff appears significantly smaller with the larger tube unit-cell size. Note that the upper acceleration limit from the structural damage threshold is also lowered with the larger unit-cells due to the lower density. It should be stressed that replacing a crystal with a nanotube lowers the cutoff of required power by three orders of magnitude, which reduces the maximum acceleration gradient from 0.36 (4 Å)–1.52 (1.5 Å) GeV/cm of crystal (Z = 30) to 0.15 GeV/cm ($a = 1.5$ Å) and 0.07 GeV/cm ($a = 2.5$ Å) of CNTs with a few nanometers of tube unit cell dimension. The analysis result shows that CNTs can significantly lower the power level required for x-ray channeling acceleration.

4. Experiment

4.1. Sample preparation

One of the well-known nanofabrication techniques of 2D CNT arrays is the anodic aluminum oxide (AAO) template process to implement periodically aligned nano-holes in a few tens of micrometer thick substrate. Straight, vertical carbon nanotubes (CNTs) can also be implanted in the porous substrate by the chemical vapor deposition (CVD) growth process (Fig. 16(a)). CNTs are grown in an AAO template by pyrolysis of C_2H_2 with argon gas at 750°C in a reactor furnace for 3–4 hours.[36, 37, 38, 39] Sub-100 μm long straight multi-wall CNTs grow along the aligned nano-pores in an AAO template, which is followed by thermal cleaning. The length of AAO-CNTs is determined by the length of the nano-pores in the substrate, i.e. the thickness of AAO membranes, which can be controlled by anodizing time. (AAO templates with thicknesses up to 100 μm and pores with diameters up to 100 nm are commercially available.) By selecting the anodizing voltage at different electrolytes during the anodizing process, AAO films with different pore sizes from a few nanometers to a few hundred nanometers can be readily obtained. The outer and inner diameters of AAO-CNTs can be tuned by pore sizes and the CVD growth conditions, respectively. Channeling of the AAO and AAO-CNT with ⁴He+ beam was demonstrated in a low energy regime (2 MeV) by a Chinese group[40]: ~ 3° and ~ 1.6° of FWHM transmission angle distributions were directly measured with CNT and AAO respectively by angle scans, accompanied with 10–20% of ion channeling. CNT samples fabricated with various tube sizes will be tested with the electron source and the laser system in the plan to find empirical relations between the beam transmission and photon-coupling efficiency and their geometrical parameters.

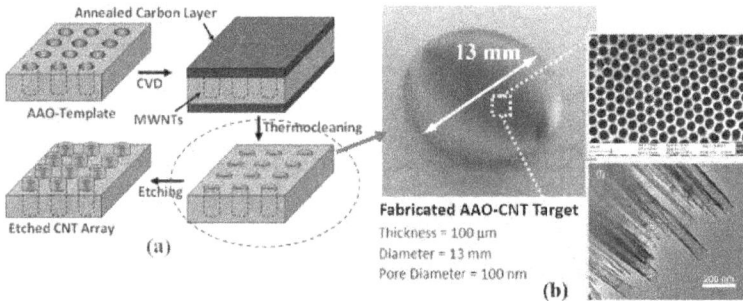

Fig. 16. (a) Flowchart of AAO-CNT fabrication process (b) fabricated AAO-CNT target (inset: SEM images of AAO-template and AAO-CNT array after chemical cleaning, Courtesy of Liu Chang[37]).

4.2. Outline of feasible proof-of-concept tests

The initial assessment of the prospective energy gains of CNT-channeled electrons was fulfilled with a PIC-based beam-driven plasma simulator combined with the beamline simulations. For the simulations, an electron linac beamline (e.g. Fermilab accelerator science and technology (FAST) facility - 50 MeV beamline from the chicane (BC1) to the imaging station (X124)) was modeled with CST and Elegant (Fig. 17(a)). The two beam profiles of the bunches with and without the modulation (modulation wavelength = λ_{mb}), which is generated by the slit-mask in the bunch compressor (BC1), are monitored at the goniometer position. The beam profiles are then manually imported to the effective CNT model.[41] A typical simulation result (Fig. 17(b)) showed that a 1 nC bunch (uncorrelated energy spread = ~ 0.01–0.015 %, λ_{mb} = 100 μm) is self-accelerated with a net energy gain

(~ 0.2 %) on the tail (witness) and an energy loss (~ 0.6 %) on the head (drive) along the 100 μm long channel with the nominal beam parameters. Our preliminary assessment with the full beamline model predicts that the FAST 50 MeV beam can produce ~ 1–2 % of maximum net gain with 3.2 nC bunch charge and 100 μm transverse beam size (circular beam, Fig. 17 (c)), corresponding to 5–10 GV/m gradient. For the simulations, the bunch charge density is about a thousand times smaller than the channel charge density (off-resonance beam-plasma coupling). However, detecting the amount of energy gain by the proposed experiment can support feasibility of TeV/m acceleration in CNT channels.

Generally, density modulations enhance energy efficiencies or power gains of coherent light sources or beam-driven accelerators. A beam, if longitudinally modulated, is more strongly coupled with accelerating or undulating structures at a resonance condition with the fundamental or higher order modes. Pre-bunched or modulated beams would improve the longitudinal beam control in energy-phase space and furthermore strongly enhance wakefield strengths or transformer ratio of beam-driven channeling accelerations. The modulated beam either maximizes the wakefield strength (when it is bunched with a plasma wavelength of an accelerating medium) or significantly increases the transformer ratio proportionally to the number of micro-bunches ($R = M \cdot R_0$, where M is the number of micro-bunches and R_0 is the transformer ratio of a single-bunch driver) with an off-resonance condition.[42]

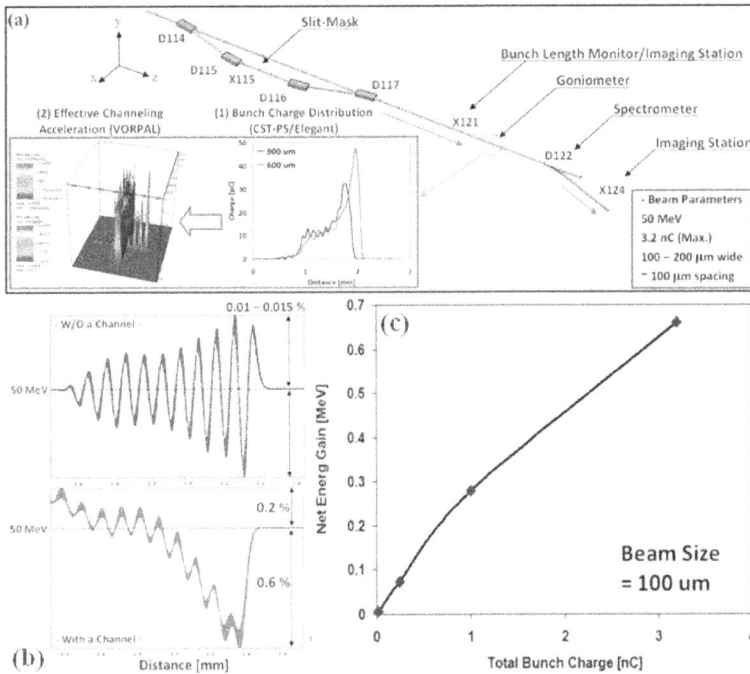

Fig. 17. (a) Electron linac beamline model (FAST). Inset is a modulated bunch charge distribution (Elegant and CST) at the goniometer position and effective CNT-channeling acceleration model (VORPAL). (b) Energy distributions of a modulated bunch (top) without and (bottom) with a channel. (c) energy gain versus bunch charge.

In principle, a beam-density modulation (or micro-bunching) corresponding to an intrinsic channel plasma frequency possibly offers the optimum beam-plasma coupling condition with maximum energy transfer efficiency and acceleration gradient. However, the crystal charge density with $\rho_p \geq 10^{19}$ cm^{-3} requires a very short modulation wavelength, $\lambda_{mb} \leq 10$ µm, which could be only produced by a short-period micro-buncher (e.g. Inverse FEL undulator). Our plan for the experiment is thus to generate a bunched beam of relatively long modulation periodicity by slit-masking the beam in the center of a magnetic chicane and then to couple a higher order mode of the modulated beam with an accelerating medium.[43] The slit-mask modulation technique is relatively easy to generate a beam modulation. At the FAST 50 MeV beamline, while a ~ 3–4 mm long photo-electron bunch passes through the bunch compressor (BC1), the slit-mask placed in the BC1 slices bunches into micro-bunch trains by imprinting the shadow of a periodic mask onto the bunch with a correlated energy spread (Fig. 18(a)). In principle, modulation strength and periodicity of the modulation can be controlled by adjusting the grid period or by the dipole magnetic field.[44] The bunch-to-bunch distance (modulation periodicity) is given as,

$$\Delta z = W \frac{\sqrt{\left(1 + h_1 R_{56}\right)^2 \sigma_{z,i}^2 + \tau^2 R_{56}^2 \sigma_{\delta i}^2}}{\eta_{x,mask} h_1 . \sigma_{z,i}} \approx W \frac{\left| \sigma_{z,i} + R_{56} \sigma_\delta \right|}{\eta_{x,mask} \sigma_\delta},$$

($\sigma_{z,i}$: initial bunch length, h_1: first order chirp = $-1/R_{56}$, $\sigma_{\delta,i}$: initial uncorrelated energy spread, R_{56}: longitudinal dispersion of BC1, and τ: energy ratio = E_{io}/E_{fo}, E_{io} and E_{fo} are the central energies before and after acceleration, respectively, and $\eta_{x,mask}$: dispersion at the mask). With nominal 50 MeV beam parameters ($\sigma_{z,i}$ = 3 ps, R_{56} = ~ − 0.192 m, E_{io} = 50 MeV), the analytic model showed that a slit-mask with slit period 900 µm and aperture width 300 µm generates ~ 100 µm spaced micro-bunches with 2.4% correlated energy spread. As shown in Fig. 18(c), the preliminary simulation data from Elegant and CST also indicated that the designed mask produces a ~ 100 µm spaced beam modulation with maximum RF chirp.

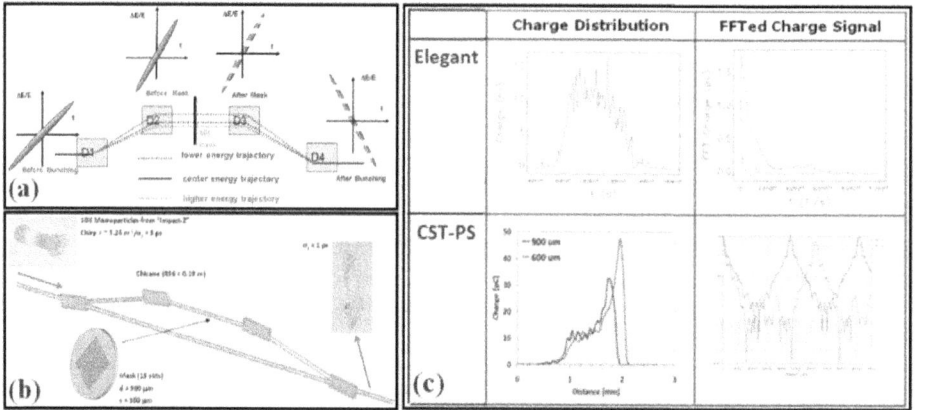

Fig. 18. (a) Conceptual drawing and (b) simulation model of a slit-mask micro-buncher (c) longitudinal charge distributions and beam signal spectra obtained by Elegant and CST simulations.

It is planned to test the CNT-channeling acceleration at the FAST 50 MeV beamline (Fig. 19(a)). A ~ 3 ps long electron bunch generated from the photo-injector is transported to the magnetic chicane (BC1). The bunch is compressed to ~ 1 ps by BC1, and the compressed beam is focused by the quadrupole triplet magnets (Q118, 119, 120). After the beam spot size is focused to ~ 100–200 µm, it is injected to a CNT target in the goniometer. The channeled electrons are transported to an electron spectrometer consisting of a dipole (D122) and a screen in the energy dispersive region following D122 at imaging station X124. Their energy distribution can be measured by the spectrometer before the beam is dumped to a shielded concrete-enclosure (beam dump).

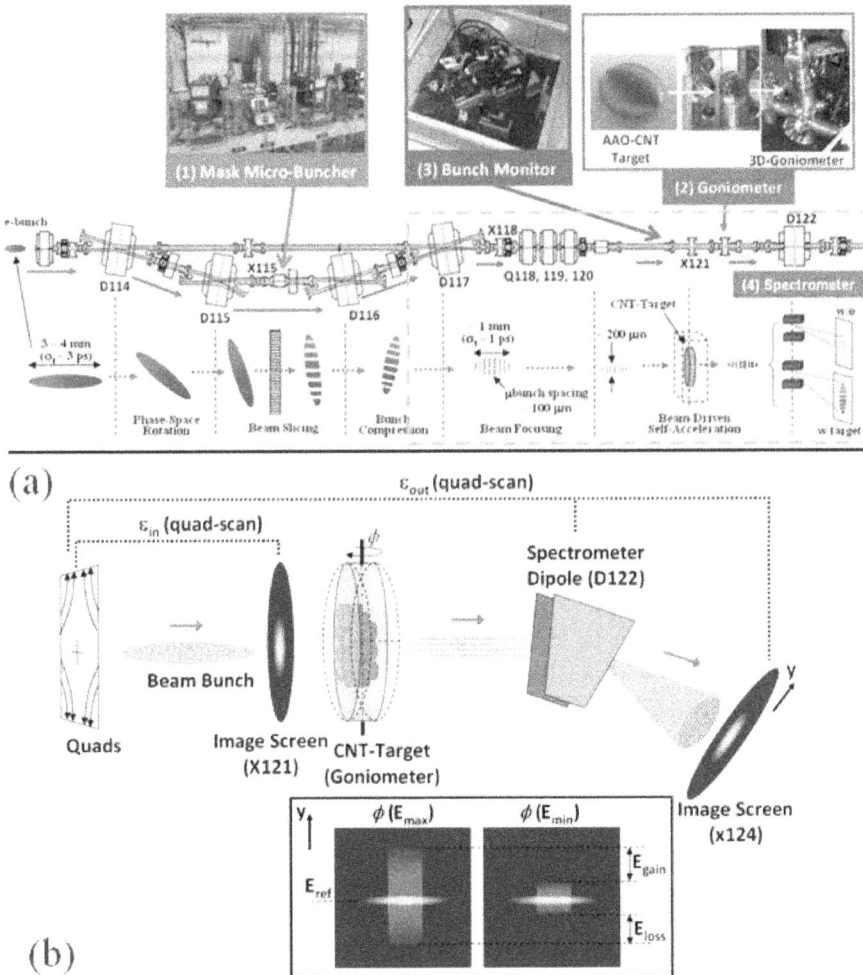

Fig. 19. (a) FAST 50 MeV beamline configuration (b) schematics for energy and emittance measurements of CNT-channeling experiment (dotted green box in (a)).

The pre-bunched beam generated by the slit-mask installed at X115 between two bending-dipoles (D115 and D116) can also be tested and the measured beam parameters can be compared with the ones of the bunched beam to check the impact of beam-modulation on channeling acceleration. Before testing the AAO-CNT target, the beam parameters can be characterized first without a target, which could be a reference for the beam-energy measurement. An experiment can be set up to check if the measured variation of the projected image on the screen due to presence of the target exceeds the nominal deviation of the image produced by the intrinsic energy spread. The initial experiment can then be followed by subsequent measurements to accurately identify net energy gains/losses and beam emittances: the channeled beam deflected by the magnetic spectrometer (D122) can be projected on the screen of the imaging station (X124). The beam-injection angle with respect to the target axis can then be scanned. A relative change of projected images from one angle to another is translated into an energy gain/loss of channeled beam (Fig. 19(b)).

5. Summary

Above we presented theoretical analysis and numerical simulations of a high gradient acceleration method for beam- and laser-driven acceleration in carbon nano-tubes (CNTs). We also analyzed parameters of CNT's for x-ray channeling acceleration with a GeV/cm level of acceleration gradient. It turns out that nanotubes can significantly lower the cutoff of the x-ray power required for the acceleration energy gain by a few orders of magnitude. Our results show that the large reduction of the required x-ray power level by employing nano-structures turns the channeling photon interactions with heavy particles, i.e. muons and protons, into a more viable approach for extremely high gradient acceleration with power levels from currently existing coherent x-ray sources.

Density modulation of high brightness beams can lead to significantly improved performance of accelerator-based coherent light sources and high energy linacs, and here we also considered a simple way for micro-bunch train generation with a masked chicane. This approach can be tested with the bunch compressor in the electron linac beamline at Fermilab's FAST facility. The linear model is derived to estimate performance of the designed masked chicane, indicating that the designed slit-mask produces $\sigma_{ms} = 33$ μm long micro-bunches spaced at ~ 100 μm in a $\sigma_t = 3$–4 ps long bunch with about 1–2 % correlated energy spread.

Numerical analysis with two simulation codes, CST-PS and Elegant, indicates that the beam modulation effectively appears with the slit period of 900 μm and 300 μm slit width. CST-PS simulations included nonlinear beam-energy distribution and space charge effects, and resulted in bunch-to-bunch distance of ~ 100 μm. Our simulations also indicate that the bunch charge density modulation would disappear when the beam is chirped with very small correlated energy spread (on-crest). The simulation results reasonably agree with theoretical analysis of the linear chicane model, which verifies a feasibility of slit-masked chicane to produce a bunch modulation on the order of 100 fs with the beam properly chirped.

Carbon nanotubes can enable efficient collimation/bending of intense beams and continuously focused acceleration in the nanochannels with exceptionally high gradients of the order of TeV/m. The CNT acceleration concept might have a great potential to advance technology for future high energy particle colliders.

Acknowledgments

This work was supported in part by the DOE Contract No. DE-AC02-07CH11359 to the Fermi Research Alliance LLC.

References

1. https://www.linearcollider.org/ILC.
2. http://clic-study.web.cern.ch/CLIC-Study/
3. https://espace2013.cern.ch/fcc/Pages/default.aspx
4. V. D. Shiltsev, *Phys.-Usp.* **55**(10), 965 (2012).
5. T. Tajima and J. M. Dawson, *Phys. Rev. Lett.* **43**(4), 267 (1979).
6. J. B. Rosenzweig, D. B. Cline, B. Cole, H. Figueroa, W. Gai, R. Konecny, J. Norem, P. Schoessow and J. Simpson, *Phys. Rev. Lett.* **61**, 98 (1988).
7. C. Joshi and T. Katsouleas, *Phys. Today* **56**(6), 47 (2003).
8. Y. Shin, *Appl. Phys. Lett.* **105**, 114106 (2014).
9. I. Y. Dodin and N. J. Fisch, *Phys. Plasmas* **15**, 103105 (2008).
10. M. A. Kumakhov and F. F. Komarov, *Energy Loss and Ion Ranges in Solids* (Gordon and Breach Science, 1979).
11. T. Tajima, and M. Cavenago, *Phys. Rev. Lett.* **59**, 1440 (1987).
12. P. Chen and R. J. Noble, *AIP. Conf. Proc.* **156**, 2122 (1987).
13. B. W. Montague and W. Schnell, *AIP Conf. Proc.* **130**, 146 (1985).
14. P. Chen and R. Noble, Channel particle acceleration by plasma waves in metals, SLAC-PUB-4187.
15. L. A. Gevorgyan, K. A. Ispiryan, and R. K. Ispiryan, *JETP Lett.* **66**, 322 (1997).
16. P. Chen and R. J. Noble, in *Relativistic Channeling*, eds. R. A. Carrigan and J. Ellison (Plenum, New York, 1987), p. 517; also *NATO ASI Ser., Ser. B* **165**, 517 (1987); SLAC-PUB-4187 1987.
17. P. Chen and R. J. Noble, *AIP Conf. Proc.* **396**, 95 (1997); also FERMILAB-CONF-97-097 1997; SLAC-PUB-7673 1997.
18. F. Zimmermann and D. H. Whittum, *Int. J. Mod. Phys. A* **13**, 2525 (1998); also SLAC-PUB-7741 1998.
19. D. S. Gemmell, *Rev. Mod. Phys.* **46**, 129 (1974).
20. J. Lindhard, *Mat.-Fys. Medd. Dan. Vid. Selsk.*, **34**, No. 14 (1965); also in *Usp. Fiz. Nauk* **99**, 249 (1969).
21. V. V. Beloshitsky, F. F. Komarov, and M. A. Kumakhov, *Phys. Rep.* **139**, 293 (1986).
22. V. M. Biryukov, Yu. A. Chesnokov, and V. I. Kotov, *Crystal Channeling and Its Application at High-energy Accelerators* (Springer, New York, 1997).
23. V. N. Baier, V. M. Katkov and V. M. Strakhovenko, *Electromagnetic Processes at High Energies in Oriented Single Crystals* (World Scientific, Singapore, 1998).

24. R. A. Carrigan Jr., J. Freudenberger, S. Fritzler, H. Genz, A. Richter, A. Ushakov, A. Zilges and J. P. F. Sellschop, *Phys. Rev. A* **68**, 062901 (2003).
25. B. Newberger, T. Tajima, F. R. Huson, W. Mackay, B. C. Covington, J. R. Payne, Z. G. Zou, N. K. Mahale and S. Ohnuma, *Proc. IEEE Part. Acc.* (IEEE, Chicago, 1989), p. 630.
26. M. Murakami, and T. Tanaka, *Appl. Phys. Lett.* **102**, 163101 (2013).
27. Z. Huang, P. Chen and R. D. Ruth. *Phys. Rev. Lett.* **74**(10), 1759 (1995).
28. P. Chen and R. J. Noble, SLAC-PUB-7402 (1998).
29. Y.-M. Shin, D. A. Still and V. Shiltsev, *Phys. Plasmas* **20**, 123106 (2013).
30. C. B. Schroeder, C. Benedetti, E. Esarey, F. J. Grüner and W. P. Leemans, *Phys. Rev. Lett.* **107**, 145002 (2011).
31. Q. Lu, R. Rao, B. Sadanadan, W. Que, A. M. Rao and P. C. Ke, *Appl. Phys. Lett.* **87**, 173102 (2005).
32. M. Takase, H. Ajiki, Y. Mizumoto, K. Komeda, M. Nara, H. Nabika, S. Yasuda, H. Ishihara and K. Murakoshi, *Nature Photonics* **7**, 550–554 (2013).
33. Z. Zhu, D. Zhu, R. Lu, Z. Xu, W. Zhang and H. Xia, *Proc. SPIE* **5974**, 597413 (2005).
34. V. M. Biryukov and S. Bellucci, *Nuclear Instruments and Methods in Physics Research B* **234**, 99–105 (2005).
35. S. Bellucci, *Nuclear Instruments and Methods in Physics Research B* **234**, 57 –77 (2005).
36. A. Larsen, Nano-scale convective heat transfer of vertically aligned carbon nanotube arrays, Project Number: MQP-JNL-CNT9.
37. P. X. Hou, C. Liu, C. Shi and H. M. Cheng, *Chinese Science Bulletin* **57**, 187 (2012).
38. T. Altalhi, M. Ginic-Markovic, N. Han, S. Clarke, D. Losic, *Membranes* **1**, 37-47 (2011); doi:10.3390/membranes1010037.
39. P. Ciambelli, L. Arurault, M. Sarno, S. Fontorbes, C. Leone, L. Datas, D. Sannino, P. Lenormand, B. Du Plouy Sle, *Nanotechnology* **22**(26), 265613 (2011).
40. Z. Zhu, D. Zhu, R. Lu, Z. Xu, W. Zhang and H. Xia, The experimental progress in studying of channeling of charged particles along nanostructure, in *International Conference on Charged and Neutral Particles Channeling Phenomena*, Proc. of SPIE Vol. 5974 (SPIE, Bellingham, WA, 2005).
41. http://www.txcorp.com/home/vsim/vsim-pa
42. E. Kallos, *Plasma Wakefield Accelerators using Multiple Electron Bunches*, PhD Dissertation (Univ. Southern California (Electrical Engineering), 2008).
43. D. C. Nguyen and B. E. Carlston, *Nuclear Instrument and Methods in Physics Research A* **375**, 597 (1996).
44. P. Muggli, V. Yakimenko, M. Babzien, E. Kallos and K. P. Kusche, *Phys. Rev. Lett.* **101**, 054801 (2008).

Channeling and Radiation Experiments at SLAC

U. Wienands

Argonne National Laboratory, 9700 South Cass Rd.,
Argonne, IL 60436, USA

S. Gessner*, M. J. Hogan, T. Markiewicz, T. Smith, J. Sheppard

SLAC National Laboratory, 2575 Sand Hill Road,
Menlo Park, CA 94025, USA

U. I. Uggerhøj, C. F. Nielsen, T. Wistisen

Department of Physics and Astronomy, Aarhus University,
Ny Munkegade 120, DK-8000, Aarhus, Denmark

E. Bagli, L. Bandiera, G. Germogli, A. Mazzolari, V. Guidi, A. Sytov

Department of Physics and Earth Sciences, University of Ferrara and
INFN Section of Ferrara, Via Saragat 1/C, I-44122 Ferrara, Italy

R. L. Holtzapple, K. McArdle, S. Tucker

California State Polytechnic University, San Luis Obispo,
CA 93407, USA

B. Benson†

Massachusetts Institute of Technology, 77 Massachusetts Ave.,
Cambridge, MA 02139, USA

The SLAC T513–E212–T523 Collaboration

Since 2014, a SLAC-Aarhus-Ferrara-CalPoly collaboration augmented by members of ANL and MIT has performed electron and positron channeling experiments using bent silicon crystals at the SLAC End Station A Test Beam as well as the FACET accelerator test facility. These experiments have revealed a remarkable channeling efficiency of about 24% under our conditions. Volume reflection is even more efficient with almost the whole beam taking part in the reflection process. A positron experiment demonstrated quasi-channeling oscillations for the first time at high beam energy. In our most recent experiment we measured the spectrum of gamma radiation for crystal orientations covering channeling and volume reflection. This series of experiments supports the development of more advanced crystalline devices capable e.g. of producing narrow-band gamma rays with electron beams or studying the interaction of the electrons with the wakefields generated in the crystal at high beam intensity.

*Now at CERN, Geneva, Switzerland
†Now at Stanford University, Stanford, CA, USA

1. Introduction

Channeling of protons in bent crystals has been thoroughly studied with the purpose of, e.g., proton extraction at accelerator facilities[1-3] and for particle-beam collimation.[4-6] Until recently, much less was known about channeling of electrons and positrons at high particle energy. Our group has been studying planar electron and positron channeling at the SLAC End Station A Test Beam (ESTB) and the Facility for Advanced aCcelerator Experiments and Tests (FACET) at beam energies up to 20 GeV. The experiments were performed using a bent, quasi-mosaic (111) Si crystal of 60 μm thickness. The (111) plane has two different lattice spacings — 0.76 and 3.2 Å effectively — which is considered of advantage for electron channeling as the highest density of channeling electrons is in the middle of the narrow channel, thus does not coincide with the "nuclear corridor" that would increase the dechanneling probability. A plot of the potential for our crystal is shown in Fig. 1 of Ref. 8.

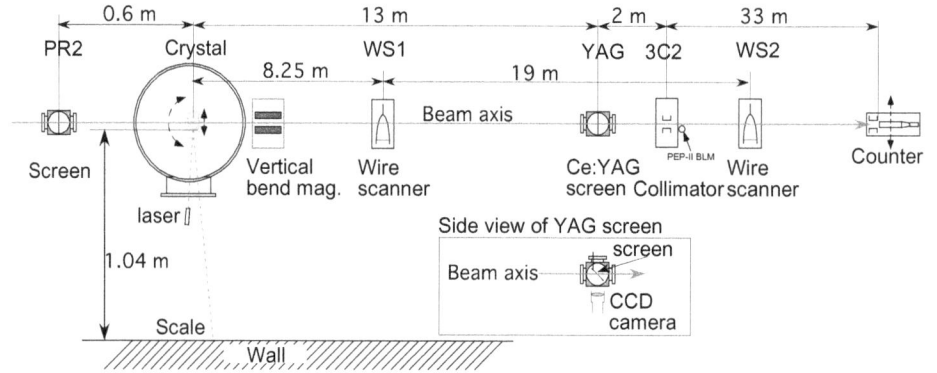

Fig. 1. Layout of the channeling and radiation experiments in the SLAC ESTB. "Counter" refers to the SciFi gamma detector.

These experiments — which have produced a body of data important for any practical application of crystals in collimation or other manipulations of electron beams at high energy — have enabled us to probe further and shift our experimental program towards the radiative aspects of channeling and volume reflection. In this overview we will first summarize the channeling data and then describe our more recent experiments investigating the radiation generated. The experimental setup is shown in Fig. 1 which shows the recent additions of a sweeper magnet and a gamma-ray detector (scintillating fibers, or SciFi for short) to allow the isolation and detection of gamma rays. Not shown is a thin scintillator paddle upstream of the SciFi counter we used as veto counter to verify the absence of charged particles when the sweeper dipole was energized. The 20-GeV positron experiment referred to below was done at FACET with a setup that was in substance the same even if the detailed detectors were different.

2. Electron Channeling

Channeling efficiency and dechanneling-length measurements were carried out with our quasi-mosaic bent crystal of 60 μm thickness and 0.15 m bending radius.[7] A summary of the data is shown in Fig. 2 together with simple model calculations. The dechanneling length turns out to be roughly independent of beam energy between 40 and 60 μm, consistent with DYNECHARM++[9] simulations but requiring a modification of the usual theoretical model for the dechanneling length. This initially unexpected result is important in energy scaling. The fitting function is

$$L_D = 15.3 \left[\frac{\mu \text{m}}{\text{GeV}} \right] \cdot E \left(1 - k_c \frac{2R_c}{R} \right) \tag{1}$$

with $k_c = 1.76$ determined by the fit. More details are given in Ref. 8. It is clear that the energy dependence of the dechanneling length for electrons is qualitatively different from that for protons.

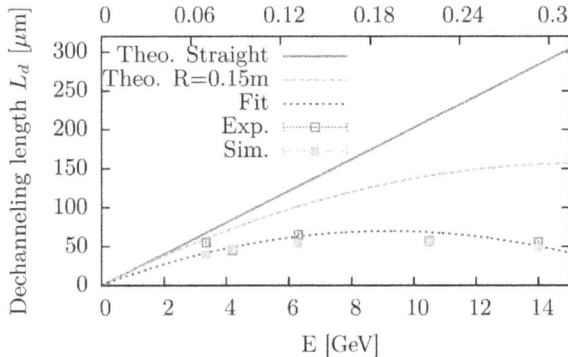

Fig. 2. Dechanneling length of electrons *vs* beam energy in our bent Si crystal with a bending radius of $\rho = 0.15$ m.

Channeling efficiency (the fraction of the incident electrons ending up in the channeling peak, as found by a fitting procedure) is up to 24%, which makes application of the channeling effect in beam-collimation systems for electrons a bit questionable. However, we found volume reflection (VR) to be effectively about 95% efficient and therefore a good candidate for collimation application.

The experiments also allowed us to assess the amount of multiple scattering, which shows itself in the "free" direction, i.e. vertically in our setup. We found that the *rms* width of the scattering angle goes up by about a factor of 2 compared to the width of multiple scattering in amorphous orientation of the crystal — see Fig. 3. The measured angular width in amorphous orientation agrees with the multiple-scattering formula[10] to within 10%. The naïve explanation for this effect is that the probability of a large-angle scatter from the nuclei goes up due to the attraction of the electrons by the nuclei. This result is of interest when compared to the measurement of the nuclear dechanneling length.[11]

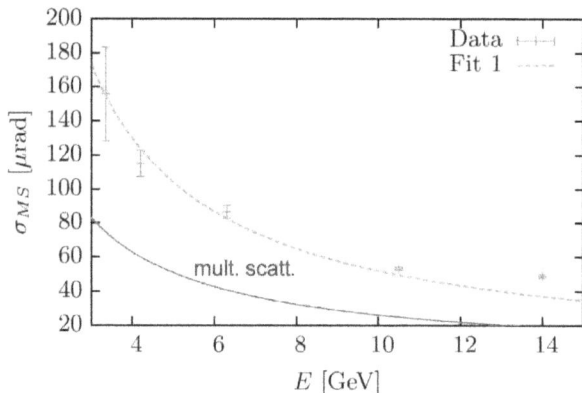

Fig. 3. Multiple scattering angle in the free, non-bending direction for channeled electrons *vs* beam energy in our bent Si crystal with a bending radius of $\rho = 0.15$ m. The multiple scattering angle was calculated using the formula in Ref. 10.

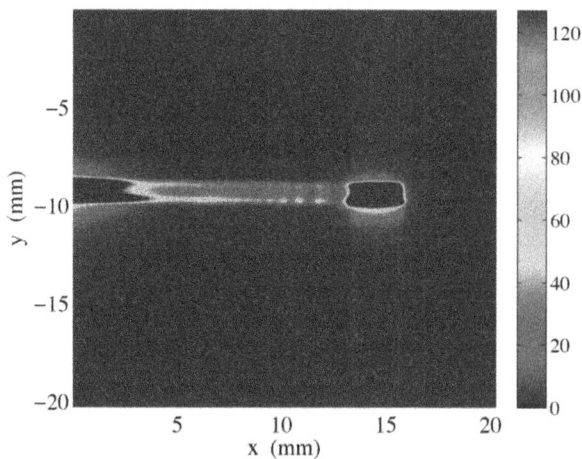

Fig. 4. Channeling of positrons at 20.35 GeV. The blob on the right is the channeled beam, on the left is a fraction of the unchanneled beam. The beamlets in between make up the dechanneling tail, broken up by quasi-channeling oscillations.

3. Positron Channeling

An experiment with positrons at 20.35 GeV was conducted at the FACET facility. The experimental setup was similar to Fig. 1. Qualitatively different than the electron data, the dechanneling tail with positrons is breaking up into small beamlets at slightly different deflection angle, see Fig. 4. This was identified as the quasi-channeling oscillations suggested by Sytov[12] to be detectable with positrons even at higher beam energy. It is a projection of the crystal structure on the screen as the dechanneling probability is always higher for electrons being close to the lower potential barrier in the periodic but also sloped potential of a bent crystal. In that publication a relation for the deflection angle was given:

$$\theta_{defl} = (\theta_b + \theta_t) - \sqrt{\frac{2d_0\,(n-1)}{R} + \frac{2d_s}{R}}\,. \tag{2}$$

Figure 5 shows the deflection angles together with the relation above using the parameters $\theta_b = 402 \pm 9\,\mu$rad (the crystal bending angle), $R = 0.15$ m (the crystal bending radius), $d_s = 3.14$ A (the plane separation) and $d_0 = 4d_s$ (a parameter), see Ref. 13.

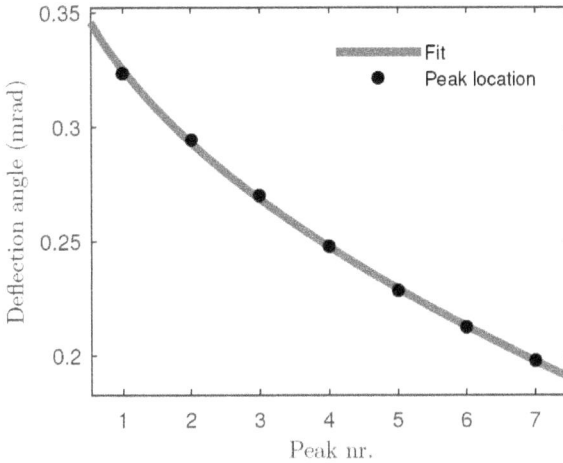

Fig. 5. Location of the quasi-channeling oscillation peaks for 20.35 GeV positrons together with a model calculation. See text for details.

4. γ radiation

Electrons (and positrons) will radiate violently if deflected with sufficiently tight bending radii. In order to detect and possibly record a spectrum of the radiation emanating from crystals, our experimental setup in the ESTB was augmented by a sweeper magnet to deflect the electrons away from the γ beam and by a scintillating-fiber calorimeter (SciFi) mounted about 45 m downstream of the crystal, both shown in Fig. 1. The full opening of the SciFi of 9 by 9 cm^2 was exposed to the incoming particles. The SciFi can be moved transversely in the horizontal plane, thus allowing to center it on the beam axis. Data was taken both with a secondary beam of an average of 10 electrons per pulse, giving less than one photon every second pulse and therefore allowing spectroscopy of the photons with little pile-up contributions. The SciFi counter is about 25 cm long, sufficient to measure spectra up to about 5 GeV photons without significant escape of the shower created in the counter. The counter was calibrated by detecting single and low-multiples of electrons using the secondary beam. The data were background-subtracted using spectra from an empty run without the crystal in the beam. Figure 6 shows a background-subtracted spectrum with the crystal rotated far out of the aligned angle to avoid coherent bremsstrahlung (CBR), it shows the expected Bethe-Heitler bremsstrahlung distribution. This spectrum also served for intensity normalization of the aligned spectra.

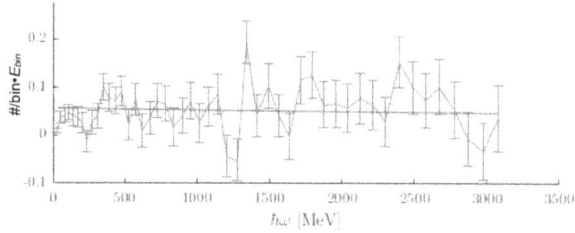

Fig. 6.   Background-corrected bremsstrahlung (Bethe-Heitler) spectrum from the crystal oriented far from channeling or CBR. The vertical scale indicates intensity/bin times energy, i.e. it is corrected for the $1/E$ behavior of bremsstrahlung.

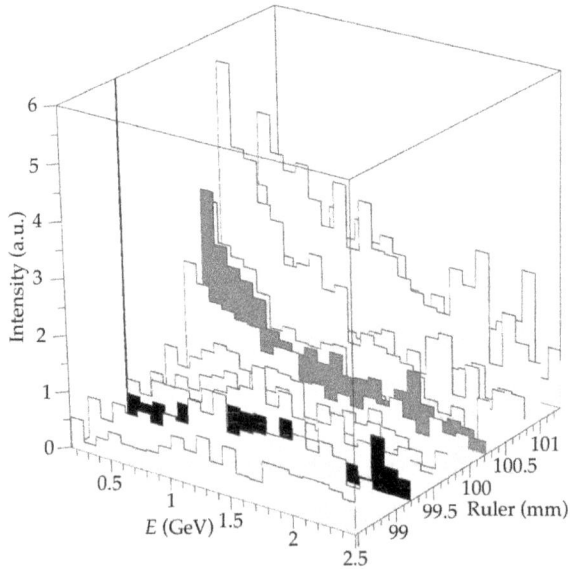

Fig. 7.   Raw energy spectra of photons for all crystal angles. The scale labelled "Ruler" is the uncalibrated crystal angle scale, the vertical scale is photon intensity per bin times the bin energy.

Crystal orientation was set with primary beam to ensure the crystal angle was near alignment. Spectra were taken at a total of 9 crystal angles covering the whole deflection triangle plus data points outside of the triangle. Figure 7 shows the raw data collected, already indicating a strong enhancement at aligned angles. Figure 8 shows a 3-d plot of the full, calibrated data set, against crystal angle and photon energy. The intensity enhancement in VR alignment is quite evident. The incoming secondary electron beam had a 1σ divergence of about 75 μrad. This exceeds the critical angle for channeling (about 60 μrad at 12.6 GeV) and therefore washes out features in the spectrum related to channeling radiation.

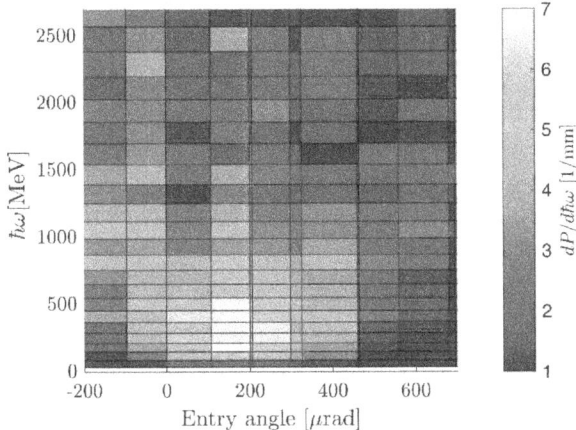

Fig. 8. Measured energy spectrum of photons *vs* crystal angle.

5. Modeling

To understand the nature of the distribution, the spectrum was compared to the results from a simulation model of the spectrum.[14] This model determines the trajectory of an electron by numeric integration of the equation of motion using the Doyle-Turner potential in an appropriate form for our crystal. With this trajectory, the Lienard-Wiechert radiation integral is evaluated using the Belkacem, Cue and Kimball (BCK) radiation integral.[15,16] This calculation is then done for 1000 particles at the different crystal orientations comparable to that of the experiment. The calculation reproduces the Bethe-Heitler spectrum at angles far enough away from the aligned orientation; the Bethe-Heitler calculation is used as an intensity calibration. Similar features are observed when comparing the experimental 3-d distribution (Fig. 8) to the simulated distribution (Fig. 9). A direct comparison of data and model calculation between 100 and 300 μrad crystal angle is shown in Fig. 10. By "tagging" the photons in the simulation model it is found that the photons associated with the high intensity peak around 100 to 300 μrad are predominantly generated by volume reflection. However, both the model as well as the data exhibit a CBR component evident by their energy-angle correlation in both the model as well as the data. The enhancement of the VR radiation over the Bether-Heitler

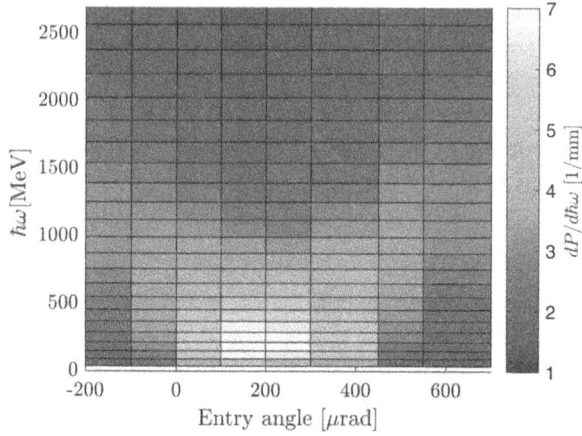

Fig. 9.   Simulation of photon energy spectrum *vs* crystal angle.

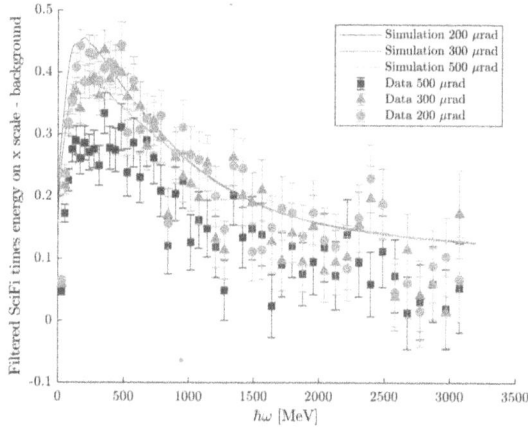

Fig. 10.   Energy spectra of photons for three crystal angles. The data are shown as points with error bars, the lines show the result of the modeling calculations.

intensity is a remarkable factor of 8 at the maximum energy and angle.

6. Future Work

With ongoing refinement of our experimental technique we will focus on radiation detection. Our results for channeling efficiency and the efficiency of the VR process as well as the γ rays from volume reflection suggest a number of follow-on experiments to further improve our understanding of the origin of the radiation as well as considering application of crystals in radiation generation. At FACET-II, experiments with much higher intensity are possible, but if spectral analysis is desired, a device like a Compton spectrometer would be needed, e.g. like the Los Alamos spectrometer.[17]

Acknowledgments

We thank the staff of the SLAC Test Facilities Dept., esp. C. Clarke, C. Hast, K. Jobe, M. Dunning and D. McCormick for their support of our experiments, without which they could not have been successful. SLAC AOSD provided us with stable beams, and we are esp. grateful to J. Nelson and T. Smith for their hard work and diligence in setting up the ESA beam and switching configurations several times for us. This work was supported in part by the US DOE under contract DE-AC02-76SF00515, by the US NSF under contracts PHYS-1068662 and PHYS-1535696 and by the Italian INFN under contract INFN-AXIAL.

References

1. W. Scandale *et al.*, *Phys. Rev. ST Accel. Beams* **11**, 063501 (2008).
2. R. A. Carrigan *et al.*, *Phys. Rev. ST Accel. Beams* **5**, 043501 (2002).
3. S. P. Møller *et al.*, *Phys. Lett. B* **256**, 91 (1991).
4. N. Mokhov *et al.*, in *Charged and Neutral Particles Channeling Phenomena* (World Scientific, Singapore, 2012), Ch. 15, p. 172.
5. N. Mokhov *et al.*, *J. Instrum.* **6**, T08005 (2011).
6. R. A. Carrigan *et al.*, in *Accelerator Physics at the Tevatron Collider* (Springer, New York, 2014), p. 187.
7. U. Wienands *et al.*, *Phys. Rev. Lett.* **114**, 034801 (2015).
8. T. Wistisen *et al.*, *Phys. Rev. Accel. Beams* **19**, 071001 (2016).
9. E. Bagli and V. Guidi, *Nucl. Instrum. Methods Phys. Res. B* **309**, 124 (2013).
10. C. Patrignani *et al.* (Particle Data Group), *Chin. Phys. C* **40**, 100001 (2016).
11. E. Bagli *et al.*, *Eur. Phys. J. C* **77**, 71 (2017).
12. A. I. Sytov *et al.*, *Eur. Phys. J. C* **76**, 77 (2015).
13. T. N. Wistisen *et al.*, *Phys. Rev. Lett.* **119**, 024801 (2017).
14. C. F. Nielsen, GPU-accelerated simulation of channeling radiation of relativistic particles, (2019), submitted to *J. Comput. Phys.*.
15. A. Belkacem, N. Cue and J. Kimball, *Phys. Lett. A* **111**, 86 (1985).
16. J. Kimball, N. Cue and A. Belkacem, *Nucl. Instrum. Methods Phys. Res. B* **13**, 1 (1986).
17. A. E. Gehring *et al.*, Los Alamos National Laboratory Report LA-UR-14-25784 (2014).

Experience with Crystals at Fermilab Accelerators

V. D. Shiltsev

Fermi National Accelerator Laboratory, MS 312,
Batavia, IL 60510, USA
shiltsev@fnal.gov

Crystals were used at Fermilab accelerators for slow extraction and halo collimation in the Tevatron collider, and for channeling radiation generation experiments at the FAST electron linac facility. Here we overview past experience and major outcomes of these studies and discuss opportunities for new crystal acceleration R&D program.

Keywords: Colliders; accelerators; crystals; collimation; halo.

1. Slow Extraction of the Tevatron Beams

Strong interplanar fields in crystals $O(10 \text{ V/Å})$ can be effectively used for various manipulations over charged particles with even the highest energies,[1] such as 1 TeV protons in the Tevatron[2] or 6.5 TeV ones in the LHC.[3]

Following the 1980's exploratory experiments with crystals in secondary particle beams at Fermilab,[5] a series of slow extraction studies at the Tevatron collider, led by R. Carrigan and organized under the umbrella of the Fermilab Experiment E853, had taken place in the 1990s.[6,7] Detailed technical description, scientific results and historical background can be found in Refs. 8 and 9.

Luminosity driven channeling extraction has been observed for the first time using a 900 GeV circulating proton beam (the highest energy at that time) at the superconducting Fermilab Tevatron[8] — see Fig. 1a). The extraction efficiency was found to be about 30%. A 150 kHz beam was obtained during luminosity driven extraction with a tolerable background rate at the collider experiments. A 900 kHz beam was obtained when the background limits were doubled.

In the follow-up E853 development,[9] the beam extraction efficiency was about 25%. Studies of time dependent effects found that the turn-to-turn structure was governed mainly by accelerator beam dynamics. Based on the results of the E853 experiment, feasibility of a parasitic 5–10 MHz proton beam from the Tevatron collider was established.

The extracted proton flux depended strongly on the crystal alignment with typical rms width of ± 2 μrad — see Figs. 1b) and 1c) - and on the source of the particle's diffusions: e.g., it scaled quadratically with the power of the transverse dipole noise heating and linearly with the proton–antiproton luminosity in the range $(0.1–1) \cdot 10^{30}$ cm^{-2}s^{-1}.

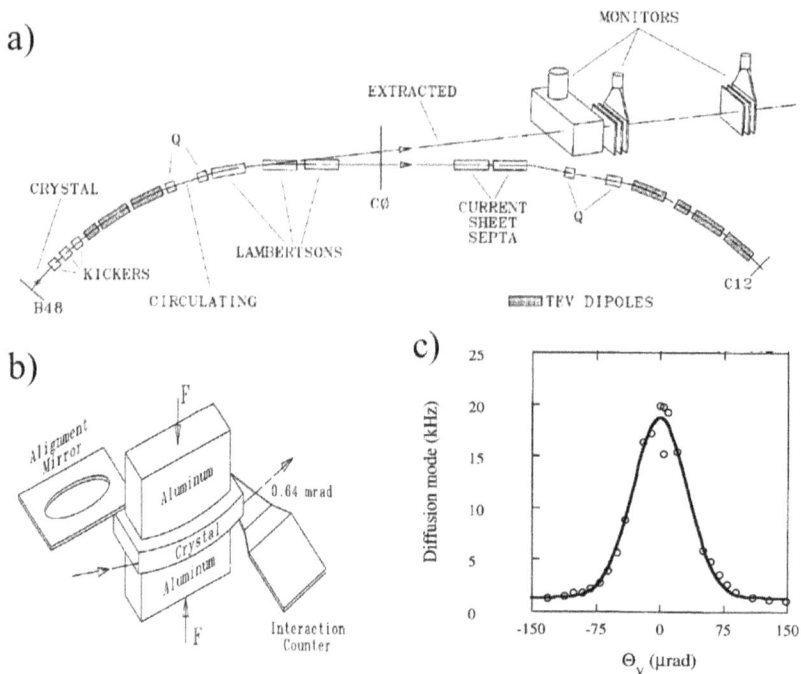

Fig. 1. Crystal assisted proton extraction from the Tevatron: a) schematic of the E853 experiment; b) 39 mm long, 3 mm high, 9 mm wide bent Si crystal cut along (111) plane; c) dependence of the extracted proton flux on angular alignment of the crystal (adapted from Refs. 8 and 9).

2. Halo Collimation by Bent Crystals in the Tevatron

The Tevatron Run II (2001–2011)[2] employed a two-stage collimation system in which a primary collimator is used to increase the betatron oscillation amplitudes of the halo particles, thereby increasing their impact parameters on secondary collimators.[10] A bent crystal can coherently direct channeled halo particles deeper into a nearby secondary absorber, thus, reducing beam losses in critical locations around the Tevatron ring and the radiation load to the downstream superconducting magnets.[11] There are several processes which can take place during the passage of protons through the crystals: a) amorphous scattering of the primary beam; b) channeling; c) *dechanneling* due to scattering in the bulk of the crystal; d) *volume reflection* off the bent planes; and e) *volume capture* of initially unchanneled particles into the channeling regime after scattering inside the crystal. Particle can be captured in the channeling regime, oscillating between two neighboring planes if it enters within crystals angular acceptance (critical angle) of:

$$\theta_c = \sqrt{\frac{2U_0}{pc}}, \tag{1}$$

where p is the particle momentum and U_0 is the crystals planar potential well depth. For 980 GeV/c protons in the Tevatron, the critical angle for (110) plane of a silicon crystal is about 7 μrad. If the particle momentum is not within the critical angle but has a tangency point with the bent planes within the crystal volume, almost all particles are deflected to the opposite direction with respect to the crystal bending - the effect is called the volume reflection (VR) which has very wide angular acceptance equal to the crystal bend angle of the order of hundreds of microradians (vs several microradians of the channeling acceptance). The drawback of the volume reflection regime is that the deflection angle is small, approximately $(1.5–2) \times \theta_c$. However, this can be overcome by using a sequence of several precisely aligned bent crystals, so that the total deflection angle is proportionally larger.

In the Tevatron beam crystal collimation experiment T980[11] both single crystals (for vertical and horizontal deflection) and multi-strip crystal assemblies (for vertical multiple VR) have been used. The bent crystal collimation of circulating beams is very different from that of extracted beams because of smaller initial impact parameter (the depth of the particle penetration at the first interaction with the crystal) and the possibility of interplay of different effects. In an accelerator such as the Tevatron several phenomena determine the impact parameter: diffusion due to scattering on the residual gas and noise in magnetic field $\sim$4 nm/turn$^{1/2}$; the RF noise results in the diffusion rate of $\sim$12 nm/turn$^{1/2}$ (hor.) and $\sim$4 nm/turn$^{1/2}$ (vert.); beam diffusion due to beam-beam or other nonlinear effects can produce up to $\sim$(10–40) nm/turn$^{1/2}$. Interactions with amorphous targets lead to the diffusion rate $\sim$200 μm/turn$^{1/2}$ for a 5 mm length of amorphous silicon, and about $\sim$4 1200 μm/turn$^{1/2}$ for a 5 mm tungsten primary target. Also, of importance transverse orbit oscillations with amplitude of about 20 μm and frequencies of 15 Hz and the 35 Hz (1300 turns) synchrotron motion of particles near the boundary of the RF bucket with amplitudes of some 1 mm (hor) and some 70 μm (vert). The resulting impact parameters are estimated to be of the order of 0.2–1 μm for transverse halo particles and 10–30 μm for the particles in the abort gaps which have leaked out of the RF buckets. All that makes the properties of the surface of the crystal, its roughness and/or the miscut angle (rather than the bulk of the crystal) pivotal for the halo collimation.

Figure 2 shows a schematic of the T-980 experimental layout, some critical hardware and results. During normal Tevatron operations, a 5-mm tungsten target scatters the proton beam halo into a 1.5-m long stainless steel secondary collimator E03, 50 m downstream of the target. For the bent crystal experiments, a goniometer containing single or multi-strip bent crystals is installed 23.7 m upstream of the E03 collimator. Scintillation counter telescopes detect secondary particles from protons interacting with the target and E03 collimator. An ionization chamber (beam loss monitor LE033) also detects secondary particles scattered from E03. A PIN diode telescope detects the secondaries scattered from the bent crystal. Under the above configuration, channeled beam is signaled by a reduction of the rate in the PIN telescope (channel LE033C) — as shown in Fig. 3. Depending on the angle between the incident particle's orbit and the Si crystal plane, observed are

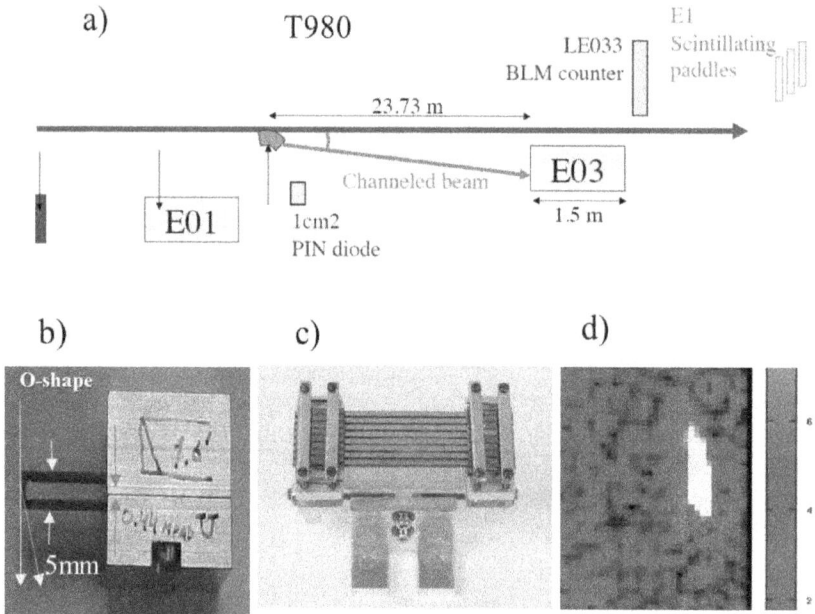

Fig. 2. Bent crystal collimation in the Tevatron: a) schematic of the T980 setup; b) *O*-shaped 5-mm silicon crystal with a bending angle of 0.44 mrad; c) eight strip crystal assembly for consecutive volume reflections; d) image of a pixel detector near E03 collimator showing the channeled beam profile (adapted from Refs. 2 and 11).

channeling dip with a width of 22±4 μrad (rms), volume reflection plateau and amorphous scattering. The width of the channeling dip is a convolution of the beam divergence, the channeling critical angle, multi-pass channeling effects and possible crystal distortions. At the bottom of the dip, the LE033C signal is about 20% of the signal at a random angular setting. This depth is a measure of the channeling efficiency of 80%.

The crystal collimation had been used during many collider stores in 2010's with peak luminosities upto $400 \cdot 10^{30}$ cm^{-2}s^{-1}. Since 2009, the system had push-pull crystal goniometers in both planes and a two-plane crystal collimation has been attempted for the first time. High-resolution pixel telescope was installed in front the downstream collimator and allowed to measure channeled and volume reflected beam profiles at those location — see Fig. 2d). Also, observed were multiple-volume reflected protons (off an 8-crystal assembly in which each strip coherently adds a 8μrad bend to the same proton) and a two-fold reduction of beam losses around the ring due to crystal channeling, specifically in the CDF and D0 collider detector regions.

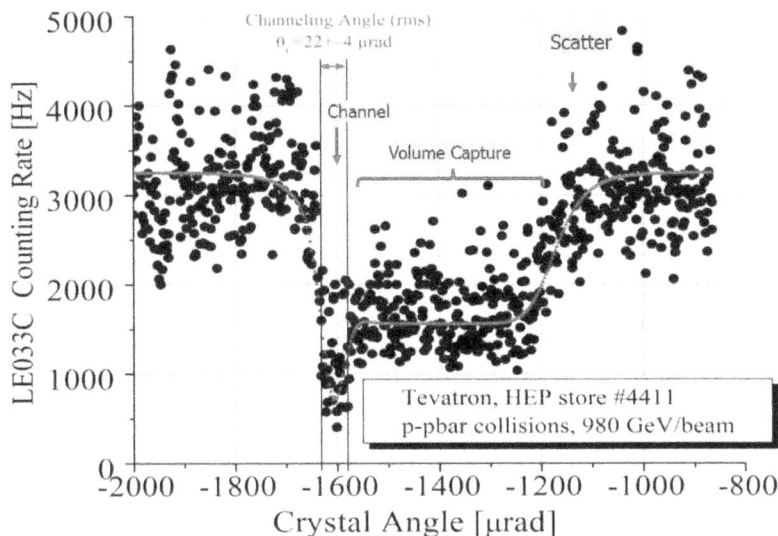

Fig. 3. The scan of the near crystal radiation counter signal (interaction probability of 980 GeV protons with Si) vs crystal alignment angle (adapted from Ref. 2).

3. Crystal Channeling Radiation Experiment at FAST

Channeling radiation (CR) can be generated when charged particles such as electrons or positrons pass through a single crystal parallel to a crystal plane or axis. The main advantage of the CR is to produce quasi-monochromatic high energy X-rays using a low energy electron beam below 100 MeV (compare with the synchrotron radiation, currently the main X-ray source that requires a few GeV electron beams to generate X-rays of tens of keV).

Recently constructed Fermilab Accelerators Science and Technology (FAST) facility[12] hosted series of studies aimed to produce high brightness X-rays using a low-emittance 50 MeV electron beam, and to demonstrate that CR can be used as a compact high-brightness X-rays source[13,14]. The FAST injector[15] (see main parameters in Table 1) which consists of a CsTe photocathode located in a 1+1/2-cell RF gun followed by two L-band (1.3 GHz) superconducting accelerating structures can generate a low emittance electron beam. The electron energy can reach up to 50 MeV downstream of the last superconducting cavity. A 160 μm thick diamond single crystal (with some 1 mrad critical angle for 43 MeV electrons) was installed and oriented so that the electron beam propagates parallel to the (110) plane of the crystal — see Fig. 4.

Many channeling spectra were summed together for 30-fC/pulse electron beam in Fig. 4d) indicating three peaks over the bremstrahlung (BS) background spectrum at 75 keV, 87 KeV and 108 keV.[16] Originally, the peaks were expected at

Table 1. Main parameters of the FAST electron beam.

Parameter	FAST 2019	ILC specs.	Comments
Beam energy: max.	301 MeV	300 MeV	100 MeV for IOTA
low-energy area	20–50 MeV		typical 34–43 MeV
electron gun	4–5 MeV		typical 4.5 MeV
Bunch charge	0.1–3.2 nC	3.2 nC	typical 0.5 nC, depends on number of bunches
Bunches per pulse	1–1000	3000	typical 100, 3 MHz rate
Pulse length	upto 1 ms	1.0 ms	typical 0.01–0.2ms
Pulse rep.rate	1 Hz	5 Hz	
Transv. emittance (n, rms)	1–5 μm	5 μm	grows with intensity
Bunch length (rms)	1.2–2.4 mm	1 mm	without compression

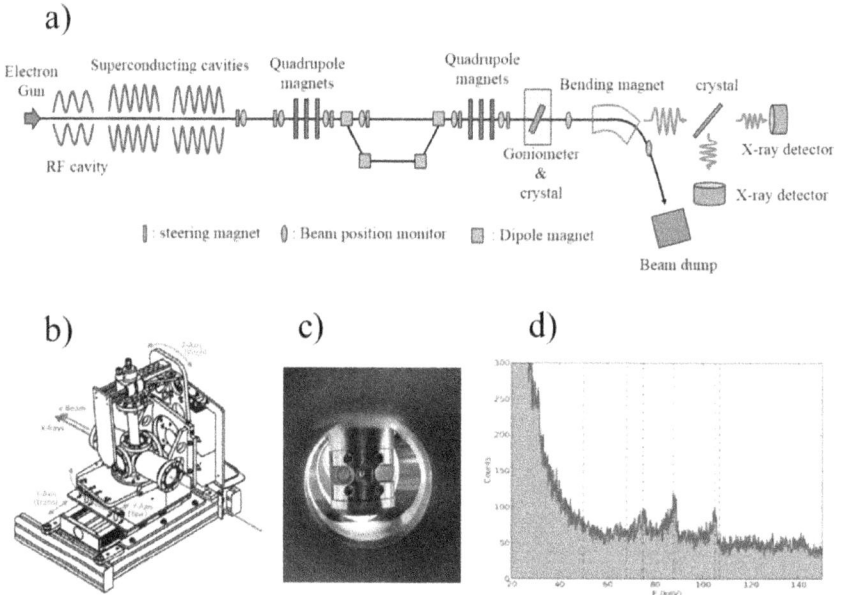

Fig. 4. Crystal channeling radiation experiment at FAST: a) schematic view of the experiment; b) crystal goniometer assembly; c) crystal goniometer assembly installed in low energy electron beamline; d) energy spectrum of emitted X-rays (adapted from Refs. 16 and 17).

51 keV, 67 keV with much higher signal above the BS background[17] and the discrepancy might be explained by the effects of nonlinear detector performance in the radiation environment, crystal morphological changes due to aging, or the electron beamline energy measurement offset.

4. Possibilities for CNT Acceleration Tests at FAST

Initial outlook for "ultimate" future energy frontier collider facility with beam energies 20–100 times the LHC energy indicates promising potential of a compact 1 PeV linear crystal collider based on acceleration of muons (instead of electrons or hadrons) in super-dense plasma of crystals was[18]. Nanostructures, such as carbon nanotubes (CNTs) can also be used offering a number of advantages over crystals for a proof-of-principle experiment (wider channels and weaker dechanneling; possibility to accept broader beams using nanotube ropes, easier 3D control of beam bending over greater lengths).[19] Simulations of the proposal of such a study at FAST[20] assume excitation of wakefields in the CNT plasma by short intense pre-modulated electron bunch — see Fig. 5. The beam density modulation (microbunching) is needed for resonant excitation of the wakefields at the CNT plasma wavelength $\lambda_p \sim 100\,\mu$m and can be arranged via, e.g., the use of a narrow vertical slit mask in the chicane section of the low-energy FAST beamline, as depicted in Fig. 5c).[21] The effect of $\sim$300 keV CNT acceleration will manifest itself as widening of the vertical image size on the beam image screen after the spectrometer magnet.

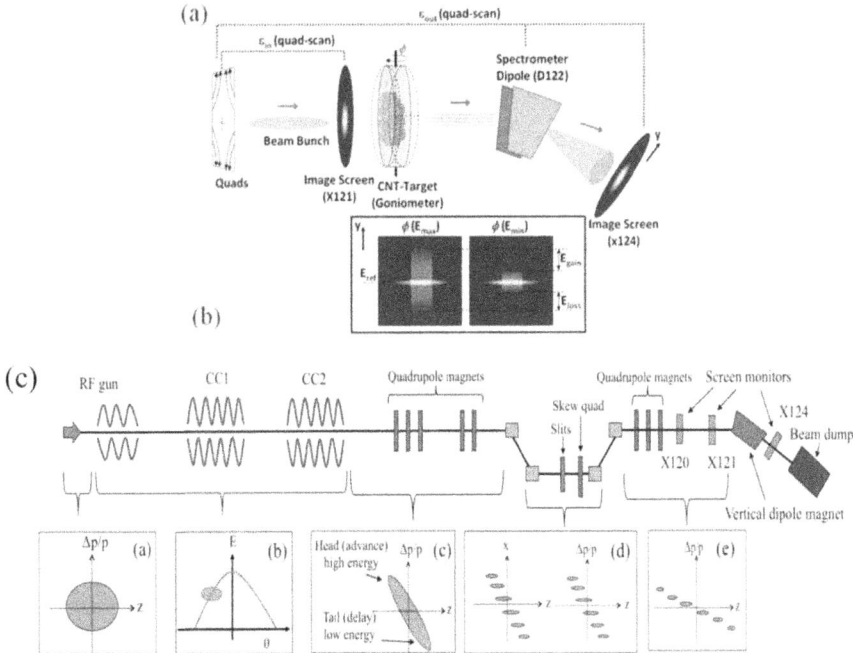

Fig. 5. Proof-of-principle CNT acceleration experiment proposal (see text): a) focusing pre-modulated electron beam on a CNT plate, b) simulated beam image on a screen after the vertical bending spectrometer magnet with(left) and without(right) proper CNT alignment; c) concept of the THz microbunching of the FAST electron bunch using vertical slit-mask in the chicane (adapted from Refs. 20 and 21).

Another possibility of using 300 MeV electron beam at FAST for effective muon production was discussed in Ref. 22 where efficiency of $e \rightarrow \mu$ conversion was estimated to be $O(10^{-8})$. FAST beams can be used for related studies of the electron beam filamentation (Weibel instability) in solid density plasmas, muon detection, calibration of theoretical models and integration of future experiments on acceleration in crystals at lager facilities, such as, e.g., FACET-II.

5. Conclusions

Fermilab has a long history of research and operational use of crystals in high energy accelerators starting with employment of them in extracted and secondary beamlines in 1980's, followed by the pioneering E853 experiment on crystal assisted proton extraction by R. Carrigan *et al.* during the Tevatron Collider Run I in 1990's and the T980 crystal collimation studies in 2005–2011 by N. Mokhov *et al..* Recently constructed FAST 50–300 MeV electron linac hosted crystal channeling radiation studies and offers opportunities for various experiments towards proof-of-principle studies of acceleration in crystals and nanostructures. Significant past experience and available hardware can be very helpful for future exploration toward pre-FACET II crystal acceleration experiment (integration and tests of detectors); particle channeling in CNTs; muon production and capture. Together with other possible experiments at FACET-II at SLAC, BELLA at LBNL and AWAKE at CERN, with high energy high-Z ions available at RHIC or LHC or with self-modulated electron beams in the SASE FEL facilities, like, e.g. the LCLS-I and -II at SLAC, these studies at FAST can provide new insights into the feasibility of the concept of beam acceleration in solid-state plasma of crystals or nanostructures like CNTs, the promise of ultra-high accelerating gradients $O(1$–$10)$ TeV/m in continuous focusing channels and, thus, prospects of the method for future high energy physics colliders.

Acknowledgments

Some materials of this summary were also presented at the *International Particle Accelerator Conference* (IPAC2010, May 23–28, 2010, Kyoto, Japan)[23] and in Refs. 2, 10 and 11. I greatly appreciate years of fruitful collaboration on the subjects of this presentation with G. Annala, R. Carrigan, A. Drozhdin, T. Johnson, A. Legan, N. Mokhov, T. Sen, R. Reilly, D. Still, R. Tesarek, J. Thangaraj, and J. Zagel (FNAL), R. Assmann, V. Previtali, S. Redaelli, and W. Scandale (CERN), Y. Chesnokov and I. Yazynin (IHEP, Protvino, Russia), V. Guidi (INFN-Ferrara, Italy), P. Piot and Y. M. Shin (NIU), Y. Ivanov (PNPI, Russia). Special thanks to Dean Still and Nikolai Mokhov for useful discussions and help with finding the materials and photos reproduced in this article.

Fermi National Accelerator Laboratory is operated by Fermi Research Alliance, LLC under Contract No. DE-AC02-07CH11359 with the United States Department of Energy.

References

1. V. Biryukov, Y. Chesnokov and V. Kotov, *Crystal Channeling and its Application at High-Energy Accelerators* (Springer, 2013).
2. V. Lebedev and V. Shiltsev (eds.), *Accelerator Physics at the Tevatron Collider* (Springer, 2014).
3. W. Scandale *et al.*, Observation of channeling for 6500 GeV/c protons in the crystal assisted collimation setup for LHC, *Phys. Lett. B* **758**, 129 (2016).
4. V. Shiltsev, High-energy particle colliders: past 20 years, next 20 years, and beyond, *Physics-Uspekhi* **55**(10), 965 (2012).
5. S .Baker *et al.*, Deflection of charged particles in the hundred GeV regime using channeling in bent single crystals, *Phys. Lett. B* **137**(1–2), 129 (1984).
6. R. Carrigan, T. Toohig and E. Tsyganov, Beam extraction from TeV accelerators using channeling in bent crystals, *Nucl. Instrum. Methods Phys. Res. B* **48**(1–4), 167 (1990).
7. R. Carrigan *et al.*, Extraction from TeV-range accelerators using bent crystal channeling, *Nucl. Instrum. Methods Phys. Res. B* **90**(1–4), 194 (1994).
8. R. Carrigan *et al.*, First observation of luminosity-driven extraction using channeling with a bent crystal, *Phys. Rev. ST Accel. Beams* **1**(2), 022801 (1998).
9. R. Carrigan *et al.*, Beam extraction studies at 900 GeV using a channeling crystal, *Phys. Rev. ST Accel. Beams* **5**(4), 043501 (2002).
10. N. Mokhov *et al.*, Tevatron beam halo collimation system: design, operational experience and new methods, *JINST* **6**(08), T08005 (2011).
11. N. Mokhov *et al.*, Crystal collimation studies at the Tevatron (T-980), *Int. J. Mod. Phys. A* **25**, 98 (2010).
12. S. Antipov *et al.*, IOTA (Integrable Optics Test Accelerator), Facility and experimental beam physics program, *JINST* **12**, T03002 (2017).
13. T. Sen and C. Lynn, Spectral brilliance of channeling radiation at the ASTA photoinjector, *Int. J. Mod. Phys. A.* **29**, 1450179 (2014).
14. D. Mihalcea *et al.*, Channeling Radiation Experiment at Fermilab ASTA, in *Proc. IPAC2015* (Richmond, VA, USA, 2015), pp.95–98.
15. D. Broemmelsiek *et al.*, Record high-gradient SRF beam acceleration at Fermilab, *New J. Phys.* **20**(11), 113018 (2018).
16. A. Halavanau *et al.*, Commissioning and First Results From Channeling Radiation At FAST, arXiv:1612.07358 (2016).
17. J. Hyun, P. Piot and T. Sen, Optics and bremsstrahlung estimates for channeling radiation experiments at FAST, arXiv:1802.06113 (2018).
18. V. Shiltsev, High-energy particle colliders: past 20 years, next 20 years, and beyond, *Physics-Uspekhi* **55**(10), 965 (2012).
19. Y. M. Shin, D. Still and V. Shiltsev, X-ray driven channeling acceleration in crystals and carbon nanotubes, *Phys. Plasmas* **20**(12), 123106 (2013).
20. Y. M. Shin, A. Lumpkin and R. Thurman-Keup. TeV/m nano-accelerator: Investigation on feasibility of CNT-channeling acceleration at Fermilab, *Nucl. Instrum. Methods Phys. Res. B* **355**, 94 (2015).
21. J. Hyun *et al.*, Micro-bunching for generating tunable narrow-band THz radiation at the FAST photoinjector, arXiv:1808.07846 (2018).
22. V. Shiltsev and S. Striganov, presented at *the 7th Annual IOTA/FAST Collaboration Meeting and Workshop on High Intensity Beams in rings, June 10–12, 2019, Fermilab*; https://indico.fnal.gov/event/20279/
23. V. Shiltsev *et al.*, Channeling radiation experiment at Fermilab ASTA, in *Proc. 1st Int. Part. Accel. Conf. (IPAC'2010)* (May 23–28, 2010, Kyoto, Japan), pp. 1242–1245.

Schemes of Laser Muon Acceleration: Ultra-short, Micron-scale Beams

Aakash A. Sahai

College of Engineering and Applied Science,
University of Colorado, Denver, CO 80204, USA
aakash.sahai@gmail.com

Toshiki Tajima

Department of Physics & Astronomy and Applied Physics,
University of California, Irvine, CA 92697, USA

Vladimir D. Shiltsev

Accelerator Division, Fermi National Accelerator Laboratory,
Batavia, IL 60510, USA

Experimentally accessible schemes of laser muon ($\mu^{\pm}$) acceleration are introduced and modeled using a novel technique of controlled laser-driven post-processing of cascade showers (or pair plasmas). The proposed schemes use propagating structures in plasma, driven as wakefields of femtosecond-scale high-intensity laser, to capture particles of divergent cascade shower of: (a) hadronic type from proton-nucleon or photo-production reactions or, (b) electromagnetic type. Apart from the direct trapping and acceleration of particles of a raw shower in laser-driven plasma, a conditioning stage is proposed to selectively focus only one of the charge states. Not only is the high gradient that is sustained by laser-driven plasma structures well suited for rapid acceleration to extend the lifetime of short-lived muons but their inherent spatiotemporal scales also make possible production of unprecedented ultrashort, micron-scale muon beams. Compact laser muon acceleration schemes hold the promise to open up new avenues for applications.

Keywords: Laser-plasma muon acceleration; muon-antimuon pair-plasma.

1. Laser Wakefield Particle Acceleration in Gaseous Plasmas

Laser-plasma electron accelerators[1] enabled by Chirped-Pulse Amplified (CPA) lasers[2] are now capable of producing several GeV[3] electron (e^-) beams in centimeter-scale gas plasmas. These widely prototyped accelerators have demonstrated propagating acceleration structures, driven as wakefields of a CPA laser pulse, that sustain average gradients of several tens of GVm^{-1} and acceleration of a few percent energy spread e^- beams in centimeter-scale gas plasmas.

Apart from this, a two-stage laser e^- accelerator[4] and a laser positron (e^+) accelerator[5] are also under active investigation.[6] These advances in acceleration techniques coupled with the rapid ongoing development of CPA lasers open up the possibility of an affordable high-energy physics (HEP) e^+-e^- collider[7] at the energy-frontier. Orders of magnitude increase in acceleration gradients is expected to enable significant reduction in the size and cost of accelerator machines that underlie a collider. Apart from high-energy physics, these compact accelerators are also expected to have a wide variety of applications in medical,[8] light-source and imaging technologies.[9]

However, laser acceleration of exotic particles remains largely unexplored. Laser acceleration of positrons in gaseous plasmas has been introduced using an innovative model of controlled interaction of positron-electron pair plasmas or particle showers with laser driven plasmas.[5] There have been recent efforts towards experimental prototyping of laser positron accelerator.[6]

In this work we introduce mechanisms of laser acceleration of muon $(\mu^{\pm})$,[10] an exotic fundamental particle which as a second generation lepton. The mechanism introduced here follows the same methodology as the laser positron accelerator[5] and is well within the reach of the experimental capabilities of existing laser-plasma and RF acceleration test facilities. We propose and analyze the short-term experimental viability of various muon production schemes and demonstrate the processing and acceleration of the generated muons using laser-driven plasmas.

In Sec. 2, scientific and technological applications of muons are reviewed and the potential of a compact muon source towards novel applications is discussed. Existing and well-studied techniques of muon production using conventional methods such as beams from traditional RF accelerators as well as techniques that utilize novel physics of uncommon interaction processes are reviewed in Sec. 3. The parameter regime for laser muon acceleration in gaseous plasmas is explored in Sec. 4 and the muon-antimuon source parameters desired for matching to laser-driven plasma acceleration structures are identified.

An evaluation of three distinct mechanisms for micron-scale muon and antimuon production, whose source properties are matched to the laser acceleration parameters as discussed Sec. 4 and that are suitable for feeding a laser muon accelerator, are presented in Sec. 5. The experimental viability of these schemes of muon production for conversion efficiency and matching to a laser-driven post-processing stage is also analyzed. Preliminary Particle-In-Cell simulations are presented in Sec. 6, to demonstrate the trapping and acceleration of muons produced as part of cascade showers using photo-production method. In Sec. 7, a mechanism for segregation of oppositely charged species of muons by selective focusing is proposed and examined using a plasma lens.

The paper concludes by summarizing various sections and a plan for future work. An important part of the future work and application of ultrashort, micron-scale muon beam is for their injection into crystal wakefield accelerators such as attosecond x-ray pulse or submicron charged particle beam driven[11-13] or submicron particle beam driven solid-state tube accelerators.[14,15] The main advantage of using muons in solid-state particle accelerators is that being second generation leptons with mass around 200 times that of first generation leptons, electrons and positrons, their synchrotron radiation losses that are $\propto (m_e/m_\mu)^4$ in high focusing fields of a plasma wave are vastly reduced. Moreover, the radiative losses (including bremsstrahlung and pair production) of muons also favorably scale as $\propto (m_e/m_\mu)^2$. Also, muons can possibly be accelerated to very high energies in a single stage continuous focusing system of crystal or CNT-based linacs.[16]

2. Significance and Applications of Muons ($\mu^\pm$)

A compact muon source is attractive due to the distinct properties that are inherently embodied in muons. These unique properties of muons in comparison with particles that have been conventionally accessible for various applications like electrons ($e^\pm$), photons (γ) or protons (p, $\bar{p}$), have not only been key to enable a wide range of technologies[17] but have also played a pivotal role in exploration of an alternative energy-frontier collider design. Being a heavier lepton, the point-like characteristic of muon despite its higher mass, $m_\mu \simeq 207 m_e$ provides precision of collision point energy over p-p or p-$\bar{p}$ collisions.[18] Moreover being heavier than $e^\pm$, muons have lower synchrotron radiation ($\propto m_\mu^{-4}$, over $e^\pm$) and radiative losses ($\propto m_\mu^{-2}$, over $e^\pm$) even at higher energies (E_μ) which enables deeper penetration depth in materials and greater stability of high E_μ storage-rings. The weak force mediated $\mu^\pm$ decay ($\tau_\mu \simeq 2.2\mu s\ E_\mu/m_\mu$)[19] has permitted neutrino flavor oscillation studies ($\nu/\bar{\nu}_{e,\mu}$) through high-intensity ($N_\nu \propto E_\mu^2$) ν-production.[20] Muon sources however currently demand many tens of meters of proton accelerators[21] under 100MVm^{-1} gradient limit.[22]

Due to lack of affordable technologies for controlled muon sources available thus far, raw cosmic $\mu^\pm$ flux from extensive air showers (1cm^{-2}min$^{-1} \gtrsim$ 1GeV with $\cos^2\theta$ fall-off from vertical) has been used in an expanding range of applications of muography[17] in nuclear threat detection,[23] archaeology,[24] geosciences[25] that require long stopping range ($E_\mu \gtrsim 1$ GeV). Slow $\mu^\pm$ with short stopping range, from lab-based sources are also widely used in material,[26] molecular[27] and medical sciences,[28] etc. through $\mu^\pm$ spin relaxation (μSR) spectroscopy.[29] Muons are also attractive for research in areas like true-muonium[30] (μ^+-μ^-) atomic physics and μ^--catalyzed fusion,[31] etc.

Development of compact and tunable muon sources with controllable E_μ spectrum, ultra-short bunch lengths and micron-scale transverse properties, as studied here, is therefore attractive not only for technological but also for HEP applications[11-15] which additionally demand high average flux and ultra-low emittance.

3. Conventional $\mu^\pm$ Sources: Hadronic Shower and Direct Production

Conventionally, production of muons in a laboratory environment has been realized using several distinct processes.

Hadronic showers,[32] produced through proton-nucleon reactions when hundreds of MeV proton beams[33,34] strike targets, predominantly contain $\pi^\pm$-mesons (pions and fractionally other mesons), with large energy and angular spread, that undergo spin-polarized decay to $\mu^{\pm}$.[32]

Hadronic $\pi^{\pm}$-$\mu^{\pm}$ showers also produced through photo-meson reaction[37] have been modeled using $\gtrsim 140$MeV electron beam undergoing bremsstrahlung in a target to produce MeV-scale $\pi^{\pm}$ that decay to $\mu^{\pm}$. At electron beam energies near $\pi^{\pm}$ mass, photo-pion production process dominates whereas contributions of other simultaneous processes towards $\pi^{\pm}$-meson production such as Bethe-Heitler (BH) $\mu^{\pm}$ pair production mediated by a nuclei and $\pi^{\pm}$ electro-production (trident-like) process mediated by virtual photons during inelastic scattering of electrons off of a nuclei, is relatively small.

Efficient $\mu^{\pm}$ production from hadronic showers demands methods for confinement of the divergent $\pi^{\pm}$-$\mu^{\pm}$ flux.[35,36] A high proton to $\mu^{\pm}$ yield requires methods[21,35] to capture and rapidly accelerate $\mu^{\pm}$ from hadronic showers to many times the rest-mass energy of muons over $\lesssim c\tau_{\mu}$[18] as well as to simultaneously cool the $\mu^{\pm}$ phase-space obtained from the divergent $\pi^{\pm}$-$\mu^{\pm}$ flux. Moreover, $\pi^{\pm}$-decay lifetime demands many meters long confinement channel ($c\tau_{\pi} \approx 8E_{\pi}/m_{\pi}\ m$).[38]

Direct $\mu^{\pm}$ production processes have also been studied. BH muon pair production using an electron beam of much higher-energy than $\pi^{\pm}$-$\mu^{\pm}$ rest mass undergoing bremsstrahlung is well established.[39] Direct $\mu^{\pm}$ production using e^{-}-photon scattering[40] requires head-on collision between GeV-scale e^{-} beam and tens of MeV photons.[41] But, scaling up the yield and energy spectrum of MeV-scale photon itself relies on a high-degree of precision to make possible Compton scattering interaction. Breit-Wheeler (BW)[42] $\mu^{\pm}$ pair production (time-reversal symmetry of $\mu^{\pm}$ annihilation), as opposed to BH pair production, requires $\gtrsim 212$ MeV center-of-mass photon-photon collision which significantly increases the complexity of simultaneous control over two hundreds of MeV photon sources (in contrast with BH process which requires one photon and a heavy nuclei).

Beam mediated direct $\mu^{\pm}$ production from photons interacting with a relativistic nuclei[43] not only requires hundred MeV-scale photons possibly being obtained using bremsstrahlung but also a highly relativistic ion beam. Nonlinear BW process (multi-photon) is more accessible using CPA lasers but suffers from ultra-low yield. Direct $\mu^{\pm}$ production using e^{+} and e^{-} annihilation in a stationary target[45] which requires $\gtrsim 43$GeV e^{+} beam and Relativistic heavy ion collisions[46] both depend on kilometer-scale machines. Resonant annihilation during ring stored e^{+} and e^{-} beams to directly produce $\mu^{\pm}$ pairs also requires hundreds of meters of rf accelerators.[47]

With the advent of advanced acceleration methods[1] laser-driven electron accelerators have been proposed for hadronic shower production using laser-plasma accelerated (LPA) multi-GeV e^{-} interacting with a target has been modeled.[48] Compact and tunable $\mu^{\pm}$ sources and corresponding laser muon acceleration schemes however still remain vastly unstudied and thus yet unrealized.

4. Laser Muon ($\mu^{\pm}$) Acceleration:
Rapid Acceleration, Ultrashort Bunches & Micron Spot-size

In this paper, compact and tunable schemes of laser muon acceleration are proposed and studied. These schemes of laser muon acceleration are based upon the interaction of controlled laser-driven plasma sustaining traveling plasma density structures with matched laser produced cascade showers which when under external confinement are also labelled as pair or exotic matter plasmas. The process of matching of cascade shower characteristics such as its energy spectra and transverse phase parameters is explored here.

4.1. *Properties of laser wakefield acceleration structures in plasmas*

The acceleration and focusing gradients (E_{plasma}) inherently sustained in laser-driven plasmas[1]:

$$E_{\text{plasma}} \simeq \sqrt{n_0(10^{20}\ \text{cm}^{-3})}\ \text{TVm}^{-1} \tag{1}$$

where n_0 is the plasma electron density, are especially well suited for trapping and acceleration of short lifetime (τ_π, τ_μ) particles. Trapping short lifetime heavier particles ($\pi^{\pm}, \mu^{\pm}$) that are predominantly produced from the source at a small fraction of the speed of light required that the velocity of the laser-driven structures is small enough to allow significant interaction time. The plasma density also plays a critical role as the group velocity of a laser (β_g^{laser}) and thus of the co-propagating acceleration structure in the plasma (β_{acc}) is:

$$\beta_g^{\text{laser}} \simeq \beta_{acc} \simeq \sqrt{1 - \frac{1}{10\pi}\ n_0(10^{20}\ \text{cm}^{-3})\ \lambda_0^2(\mu m)\ r_e(\text{fm})} \tag{2}$$

where, λ_0 is the laser wavelength and r_e is the classical electron radius, 2.818 fm.

Rapid acceleration within the lifetime of unstable particles to many times their rest-mass energy, using high-gradient laser-driven acceleration structures, extends their lifetime proportional to the Lorentz factor acquired through acceleration. This rapid extension of lifetime enabled by high-gradient laser acceleration thus increases the efficacy of unstable particles for applications.

While the high focusing fields are quite effective for initial trapping, muons being a heavier lepton undergo smaller transverse oscillations driven radiation losses under off-axis displacement relative to electrons interacting with equivalent fields.

Moreover, the natural dimension of laser acceleration structures in plasma is of the order of plasma wavelength (λ_{plasma}),

$$\lambda_{\text{plasma}} \simeq \frac{3.3}{\sqrt{n_0(10^{20}\ \text{cm}^{-3})}}\ \mu m \simeq L_{\text{beam}}, r_{\text{beam}} \tag{3}$$

which makes possible for the spot-size of the accelerated particle beams (r_{beam}) to be micron scale in plasma. Similarly, the bunch-lengths (L_{beam}) when using laser accelerators is also of the order of the plasma wavelength which has a range of a few to tens of microns. Thus, the muon beam when accelerated within a plasma acceleration structure is not only micron-scale transversely but also has an ultra-short bunch-length. This inherently micron-scale dimensionality of the acceleration structures makes possible unprecedented energy density of exotic particle beams when accelerated using laser accelerator schemes, as proposed here, if a sufficient number of exotic particles are effectively produced and trapped.

For a Ti:Sapphire CPA laser with characteristic center wavelength, $\lambda_0 = 0.8\,\mu$m, laser group velocity (β_g^{laser}, Eq. (2)) and plasma wavelength ($\lambda_{\mathrm{plasma}}$, Eq. (3)) with plasma density is shown in Fig. 1.

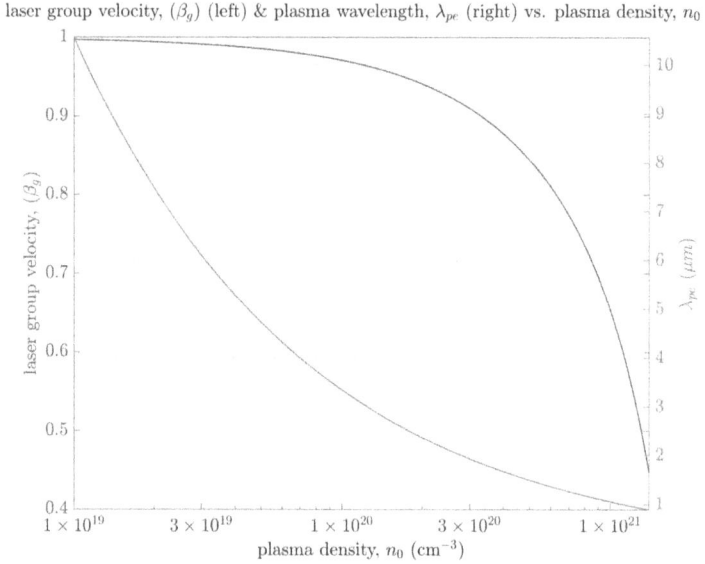

laser group velocity, (β_g) (left) & plasma wavelength, λ_{pe} (right) vs. plasma density, n_0

Fig. 1. Laser group velocity (β_g^{laser}, Eq. (2)) and plasma wavelength ($\lambda_{\mathrm{plasma}}$, Eq. (3)) with plasma density for a Ti:Sapphire (active medium) CPA laser with central wavelength, $\lambda_0 = 0.8\,\mu$m.

4.2. Matching of cascade showers with laser-driven plasmas

As evident from the micron-scale sizes of acceleration structure in laser-driven plasma it is however important that the exotic particle source that couples the particles into the acceleration structure is precisely controllable. The source should have the ability to constrain the produced exotic particles within a small transverse spot-size in addition to producing an energy spectrum which optimizes the capture efficiency through longitudinal interaction dynamics.

Therefore, it is quite critical to understand various limits under which mechanisms that produce cascade shower can deliver a micron-scale spot-size to match with the transverse size of the laser acceleration structures or at least be of the same order of size. For further enhancement of the efficiency of capture from these exotic particle sources it is essential to understand and control its energy spectra and transverse phase-space properties by varying the beam and target properties. Through detailed characterization of shower properties over the drive beam and target property parameter-space the laser post-processing stage can be optimized to match with the characteristics of the shower.

The exotic particles produced from a controlled source are post-processed using a laser-driven plasma acceleration structure. The laser-driven plasma acceleration structure may be a plasma wave or a slowly propagating charge separation structure. By varying the laser-plasma properties of the post-processing stage the proposed schemes seek to enable a match with the pair plasma properties.

In this paper, schemes of laser muon acceleration are proposed, as listed below, and examined:

(1) Laser-driven post-processing of electron and/or positron beam driven hadronic shower to trap and accelerate muon as well as pion pairs.

(2) Laser-driven post-processing of proton beam driven hadronic showers to trap and accelerate muon as well as pion pairs.

(3) Laser-driven positron-electron storage ring for tunable muon pair production through positron-electron annihilation.

Apart from the direct interaction between cascade showers and laser-driven plasma wave, it is also possible to segregate a species of one charge state before the interactions. To enable this indirect interaction with a conditioned cascade shower, a plasma lens[52] is proposed to be inserted between the production and post-processing stage. In the conditioning state, the discharge current direction (sign of the external voltage) decides the sign of the particle species that will get focused by the plasma lens.

5. Matched Muon ($\mu^{\pm}$) Production for Laser-Plasma Muon Acceleration

In this section, we present the underlying mechanisms and corresponding analytical evaluation of muon-antimuon production schemes that produce muon phase-space that are likely to be matched with the laser muon accelerator properties outlined in Sec. 4.

5.1. *Scheme I:*
Laser-driven plasma based post-processing of photo-produced hadronic shower ($e^{\pm}$-beam driven)

Electron and positron beams undergo bremsstrahlung radiation loss when propagating in materials as they experience change in velocity due to the electromagnetic force of the material nuclei in their propagation path. The energy of the radiation emitted depends inversely on square of the particle mass undergoing bremsstrahlung. The rate of energy loss of the particle is known to be directly proportional to the particle energy ($dE/dx = E/X_0$, where, E is the nominal energy of an arbitrary particle undergoing bremsstrahlung radiation, x is the coordinate along penetration in the material and X_0 is the radiation length of the material.

From its definition, X_0 is the penetration depth at which the particle energy reduces to $1/e$ of its initial value). The photon energy spectrum follows the Bethe-Heitler (BH) function, $d\sigma_\gamma/d\varepsilon_\gamma = \frac{4}{3}\frac{1}{X_0} F(\varepsilon_\gamma, E_\square) \varepsilon_\gamma^{-1}$, where $F(\varepsilon_\gamma) \simeq 1 - \frac{\varepsilon_\gamma}{E_\square} + \frac{3}{4}\left(\frac{\varepsilon_\gamma}{E_\square}\right)^2$. It is to be noted that as in the discussion in Tsai (1974),[39] BH formalism does not include several effects attributed to the nuclei in the material such as screening of nuclear field by atomic electrons and the shape of the nuclei, etc.

Bremsstrahlung radiation from an electron or positron beam with energies much higher than the rest-mass energy of muons and pions ($E_{e\pm} \gg 140$ MeV) trigger photo-production reactions in the presence of the nuclei in materials:

(R.1) photo-meson reaction:
$$\gamma + p \to \pi^+ + n$$
$$\gamma + n \to \pi^- + p$$
$$\gamma + Z_1 \to \pi^\square + Z_2 \text{, and}$$

(R.2) Bethe-Heitler muon pair-production reaction:
$$\gamma + Z \to \mu^+\mu^- + Z \text{.}$$

These reactions result in the production of $\mu^\pm(\text{-}\pi^\pm)$ flux in addition to the $e^\pm$ electromagnetic shower (primarily through the BH positron-electron pair-production process). The photo-meson reaction in (R.1), that produces $\pi^\pm$ flux through bremsstrahlung photon interaction with the nuclei has a differential cross section which is at least one order of magnitude higher than that of the BH muon pair-production. But, for centimeter-scale thick targets it is expected that the $\pi^\pm$ flux component gets suppressed and de-collimated due to absorption and scattering off of the nuclei after pions are produced. On the contrary, thinner targets while increasing the $\pi^\pm$ flux result in the suppression of the $\mu^\pm$ flux.

These photo-production reactions, (R.1) and (R.2), dominate the production of hadronic $\mu^\pm$ showers. The inelastic electron scattering reaction, $e + Z \to e' + Z + \pi^\square$ has a differential cross-section which is less than about 1% of the cross section of above reactions when the electron energy is much higher than the threshold energy of ~ 140 MeV. This considerably smaller differential cross-section for the inelastic scattering process in the eZ process is due to the extra electromagnetic vertex associated with a virtual photon, $d\sigma_{eZ} \propto \alpha^4 Z^2$ as compared to $d\sigma_{\gamma Z} \propto \alpha^3 Z^2$.

The cross-section of BH muon pair-production has been estimated over a wide range of parameters. In the case where the electron or positron beam that is used is ultra-relativistic ($\gamma_{e\pm} \gg 1$) such that the typical bremsstrahlung photon energy, ε_γ is much higher than muon pair rest-mass $2\ m_\mu c^2$, then the integrated cross-section of the photo-production reaction, $\gamma + Z \to \mu^+\mu^- + Z$ can be simplified.[48]

$$\varepsilon_\gamma \gg 2m_\mu c^2$$

$$\varepsilon_\gamma \sim E_{e\pm} \gtrsim 3\,\text{GeV (for validity of below BH cross-section)}$$

$$\sigma_{\gamma Z_1 \to \mu^+ \mu^- Z_2} \simeq \frac{28}{9} Z^2 \alpha \, r_0^{\mu\,2} \left(\ln \frac{2\varepsilon_\gamma}{m_\mu c^2} - \frac{109}{42} \right) \tag{4}$$

$$\sigma_{\gamma Z_1 \to \mu^+ \mu^- Z_2} \simeq 10^{-31}\,\text{m}^{-2} = 0.5\,\text{milli-barn} \quad (\varepsilon_\gamma \sim 200\,\text{MeV}, Z \sim 79)$$

where,

- r_0^μ the classical muon radius ($r_0^\mu = 1.36 \times 10^{-17}\,\text{m}$), $r_0^\mu = r_0^e \times m_e/m_\mu$ where $r_0^e = r_0 = e^2 m_e^{-1} c^{-2}$ (in cgs) is the classical electron radius ($r_0^e = 2.82 \times 10^{-15}\,\text{m}$)
- Z is the atomic number of the material nuclei
- α is the fine structure constant $= v_B/c = e^2\hbar^{-1}c^{-1} = r_0/\lambda_c = e^2(m_e c^2)^{-1}(\hbar m_e^{-1}c^{-1})$(cgs), where v_B is the velocity of the first orbit of a Bohr atom, λ_c is the reduced Compton wavelength
- ε_γ is energy of bremsstrahlung photon.

The inefficiency of the muon pair BH photo-production process using an electron beam from a conventional rf accelerator relative to the positron-electron BH pair production process is well known[49] and is due to the smaller cross-section of the muon pair-production process by a factor of $(m_e/m_\mu)^2 \sim 1/(207)^2$.

The BH muon pair-production event rate $\mathcal{R}_{\gamma Z_1 \to \mu^+ \mu^- Z_2} \equiv dN_{\mu\pm}dt^{-1}$ can be estimated using $\mathcal{R}_{\gamma Z_1 \to \mu^+ \mu^- Z_2} \equiv \frac{dN_{\mu\pm}}{dt} = \mathcal{L} \times \sigma_{\gamma Z_1 \to \mu^+ \mu^- Z_2}$.

$$\mathcal{L} = \frac{N_{\text{beam}}}{\sigma_{z-\text{beam}}/c}\, n_{\text{target}}\, T_{\text{target}}$$

$$\mathcal{R}_{\gamma Z_1 \to \mu^+ \mu^- Z_2} \,(\text{in 50 fs}) = \frac{1\text{nC}}{e}\, 5.9 \times 10^{28}\,\text{m}^{-3}\, 1\,\text{cm}\, 0.5\,\text{milli-barn} \tag{5}$$

$$\mathcal{R}_{\gamma Z_1 \to \mu^+ \mu^- Z_2} \,(\text{in 50 fs}) \simeq 10^5\,\text{pairs (1 nC, 50 fs, } \sigma_r \sim 20\,\mu\text{m)}$$

where,

- N_{beam}, $\sigma_{z-\text{beam}}$ are the number of particles and the bunch length of the beam, respectively
- n_{target}, T_{target} are the number density and the thickness of the target.

The estimated number of muon pairs per 10 GeV electron is thus between $10^{-4} - 10^{-5}/e$. For nominal FACET-II parameters[50] it is estimated that 10^5 pairs can be produced when electron beam with $N_{\text{beam}} = \frac{1\text{nC}}{e} \simeq 6.24 \times 10^9$ in 50 fs bunch length is incident on Gold target (Au) with number density $n_{\text{target}} = 5.9 \times 10^{28}\,\text{m}^{-3}$ and target thickness, $T_{\text{target}} = 10^{-2}\,\text{m}$. The cross-section of BH muon pair production is 0.5 milli-barn as estimated in Eq. (4).

The energy spectrum of the muon pairs photo-produced by (R.2) is exponential and it peaks slightly above $2m_\mu c^2$. This implies that a large number of muons have

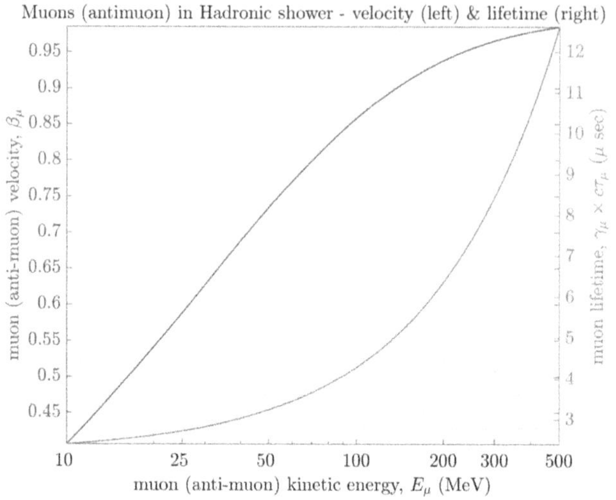

Fig. 2.   Photo-produced Hadronic shower with an exponential velocity distribution has the shown muon velocity (β_μ) and corresponding lifetime ($\gamma_\mu \tau_\mu$) over a range of muon kinetic energy.

a low velocity. The angle of propagation of the muons is directly proportional to their relativistic momentum or the Lorentz factor.

In Fig. 2, the properties of muons and antimuons contained a photo-produced hadronic shower are captured. The left-hand axis shows the muon (and antimuon) velocity while the right-hand axis their corresponding lifetime.

The energy spectrum and transverse phase-space of the particles in a hadronic shower are therefore unconstrained. In the laser muon acceleration scheme introduced here, the photo-produced hadronic shower which primarily comprises of muon pairs is coupled into a laser-driven slowly propagating acceleration structure in the plasma. This slowly propagating laser acceleration structure traps the charged particles and accelerates them.

From a comparison of Fig. 2 on hadronic shower muon properties and Fig. 1 on laser group velocity it is apparent that to trap muons with greater than 10MeV kinetic energy and velocities around 0.5c, it is necessary to use plasma densities as high 10^{20} cm^{-3}.

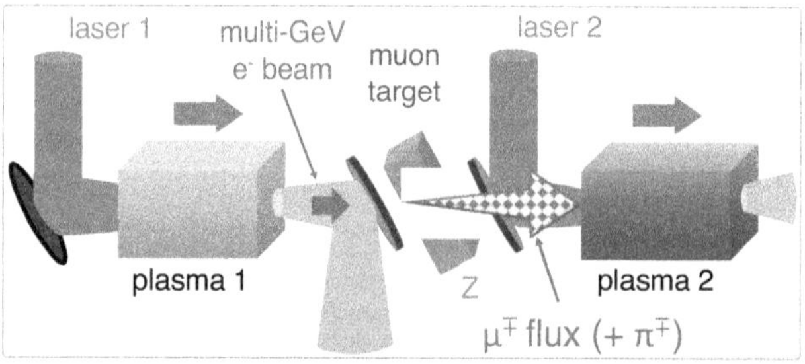

Fig. 3.   Schematic of laser-plasma post-processing of photo-produced (photo-meson and BH muon pair-production) $\pi^\pm$-$\mu^\pm$ Hadronic shower driven by $e^\pm$ beam in a target.

5.2. *Scheme II:*
Laser-driven plasma based post-processing of proton beam driven hadronic shower

When proton beam is shot onto a target, direct interaction of protons with atomic nuclei dominates the interaction. The resulting proton-nucleon reaction result in the production of pions mediated by the strong force. These reactions occur between the high-energy protons in the beam and the nucleons that constitute the atomic nuclei:

(R.3) protons, $p + p \to \pi^+ + p + p$

(R.4) neutrons, $p + n \to \pi^- + p + p$.

Thus, the hadronic shower produced is primarily a pion shower. The threshold proton beam energy required can be estimated from momentum four-vector of the interaction:

$$E_p^{\text{th}}\Big|_{\pi^\pm} = \left[\frac{1}{2}\left(2 + \frac{m_{\pi^\pm}}{m_p}\right)^2 - 2\right] m_p c^2 \simeq 0.31 \, m_p c^2 \sim 290 \text{ MeV} . \quad (6)$$

The cross-section of proton-nucleon reactions which is dictated by strong interactions is,

$$\sigma_{pp}(E_p) \simeq 40 \times 10^{-27} \text{ cm}^{-2} = 40 \text{ mb} \quad (1 \text{ barn} = 10^{-24} \text{ cm}^{-2}),$$
$$\sigma_{pZ}(Z, A) \simeq \sigma_{pp} \times A^{0.7} \quad [\sigma_{pZ}(A > 100) \simeq 1 \text{ barn}]. \quad (7)$$

The event-rate of pion production is calculated for a hypothetical ultra-short (< 1 ps) 500 MeV proton bunch with 1 nC charge ($N_{\text{beam}} = 6.24 \times 10^9$ protons) incident on a 1 cm thick Tungsten ($A_{\text{W}}(Z = 74) \simeq 184$) target of number density $n_{\text{target}} = 6.3 \times 10^{22} \text{ cm}^{-3}$ using below,

$$\mathcal{L} = \frac{N_{\text{beam}}}{\sigma_{z-\text{beam}}/c} \, n_{\text{target}} \, T_{\text{target}}$$
$$\mathcal{R}_{pZ \to \pi^\pm pZ} = \frac{dN_{\mu^\pm}}{dt} = \mathcal{L} \times \sigma_{pZ} \quad (8)$$
$$= \frac{1\text{nC}}{e} \, 6.3 \times 10^{28} \, \text{m}^{-3} \, 1 \, \text{cm} \, 1.54 \text{ barn}$$
$$= 6.1 \times 10^8 \, \pi^\pm .$$

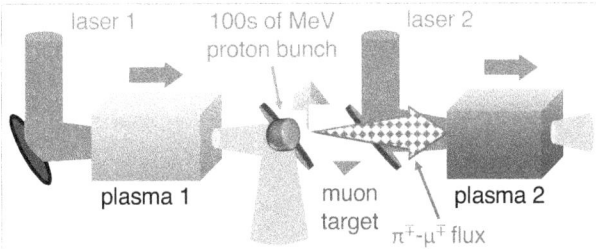

Fig. 4. Schematic of laser-plasma post-processing of proton-neutron reaction based $\pi^\pm$-$\mu^\pm$ hadronic shower driven by proton beam in a target.

This pion production process is thus quite efficient relative to the photo-production process as it produces 0.1 pion per proton. It is also however well known that 0.5 GeV proton beams that can be focused down to micron-scale spot-size are not available in ultra-short pico-second scale bunches with 1nC scale charge. Therefore, this scheme relies on a possible yet currently non-existent proton beam.

The energy spectrum and transverse phase-space of the shower pions is not usable in any real applications. This demands capture and storage of pions in a ring before they predominantly decay to muons.

In this laser muon acceleration scheme, the charged pions in the proton beam driven hadronic shower are coupled into a slowly propagating acceleration structure in the plasma. This slowly propagating laser-driven plasma acceleration structure traps the charged pions (thus, does not trap π^0) and accelerates them to high energies. Relativistic pions decay in a small forward angle and thus acceleration of pions is essential to increase the capture efficiency of muons produced from pion decay. However, the meters long pion confinement channel needed after the laser acceleration stage does not allow for a compact design.

5.3. Scheme III:
Laser-plasma positron electron mini-collider storage ring

Standalone laser-plasma electron accelerators have been shown to produce multi-GeV beams which undergo bremsstrahlung in a metal target which results in BH pair production of positrons and electrons. The resulting particle shower is post-processed using a laser-driven plasma stage to trap and accelerate a positron and electron dual bunch beam. The proof-of-principle of a positron laser-plasma accelerator has been demonstrated[5] and is currently under active investigation.[6]

These beams are stored in a mini-Collider ring where at the interaction or collision point muon pairs are produced close to their resonance from electron positron annihilation mediated by a virtual photon of the collision point energy. Energy asymmetry between the positron and electron beams is preferable as the produced muon pairs then have an initial kinetic energy and can thus be injected into a subsequent acceleration stage.

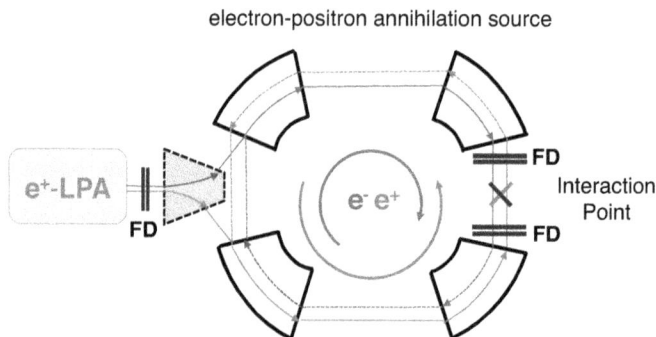

Fig. 5. Schematic of a muon source using mini Collider based upon laser-plasma positron electron storage ring.

Positron-electron annihilation allows access to the frontiers of center-of-mass energy and have thus been tools for discovering new physics. Advances in accelerator physics strive to make these tools compact and affordable. However, in this scheme a mini-Collider with a tunable energy symmetry between the colliding electron and positron beam energy enables the production of tunable energy muon beams.

During the collision of unpolarized spin electron and positron beams, the annihilation differential cross-section of muon pair production (which exhibits the typical QED $1/s$ dependence of the cross-section) is,

$$\frac{d\sigma_{e^\pm \to \mu^\pm}}{d\Omega} \simeq \frac{e^4}{64\pi^2} \frac{\hbar^2 c^2}{s} \frac{\sqrt{s - 4m_\mu^2}}{\sqrt{s - 4m_e^2}}$$
$$\times \left(1 + 4\frac{(m_e^2 + m_\mu^2)}{s} + \left(1 - \frac{4m_e^2}{s}\right)\left(1 - \frac{4m_\mu^2}{s}\right)\cos^2\theta\right) \quad (9)$$

$$\frac{d\sigma_{e^\pm \to \mu^\pm}}{d\Omega} \simeq \frac{e^4}{64\pi^2} \frac{\hbar^2 c^2}{s} (1 + \cos^2\theta) \quad (\text{under, } \sqrt{s} \gg m_\mu > m_e)$$

and the integrated cross-section of electron-positron annihilation during collision to muon-antimuon pair is,

$$\sigma_{e^\pm \to \mu^\pm} = \int \left(\frac{d\sigma_{e^\pm \to \mu^\pm}}{d\Omega}\right) d\Omega = \frac{4\pi}{3} \alpha^2 \frac{\hbar^2 c^2}{s} = \frac{87 \text{ nbarns}}{s \text{ (in GeV}^2)} \quad (10)$$

where,

- $\sigma_{e^\pm \to \mu^\pm}$ is the cross-section of the reaction electron-positron collision to muon-antimuon pair production
- Ω is the solid angle in the real space
- m_e, m_μ are electron and muon mass respectively
- s is the norm of the summed momentum four-vectors of electron ($\mathbf{p}$) and positron ($\tilde{\mathbf{p}}$) beam at the point of collision ($\|\mathbf{p} + \tilde{\mathbf{p}}\|^2$),
  it is also the norm of the summed muon (k) and antimuon ($\tilde{k}$) momentum four-vectors at the point of collision ($\|\mathbf{k} + \tilde{\mathbf{k}}\|^2$)
- θ is the angle between $\mathbf{p}$ and $\mathbf{k}$
- α is the fine structure constant.

The event rate ($\mathcal{R}_{e^\pm \to \mu^\pm} \equiv dN_{\mu^\pm} dt^{-1}$) muon-antimuon pair-production in positron-electron beam (assumed to have a Gaussian spatio-temporal profile) collision which depends on the luminosity ($\mathcal{L}$, cm^{-2}s^{-1}) parameter is therefore, and the integrated cross-section of electron-positron annihilation during collision to muon-antimuon pair is,

$$\mathcal{L} = \frac{N^{e^+} N^{e^-}}{4\pi \, \sigma_r^{e^+} \sigma_r^{e^-}} \, \hat{S} \, f_{rep} \quad (\text{equal bunchlengths}, \sigma_s^{e^+} \sigma_s^{e^-})$$

$$\mathcal{R}_{e^\pm \rightarrow \mu^\pm}(\text{per collision}) \equiv \frac{dN_{\mu^\pm}}{dt} = \mathcal{L} \times \sigma_{e^\pm \rightarrow \mu^\pm} = \frac{1}{s} \frac{\alpha^2}{3} \frac{N^{e^+} N^{e^-}}{\sigma_r^{e^+} \sigma_r^{e^-}} \, \hat{S} \quad (11)$$

$$= \frac{7}{s \, (\text{in GeV}^2)} \frac{N^{e^+} N^{e^-}}{\sigma_r^{e^+} \sigma_r^{e^-}} \, \hat{S}$$

where,

- $\sigma_r^{e^+}$, $\sigma_r^{e^-}$ is the radial waist-size of radially symmetric positron and electron bunches of spatio-temporal Gaussian profile, respectively.
- N^{e^+}, N^{e^-} is the number of particles in Gaussian positron and electron bunches, respectively.
- $\hat{S}$ is the Luminosity reduction factor due to several practical considerations such as crossing angles, Non-Gaussian profiles, Hourglass effect due to tight focusing, collision offsets, etc.
- f_{rep} is the number of collision per second.

For this scheme, we assume that laser-plasma accelerator produced positron and electron bunches of 200 pC ($N^{e^+} \sim N^{e^-} \simeq 1.25 \times 10^9$ particles per bunch) each are coupled to a storage ring and are made to collide at a collision point with tightly focused beam waist-size of $\sigma_r^{e^+} \simeq \sigma_r^{e^-} \simeq 10^{-10}$ m $= 0.1$ nm (difficult, if not impossible). Assuming that the electron beam energy is $E_{e^-} = 150$ MeV and that of the positron beam is $E_{e^+} = 100$ MeV then the number of muon pairs produced is only about $\simeq 1500$ with an initial kinetic energy of few 10s of MeV.

So, while the mini collider-based scheme produces beams of small spot-size, bunch length and transverse emittance, it is quite ineffective at scaling the number of muons at each collision event.

6. Particle In Cell Simulation of Scheme 1: Controlled Interaction of μ^+-μ^- Pair-Plasma with Laser-Driven Plasma

Multi-dimensional PIC simulations are used to validate the laser muon acceleration schemes outlined above, especially with relevance to the Scheme I (Subsec. 5.1). The PIC simulation reported below use the open-source EPOCH code.[51] In this section, $2\frac{1}{2}$D simulations adjusted to match 3D simulations are presented for preliminary evaluation of trapping and acceleration of muons. In these simulations, a 2D cartesian grid which resolves $\lambda_0 = 0.8\,\mu$m with 20 cells in the longitudinal and 15 cells in the transverse direction tracks a linearly-polarized laser pulse at its group velocity.

The photo-produced hadronic particle shower driven by multi GeV electron beam shows a peak in its energy spectrum around muon energy of 200 MeV.[48]

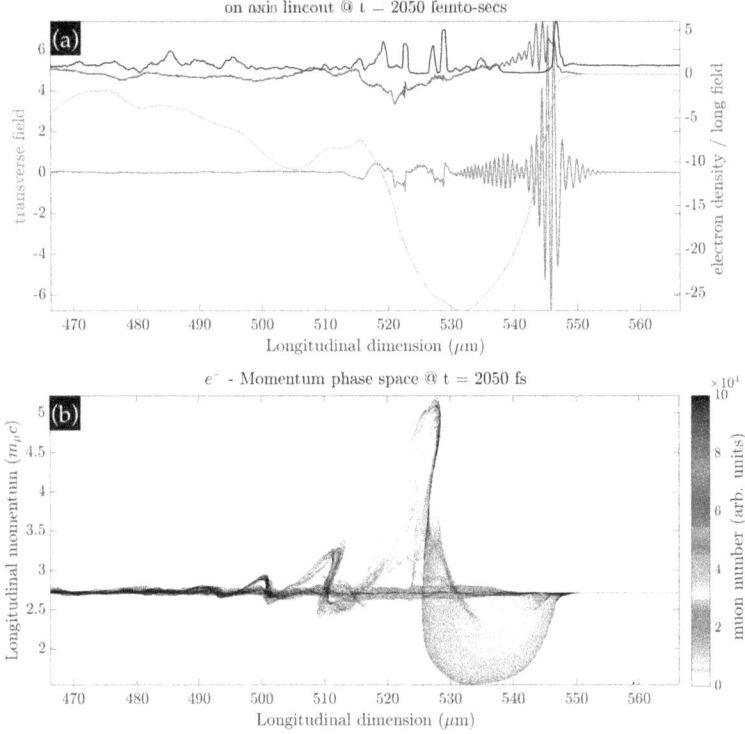

Fig. 6. On-axis lineout of laser-plasma interaction parameters (in a) and muon (μ^-) momentum phase space (in b), with muon longitudinal momentum shown along the longitudinal dimension at around 2 ps and about 0.6 mm of plasma length. The initialized 200 MeV muons get trapped and accelerated in laser plasma acceleration structures driven using a 1 J, 30 fs laser focused to a focal spot-size of $w_0 = 5\,\mu m$ interacting with a laser-ionized $n_0 = 2 \times 10^{19}$ cm^{-3} plasma.

In the simulation results presented in this section, the hadronic shower (made only of μ^- particles) is initialized with transverse size of $\sigma_r = 20\,\mu$m and the shower longitudinally spans the entire simulations box. Each particle species is initialized with 4 particles per cell. Absorbing boundary conditions are used for both fields and particles. A 1J laser with a Gaussian envelope of 30 fs pulse length is focussed to a spot-size of $w_0 = 5\,\mu$m at the plasma and propagates in 50 μm of free-space before it impinges on a fixed-ion plasma.

The preliminary $2\frac{1}{2}$D PIC simulations carried out as described above provide good understanding of the process of trapping and acceleration of muons contained within the hadronic shower or muon-antimuon pair-plasma. Below we present a few $2\frac{1}{2}$D PIC simulation snapshots (Figs. 6–10) to establish the viability of laser muon acceleration for a pre-ionized $n_0 = 2 \times 10^{19}$ cm^{-3} plasma.

The $2\frac{1}{2}$D PIC simulation snapshots presented as evidence of laser muon acceleration are as follows:

(1) Fig. 6(a) shows the on-axis lineout of various laser-plasma parameters (laser transverse field is normalized to $m_e c \omega_0 e^{-1}$, longitudinal plasma field is normalized to $m_e c \omega_{pe} e^{-1}$, plasma electron density, n_0(initial), plasma potential

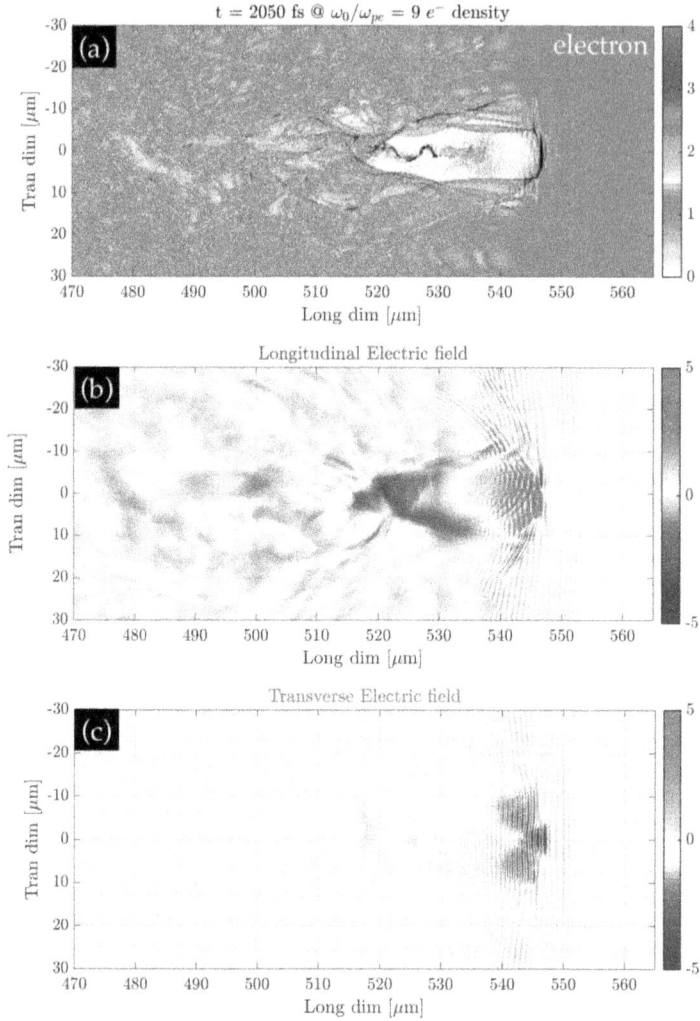

Fig. 7. Laser-plasma interaction characteristics (normalized) of laser muon accelerator corresponding to the on-axis lineout in Fig. 6(a) at around 2 ps in a $n_0 = 2 \times 10^{19} \, cm^{-3}$ plasma: (a) plasma electron density of the acceleration structure, (b) longitudinal electric field associated with the plasma acceleration structure, (c) the transverse field of the evolving laser pulse.

is normalized to $m_e c^2 e^{-1}$). Fig. 6(b) shows corresponding muon longitudinal momentum phase-space with longitudinal muon momentum along the y-axis and longitudinal dimension along the x-axis. From this phase-space, it is clear that the muons gain around 200 MeV over 0.6 mm.

(2) Fig. 7 shows the 2D real-space simulation snapshot of: plasma electron density of the acceleration structure (in a), plasma longitudinal field (in b) and laser transverse field (in c) corresponding to the time snapshot in Fig.6.

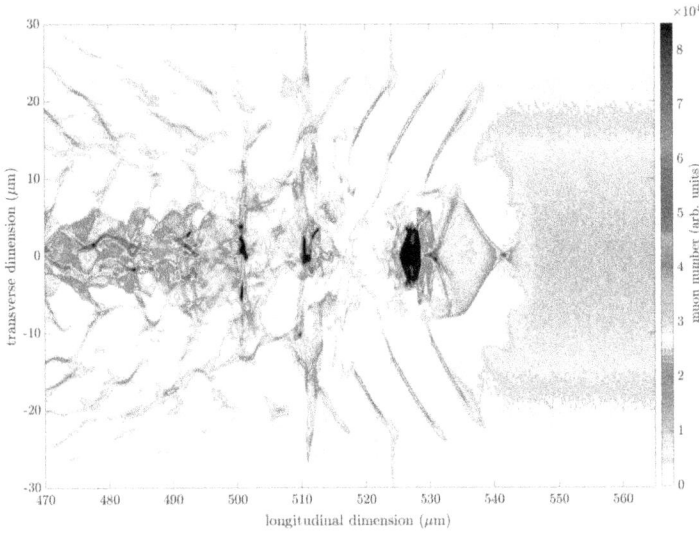

Fig. 8. Trapped and accelerated muon (μ^-) bunch in real space with micron-scale transverse and longitudinal dimensions at around 2 ps of laser-plasma interaction corresponding with the laser muon acceleration snapshot presented in Fig. 7 in a $n_0 = 2 \times 10^{19} \, \text{cm}^{-3}$ plasma.

(3) Fig. 8 shows the 2D real-space simulation snapshot of trapped and accelerated muon bunch density.

(4) Fig. 9 shows the muon momentum phase-space of longitudinal momentum along the y-axis against transverse real-space along the x-axis. From this snapshot it can be inferred that muons have small amplitude transverse oscillations as they gain energy. Additionally, the accelerated beam transverse spot-size can be inferred to be $< 10 \, \mu\text{m}$.

(5) Fig. 10 shows the muon momentum phase-space of longitudinal momentum along the y-axis against transverse momentum along the x-axis. This snapshot further reinforces the transverse dynamics of muons as they gain energy. Moreover, it shows that $< 10 \, \text{mrad}$ opening angles are likely from a laser muon accelerator.

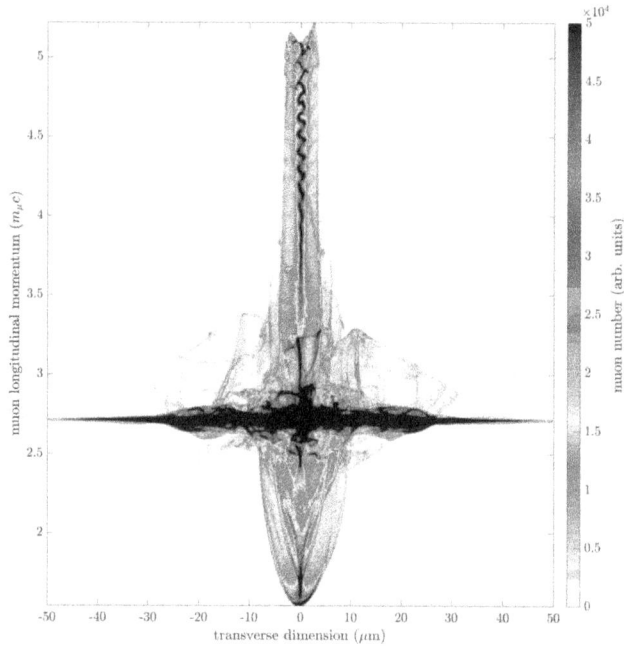

Fig. 9. Muon (μ^-) momentum phase space, with muon longitudinal momentum along the y-axis shown against the transverse real-space dimension along the x-axis at around 2 ps and about 0.6 mm of laser propagation in plasma in correspondence with the snapshots presented in above figures.

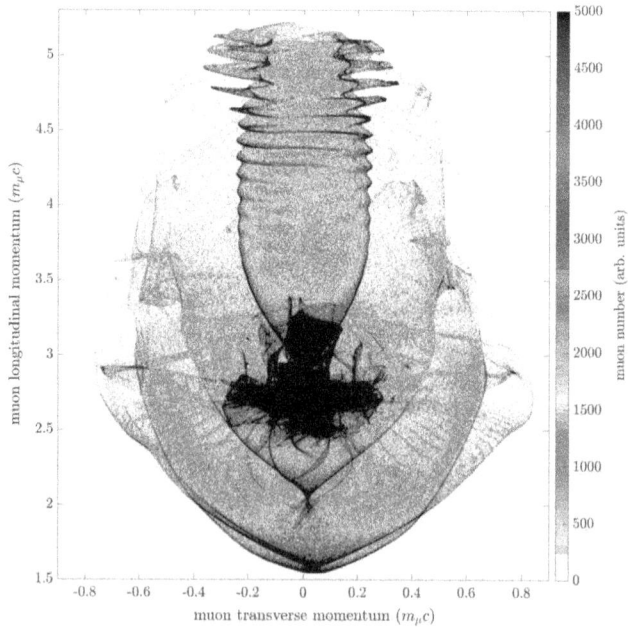

Fig. 10. Muon (μ^-) momentum phase space, with muon longitudinal momentum along the y-axis shown against the transverse momentum along the x-axis at around 2 ps and about 0.6 mm of laser propagation in plasma in correspondence with the snapshots presented in above figures.

From analysis of the simulation snapshots it can be inferred that acceleration of ultrashort, micron-scale muon beams is viable using laser muon accelerator even with a few tens of TW peak power CPA laser.

7. Conditioning-Stage Using a Plasma Lens: Charge Dependent Focusing of Oppositely Charged Particles

The proposed conditioning step involves the use of a discharge plasma lens to segregate oppositely charged species of a particle-shower or an oppositely charged dual bunch configuration by selectively focusing one charge sign of the particles.

Cascade shower (electromagnetic or hadronic) that contains both the oppositely charged particle species such as electron-positron $(e^{\pm})$ pair or muon-antimuon $(\mu^{\pm})$ pair, is produced for instance using an electron and/or positron beam or a proton beam. These beams are themselves possibly obtained using a laser-plasma accelerator. In the plasma lens based conditioning stage, the oppositely charged species of the cascade shower are selectively segregated due to only one of the charge sign being focused in the device while the other is de-focused.

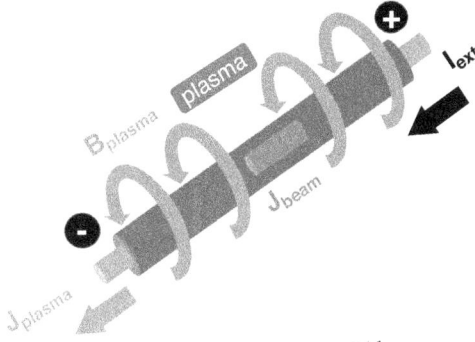

$$B_\phi(r < R_{\text{lens}}) = C_1 \, J_0 \, \frac{r^{\alpha+1}}{R^\beta}$$

$$\frac{\partial v_r}{\partial z} + k_{\text{lens}}^2 \, r = \frac{\partial^2 r}{\partial z^2} + k_{\text{lens}}^2 \, r = 0$$

$$k_{\text{lens}}^2 = \pm e \, C_2 \, J_0 \, \frac{r^\alpha}{R^\beta} \, \frac{1}{p_z}$$

Fig. 11. Schematic of the conditioning stage where a discharge plasma lens[52] is used to segregate species of opposite charge sign by phasing the discharge current direction (if RF voltage is applied).

The plasma lens is based upon a discharge plasma where a high RF terminal voltage is applied to sustain the discharge. By appropriately phasing the entrance phase of the cascade shower and the RF voltage phase it is possible to choose the charge sign to be focused.

With the choice of amplitude and polarity of the externally injected current, discharge plasma dimensions, gas pressure and gas type it is possible to control the acceptance and focusing properties of the cascade shower processing device.

In the 1965 [52] BNL work on using plasma discharges as active plasma lens it was also found that the focusing strength of an active plasma lens was directly proportional to the radial distance from the axis (particles away from the axis experience higher focusing force) and inversely proportional to the momentum of the cascade shower particle (lower energy particles experience higher focusing force).

These focusing characteristics of the discharge plasma lens can be understood if the problem is considered in a cylindrical coordinate system (r, z, ϕ) as depicted in Fig. 11. The azimuthal field in plasma ($B_{\text{plasma}} = B_{\phi}$) is excited due to the plasma current ($J_{\text{plasma}} = J_0$) which is driven by externally injected current, I_{ext} along the z direction. This azimuthal magnetic field due to the plasma current exerts a force on longitudinally (along z) propagating charged particle beam. The Lorentz force on the charged beam particle injected along z interacting with a magnetic field oriented along is in the radial, r direction. The equation of radial motion of each charged particle in the beam is governed by the lens equation above.

Cascade showers comprising of oppositely charged particle species generated by the decay of high-energy gamma-ray photons in a metallic target have energy spectra which has a concentration of particles at non-relativistic energies. However, due to the randomized decay of high-energy gamma-rays, the divergence angle of the particle shower particles can be large. A plasma lens that is located right next to a metallic converter target can also capture the divergent charged particles of the cascade shower.

Depending upon the direction of the discharge plasma current one of the charged particle species is focused and the other is defocused. This leads to the segregation of the oppositely charged particle species of the cascade shower. Therefore, at the output of the assembly of particle-shower target and plasma lens one of the charged particle species is detectable whereas the species with the opposite sign of charge is excessively defocussed and thus diluted.

8. Discussion and Future Work

In this paper, experimentally viable and affordable laser muon sources have been introduced in consideration of the numerous possibilities offered for a wide-range of applications. This paper proposes and estimates the experimental viability of compact and tunable schemes of laser muon acceleration in gas plasmas using an innovative technique of post-processing of muon-antimuon cascade hadronic showers or pair plasmas through their controlled interaction with laser-driven plasmas.

The laser muon acceleration schemes introduced and investigated here are designed in consideration of being well within the reach of experimental verification using existing experimental facilities as reflected in the choice of experimental setup and laser, plasma and beam parameters. Although the first-stage of the muon acceleration schemes presented here rely on laser accelerated particles such as multi GeV electrons, for proof-of-principle experiments may be based on more controlled and reliable 10 GeV scale electron beams from rf accelerators, in the short-term.

In the short term, the scheme I (Subsec. 5.1) which uses direct photo-production of muon pairs, although being limited in overall conversion efficiency, is found to be most suitable for the development of an experimental prototype of a laser muon accelerator that produces ultrashort and micron-scale muon beam. A few facilities, like BELLA at Berkeley and FACET-II at Stanford, that offer collocated CPA laser and micron-scale e^- and/or e^+ beams of many 10 s to 100 s of pC charge at 10 GeV scale beam energy, can be utilized for this prototyping effort.

Preliminary PIC simulations presented in Sec. 6 demonstrate the potential to trap and accelerate muons in gas plasmas. In these simulations, we observe several 100MeV gain in muon energy in less than a millimeter. Further modeling using analysis, particle-tracking and particle-in-cell simulations will be carried out to accurately estimate the properties of the accelerated muon beams within the reach of an experimentally viable laser muon accelerator prototype. In future work, we thus propose to extend the preliminary Particle-In-Cell based modeling of laser muon acceleration schemes reported here along with substantially more detailed modeling of muon photo-production using micron-scale electron or positron beams.

We will model the possibilities of tunable and spectrally controlled acceleration of muon and antimuon beams, dual bunch muon-antimuon beams, spatio-temporally overlapped electron and muon bunch beams, increasing the total trapped muon beam charge, segregation of oppositely charged muons prior to the acceleration stage etc. Moreover, our experimental prototyping effort will work hand-in-hand with theoretical modeling to better understand the expected muon cascade shower or pair-plasma properties and its interaction with laser-driven plasmas.

Compact and tunable production and acceleration of unprecedented ultra-short (femtosecond to attosecond) micro-scale spot muon beams is essential for an advanced acceleration program using muons and suited for acceleration mechanisms with inherently micron to nanometer spatial scale. Moreover ultrashort, micron-scale muon beams can be injected into crystal wakefield accelerators such as attosecond x-ray pulse driven [11-13] or sub-micron particle beam driven solid-state tube accelerators [14,15] in order to minimize the synchrotron radiation losses ($\propto m_\mu^{-4}$) as well as the radiative losses ($\propto m_\mu^{-2}$) in comparison with electrons and positrons. The significance of a compact tunable high-energy muon source also importantly lies in its multitude of technological, security and medical applications.

Acknowledgments

A. A. S. was supported by the College of Engineering and Applied Science, University of Colorado, Denver. V. D. S. was supported by Fermi National Accelerator Laboratory, which is operated by the Fermi Research Alliance, LLC under Contract No. DE-AC02-07CH11359 with the United States Department of Energy. This work utilized the RMACC Summit supercomputer through the XSEDE program, which is supported by the National Science Foundation (awards ACI-1532235 and ACI-1532236), the University of Colorado Boulder, and Colorado State University. The

Summit supercomputer is a joint effort of the University of Colorado Boulder and Colorado State University.[53]

References

1. T. Tajima and J. M. Dawson, Laser electron accelerator, *Phys. Rev. Lett.* **43**, 267 (1979).
2. D. Strickland and G. Mourou, Compression of amplified chirped optical pulses, *Opt. Commun.* **56**(3), 219 (1985).
3. W. P. Leemans *et al.*, Multi-GeV electron beams from capillary-discharge-guided sub-Petawatt laser pulses in the self-trapping regime, *Phys. Rev. Lett.* **113**, 245002 (2014); X. Wang *et al.*, Quasi-monoenergetic laser-plasma acceleration of electrons to 2 GeV, *Nat. Commun.* **4**, 1988 (2013).
4. H. T. Kim *et al.*, Enhancement of electron energy to the multi-GeV regime by a dual-stage laser-wakefield accelerator pumped by Petawatt laser pulses, *Phys. Rev. Lett.* **111**, 165002 (2013); S. Steinke *et al.*, Multistage coupling of independent laser-plasma accelerators, *Nature* **530**, 190–193 (2016).
5. A. A. Sahai *et al.*, Quasimonoenergetic laser plasma positron accelerator using particle-shower plasma-wave interactions, *Phys. Rev. Accel. Beams* **21**, 081301 (2018).
6. A. A. Sahai *et al.*, Laser positron accelerator: Proof-of-Principle experimental effort and novel applications, *LaserNetUS Expt. proposal*.
7. M. Xie *et al.*, Studies of laser-driven 5 TeV colliders in strong quantum e^+e^- beamstrahlung regime, *AIP Conf. Proc.* **398**, 233 (1997); W. Leemans and E. Esarey, Laser-driven plasma-wave electron accelerators, *Phys. Today* **62**(3), 44–49 (2009).
8. M. Uesaka and K. Koyama, Advanced accelerators for medical applications, *Rev. Accel. Sci. Tech.* **9**, 235–260 (2016) .
9. F. Albert and A. G. R.Thomas, Applications of laser wakefield accelerator-based light sources, *Plasma Phys. Control. Fusion* **58**(10), 103001 (2016).
10. H. Yukawa, On the interaction of elementary particles. I, *Proc. of the Physico-Mathematical Society of Japan* **17**, 48–57 (1935); S. H. Neddermeyer and C. D. Anderson, Note on the nature of cosmic-ray particles, *Phys. Rev.* **51**, 884–886 (1937).
11. T. Tajima and M. Cavenago, Crystal x-ray accelerator, *Phys. Rev. Lett.* **59**, 1440 (1987), doi:10.1103/PhysRevLett.59.1440.
12. S. Hakimi *et al.*, Wakefield in solid state plasma with the ionic lattice force, *Phys. Plasmas* **25**, 023112 (2018).
13. S. Hakimi *et al.*, X-ray laser wakefield acceleration in a nanotube, in *Proc. of the Workshop on Acceleration in Crystals and Nanostructures (XTALS 2019), 24–25 June 2019, Fermilab*.
14. P. Chen and R. J. Noble, A solid state accelerator, *AIP Conf. Proc.* **156**, 222 (1987); P. Chen and R. J. Noble, Crystal channel collider: Ultra-high energy and luminosity in the next century, *AIP Conf. Proc.* **398**, 273 (1997), [SLAC-PUB-7402 (1998)].
15. A. A. Sahai *et al.*, Solid-state tube wakefield accelerator using surface waves in crystals, in *Proc. of the Workshop on Acceleration in Crystals and Nanostructures (XTALS 2019), 24–25 June 2019, Fermilab*.
16. V. D. Shiltsev, High-energy particle colliders: Past 20 years, next 20 years, and beyond, *Physics-Uspekhi* **55**(10), 265 (2012).
17. K. Nagamine, *Introductory Muon Science* (Cambridge University Press, 2003), ISBN: 9780511470776, DOI: 10.1017/CBO9780511470776.
18. R. B. Palmer *et al.*, *AIP Conf. Proc.* **372**, 3 (1996), [arXiv:acc-phys/9602001]; Muon Collider Collaboration, Status of muon collider research and development and future plans, *Phys. Rev. ST Accel. Beams* **2**, 081001 (1999).

19. E. J. Williams and G. E. Roberts, Evidence for transformation of mesotrons into electrons, *Nature* **145**, 102 (1940).

20. D. G. Kosharev, Proposal for a decay ring to produce intense secondary particle beams at the SPS, CERN Report No. CERN/ISR-DI/74-62 (1974); S. Geer, Neutrino beams from muon storage rings: Characteristics and physics potential, *Phys. Rev. D* **57**, 6989 (1998).

21. A. Moretti *et al.*, Effects of high solenoidal magnetic fields on rf accelerating cavities, *Phys. Rev. ST Accel. Beams* **8**, 072001 (2005); R. B. Palmer *et al.*, rf breakdown with external magnetic fields in 201 and 805 MHz cavities, *Phys. Rev. ST Accel. Beams* **12**, 031002 (2009).

22. R. Wideroe, Uber ein neues Prinzip zur Herstellung hoher Spannungen, *Archiv fur Elektrotechnik* **21**(4), 387 (1928).

23. K. N. Borozdin *et al.*, Surveillance: Radiographic imaging with cosmic-ray muons, *Nature* **422**, 277 (2003).

24. K. Morishima *et al.*, Discovery of a big void in Khufuś Pyramid by observation of cosmic-ray muons, *Nature* **552**, 386–390 (2017); L. W. Alvarez *et al.*, Search for hidden chambers in the pyramids, *Science* **167**(3919), 832–839 (1970).

25. N. Lesparre *et al.*, Geophysical muon imaging: feasibility and limits, *Geophys. J. Int.* **183**(3), 1348–1361 (2010); K. Jourde *et al.*, Muon dynamic radiography of density changes induced by hydrothermal activity at the La Soufriere of Guadeloupe volcano, *Sci. Rep.* **6**, 33406 (2016).

26. T. Yamazaki *et al.*, Negative moun spin rotation, *Phys. Scr.* **11**, 133 (1975); J. H. Brewer *et al.*, mu+SR spectroscopy: The positive muon as a magnetic probe in solids, *Phys. Scr.* **11**, 144 (1975).

27. K. Barnabas *et al.*, Contrasts between uracil and thymine in reaction with hydrogen isotopes in water, *J. Phys. Chem.* **95**(24), 10204–10207 (1991); R. H. Scheicher *et al.*, First-principles study of muonium in A- and B-form DNA, *Physica B* **374–375**, 448–450 (2006).

28. L. Bossoni *et al.*, Human-brain ferritin studied by muon spin rotation: a pilot study, *J. Phys. Condens. Matter* **29**, 415801 (2017); K. Nagamine *et al.*, Probing magnetism in human blood by muon spin relaxation, *Physica B* **374–375**, 444–447 (2006).

29. E. Fermi and E. Teller, The capture of negative mesotrons in matter, *Phys. Rev.* **72**, 399 (1947); R. L. Garwin, L. M. Lederman and M. Weinrich, Observations of the failure of conservation of parity and charge conjugation in meson decays: The magnetic moment of the free muon, *Phys. Rev.* **105**, 1415 (1957).

30. S. J. Brodsky and R. F. Lebed, Production of the smallest QED atom: True muonium (μ^+-μ^-), *Phys. Rev. Lett.* **102**, 213401 (2009); S. G. Karshenboim *et al.*, Next-to-leading and higher order corrections to the decay rate of dimuonium, *Phys. Lett. B* **424**, 397–404 (1998)

31. J. D. Jackson, Catalysis of nuclear reactions between hydrogen isotopes by mu-minus mesons, *Phys. Rev.* **106**, 330 (1957); L. W. Alvarez *et al.*, Catalysis of nuclear reactions by mu-minus mesons, *Phys. Rev.* **105**, 1127 (1957).

32. D. M. Bose, B. Choudhuri and M. Sinha, Cosmic-ray meson spectra, *Phys. Rev.* **65**, 341 (1944); C. M. G. Lattes *et al.*, Processes involving charged mesons, *Nature* **159**, 694–697 (1947).

33. E. Fermi, High energy nuclear events, *Prog. Theor. Phys.* **5**(4), 570–583 (1950); D. S. Kothari, Fermi's thermodynamic theory of the production of pions, *Nature* **173**, 590 (1954).

34. R. Cywinski *et al.*, Towards a dedicated high-intensity muon facility, *Physica B* **404**, 1024–1027 (2009); E. Cartlidge, Muon users consider going it alone, *Phys. World*

$\mathbf{19}$(12), 13 (2006).

35. V. I. Balbekov and N. V. Mokhov, Low budget muon source, in *Proc. of Particle Acc. Conf.(PAC), Chicago, TPAH144, IL, USA (2001)*; H. Miyadera, A. J. Jason and S. S. Kurennoy, Simulation of large acceptance linac for muon, in *Proc. of PAC, FR5REP071, Vancouver, BC, Canada (2009)*.

36. H. K. Sayed and J. S. Berg, Optimized capture section for a muon accelerator front end, *Phys. Rev. ST Accel. Beams* $\mathbf{17}$(7), 070102 (2014).

37. E. M. McMillan, J. M. Peterson and R. S. White, Production of mesons by X-rays, *Science* $\mathbf{110}$(2866), 579–583 (1949); G. F. Chew *et al.*, relativistic dispersion relation approach to photomeson production, *Phys. Rev.* $\mathbf{106}$, 1345 (1957).

38. K. Nagamine *et al.*, Compact muon source with electron accelerator for a mobile mSR facility, *Physica B* $\mathbf{404}$, 1020–1023 (2009).

39. Y.-S. Tsai, Pair production and bremsstrahlung of charged leptons, *Rev. Mod. Phys.* $\mathbf{46}$, 815 (1974).

40. H. Athar, G. L. Lin and J. J. Tseng, Muon pair production by electron-photon scatterings, *Phys. Rev. D* $\mathbf{64}$, 071302(R) (2001).

41. L. Serafini *et al.*, A muon source based on plasma accelerators, *Nucl. Instrum. Methods Phys. Res. A* $\mathbf{909}$, 309–313 (2018).

42. G. Breit and J. A. Wheeler, Collision of two light quanta, *Phys. Rev.* $\mathbf{46}$, 1087 (1934).

43. C. Muller, C. Deneke and C. H. Keitel, Muon-pair creation by two x-ray laser photons in the field of an atomic nucleus, *Phys. Rev. Lett.* $\mathbf{101}$, 060402 (2008).

44. C. Muller *et al.*, Lepton pair production in high-frequency laser fields, *Laser Phys.* $\mathbf{19}$, 791-796 (2009).

45. M. Antonelli *et al.*, Novel proposal for a low emittance muon beam using positron beam on target, *Nucl. Instrum. Methods Phys. Res. A* $\mathbf{807}$, 101–107 (2016); H. Burkhardt, S. R. Kelner and R. P. Kokoulin, Production of muon pairs in annihilation of high-energy positrons with resting electrons, CERN-AB-2003-002 (2003).

46. I. F. Ginzburg *et al.*, Production of bound $\mu+\mu-$ systems in relativistic heavy ion collisions, *Phys. Rev. C* $\mathbf{58}$, 3565 (1998).

47. D. M. Kaplan, T. Hart and P. Allport, Producing an intense, cool muon beam via e^+e^- annihilation, arXiv:0707.1546.

48. A. I. Titov, B. Kampfer and H. Takabe, Dimuon production by laser-wakefield accelerated electrons., *Phys. Rev. ST Accel. Beams* $\mathbf{12}$, 111301 (2009); B. S. Rao *et al.*, Bright muon source driven by GeV electron beams from a compact laser wakefield accelerator, arXiv:1804.03886.

49. W. A. Barletta and A. M. Sessler, Characteristics of a high energy $\mu^+\mu^-$ collider based on electro-production of muons, *Nucl. Instrum. Methods Phys. Res. A* $\mathbf{350}$, 36 (1994).

50. V. Yakimenko, Ultimate beams at FACET-II, presented at the Workshop on Beam Acceleration in Crystals and Nanostructures (XTALS 2019), 24–25 June 2019, Fermilab; V. Yakimenko *et al.*, Prospect of studying nonperturbative QED with beam-beam collisions, *Phys. Rev. Lett.* $\mathbf{122}$, 190404 (2019).

51. T. D. Arber *et al.*, Contemporary particle-in-cell approach to laser-plasma modelling, *Plasma Phys. Control. Fusion* $\mathbf{57}$, 113001 (2015).

52. E. B. Forsyth, L. M. Lederman and J. Sunderland, The Brookhaven-Columbia plasma lens, *IEEE Trans. Nucl. Sci.* $\mathbf{12}$, 872 (1965).

53. J. Towns *et al.*, XSEDE: Accelerating scientific discovery, *Comput. Sci. Eng.* $\mathbf{16}$, 62–74 (2014); J. Anderson *et al.*, in Deploying RMACC summit: An HPC resource for the Rocky mountain region, *Proc. of PEARC17, New Orleans, LA, USA, July 09-13,* (2017).

Solid-state Tube Wakefield Accelerator
Using Surface Waves in Crystals

Aakash A. Sahai

College of Engineering and Applied Science,
University of Colorado, Denver, CO 80204, USA
aakash.sahai@gmail.com

Toshiki Tajima, Peter Taborek

Department of Physics & Astronomy and Applied Physics,
University of California, Irvine, CA 92697, USA

Vladimir D. Shiltsev

Accelerator Research Department,
Fermi National Accelerator Laboratory, Batavia, IL 60510, USA

Solid-state or crystal acceleration has for long been regarded as an attractive frontier in advanced particle acceleration. However, experimental investigations of solid-state acceleration mechanisms which offer TVm^{-1} acceleration gradients have been hampered by several technological constraints. The primary constraint has been the unavailability of attosecond particle or photon sources suitable for excitation of collective modes in bulk crystals. Secondly, there are significant difficulties with direct high-intensity irradiation of bulk solids, such as beam instabilities due to crystal imperfections and collisions etc.

Recent advances in ultrafast technology with the advent of submicron long electron bunches and thin-film compressed attosecond x-ray pulses have now made accessible ultrafast sources that are nearly the same order of magnitude in dimensions and energy density as the scales of collective electron oscillations in crystals. Moreover, nanotechnology enabled growth of crystal tube structures not only mitigates the direct high-intensity irradiation of materials, with the most intense part of the ultrafast source propagating within the tube but also enables a high degree of control over the crystal properties.

In this work, we model an experimentally practicable solid-state acceleration mechanism using collective electron oscillations in crystals that sustain propagating surface waves. These surface waves are driven in the wake of a submicron long particle beam, ideally also of submicron transverse dimensions, in tube shaped nanostructured crystals with tube wall densities, $n_{tube} \sim 10^{22-24}\,cm^{-3}$. Particle-In-Cell (PIC) simulations carried out under experimental constraints demonstrate the possibility of accessing average acceleration gradients of several TVm^{-1} using the solid-state tube wakefield acceleration regime. Furthermore, our modeling demonstrates the possibility that as the surface oscillations and resultantly the surface wave transitions into a nonlinear or "crunch-in" regime under $n_{beam}/n_{tube} \gtrsim 0.05$, not only does the average gradient increase but strong transverse focusing fields extend down to the tube axis. This work thus demonstrates the near-term experimental realizability of Solid-State Tube Wakefield Accelerator (SOTWA).

The ongoing progress in nanoengineering and attosecond source technology thereby now offers the potential to experimentally realize the promise of solid-state or crystal acceleration, opening up unprecedented pathways in miniaturization of accelerators.

1. Introduction

Particle acceleration techniques using collective charge density oscillations in crystals have been known to be an attractive possibility for the past many decades.[1]

1.1. *Solid-state acceleration using wakefields in crystal plasmas: Attosecond sources and crystal tubes*

In solid-state or crystal acceleration mechanisms a charged particle beam gains energy by extracting the electromagnetic field energy of collective electron oscillation modes excited in crystals. These solid-state collective oscillations are known to sustain propagating charge density waves of high energy densities. These collective oscillations and the associated waves can be efficiently excited as wakes of pulsed sources of particles or photons with pulse dimensions that are resonant with the scales of collective oscillations in solid-state. However, the theoretically modeled solid-state acceleration gradients[2,3] which are known to be orders of magnitude higher than the time-tested radio-frequency technology as well as the emerging gaseous plasma acceleration[4,5] techniques, are yet to be experimentally verified and further studied.

Experimental verification of solid-state acceleration mechanisms has been so far hampered by several technological challenges such as unavailability of pulsed particle and photon sources that are resonant with the collective oscillations in crystals. However, technological advances in intense particle and photon pulsed ultrafast source compression technologies have continued to drive the pulse dimensions towards ever shorter time and spatial scales. These technological advancements in ultrafast source compression techniques have made scales required to resonantly excite collective electron modes for solid-state acceleration mechanisms experimentally accessible. Especially, recent breakthroughs in attosecond scale photon[6] and particle[7] bunch ultrafast source technologies have opened up the potential for experimental realization of long-sought solid state acceleration.[1-3]

Although attosecond source technologies provide an effective means for resonant excitation of collective modes in solid-state crystal media, there still exist other technological barriers. In addition to the barriers due to the scarcity of attosecond sources, accessing solid-state gradients has also been impeded by difficulties with direct irradiation of solids at high intensities using particle or photon beam. Advances in nano-structured materials and nanoengineering of tube-like structures in crystals, such as nanotubes, however now offer the possibility of overcoming these difficulties with direct interaction of a crystal with high-intensity sources. Direct interaction of a high-intensity particle beam is reported to undergo severe filamentation due to the deformities in the crystal structure.[8] Not only is a filamented beam detrimental to driving a coherent wake but it also leads to a severely uncontrolled interaction and energy dissipation.

Solid-state plasmas with electron densities $n_0 \sim 10^{22-24}\,\mathrm{cm}^{-3}$, sustain electron oscillations at superoptical time, $177(n_0[10^{22}\,\mathrm{cm}^{-3}])^{-1/2}$ attosec ($\sim \omega_{pe}^{*-1}$

where $\omega_{pe} = (n_0 e^2 \epsilon_0^{-1} m^{-1})^{1/2}$ and m_e the electron mass[9]) and spatial scales, $330(n_0[10^{22}\,\mathrm{cm}^{-3}])^{-1/2}$ nm ($\sim \lambda_{\mathrm{pe}}$). Electron modes at such scales offer Tajima-Dawson (wavebreaking) acceleration gradients[4] of the order of, $E_{wb} \simeq 9.6(n_0[10^{22}\,\mathrm{cm}^{-3}])^{-1/2}\,\mathrm{TVm}^{-1}$. By coupling with these superoptical scales, sub-micron particle bunches (e.g., $\sigma_z < 1\,\mu m$[10]) or intense keV photon lasers[11] make excitation of unprecedented TVm^{-1} average gradients experimentally feasible.

1.2. *Progress of wakefields acceleration in gaseous plasmas: Femtosecond sources*

Over the past few decades, access to tens of femtosecond chirped pulse amplified[12] $0.8\mu m$ wavelength lasers (with few femtosec single cycle) and particle bunches has enabled experimental verification of advanced particle acceleration techniques that use collective electron oscillations in gaseous plasmas.[4,5] Whereas lasers have been compressed to few cycle long pulses using the innovative chirped pulse amplification technique,[12] ultrashort particle bunches have been obtained via phase-space gymnastics[13] or self-modulation in plasma.[14] Both these ultrafast source technologies have enabled successful gaseous plasma acceleration experiments with many GVm^{-1} gradients.[15,16] These experiments have used micron-scale charge-density waves in gaseous bulk plasma.

Control over bulk plasma waves in homogeneous gases by femtosecond-scale sources[4,5] has lead to the successful demonstration of gaseous plasma wakefield acceleration techniques. Numerous advantages of these techniques over conventional radio-frequency acceleration techniques has now lead to them being enhanced and fine-tuned for real-world applications using commercially available femtosecond sources. Some of these enhancements include control over: (a) wakefield profile distortions from ion motion,[17] (b) dark current injection and acceleration due to secondary ionization,[18] (c) accelerated beam emittance growth due to scattering off of plasma ions,[19] (d) positron defocusing by the bared ions, (e) repetition rate constraints due energy coupling to long-lived ion modes,[20] etc.

Non-homogeneous plasmas of specific shapes have been proposed to address many of the above enhancements of gaseous bulk plasma acceleration. The earliest shaped plasma proposal[21] for a fiber accelerator sought to keep high-intensity laser pulses continuously focussed.[22] Utilizing this shaped plasma proposal, gaseous plasma fibers that are excited using mechanisms such as laser-heated capillary,[23] etc. are now regularly used for plasma fiber guided laser-plasma acceleration. Mechanism of beam-driven shaped gaseous hollow plasma (later labeled hollow-channel) acceleration has also been studied.[25] Experiments on beam-driven gaseous hollow plasma using intense positron beams have observed $\sim 200\,\mathrm{MVm}^{-1}$ peak gradients ($\sim 0.01 E_{wb}$).[26] Access to higher gradients in shaped gaseous hollow plasmas is currently under active research. Active areas of research include technological difficulties in shaping a desired channel in gaseous plasmas apart from challenges due to the absence of any focusing force[25] such as control of higher-order transverse wakes excited by the drive beam due to its misalignment from channel axis[27] and beam-breakup resulting from these transverse wakes. Recent results have demonstrated that beam breakup may be controllable via further shaping of the gaseous hollow plasmas.[28]

In this paper, we introduce and model a regime of experimentally realizable solid-state acceleration that uses charge density waves of submicron scale lengths in nanotube shaped solid-state plasmas. The acceleration modes in this regime of solid-state tube wakefield acceleration take advantage of the developments in nano-fabrication as well as submicron particle or attosecond photon pulsed source technology. We show using analytical and computationally modeling that the crystal tube surface electron oscillations sustain an electrostatic "crunch-in" mode.[11,20,29] This electrostatic mode supports electromagnetic surface wave modes with phase velocity close to the driver velocity and on-axis longitudinal electric fields that approach the Tajima-Dawson gradient of the tube wall electron density. This high phase velocity surface wave mode supported by excitation of tube wall electron oscillations makes solid-state tube wakefield accelerator regime quite effective.

Although significantly different from traveling wave modes supported by electron oscillations in solid-state, a similar electron oscillation mode of gaseous plasma hollow channels has been computationally observed in a few previous works. However, neither its structural and electromagnetic properties nor its acceleration characteristics have been extensively modeled. In gaseous plasmas the "crunch-in" like mode has been observed in simulation works that have used experimentally feasible parameter regime such as a laser-driven shaping of a hollow plasma proposal,[30] a proton beam driven shaped hollow plasma acceleration proposal in externally magnetized plasma[31] and a electron or positron beam driven shaped hollow plasma.[20]

An important recent work[11,40] has recently studied and modeled the excitation of modes in crystal tubes using attosecond keV photon x-ray pulses. This work on modeling of x-ray wakefield tube accelerator has demonstrated the potential of using x-ray wakefield acceleration mechanism in tubes for sustaining many TV-cm^{-1} gradients. With the advent of a few cycle high-intensity x-ray laser using thin-film compression technique, the crystal x-ray wakefield acceleration mechanism has the potential to further advance the progress made by the Ti:Sapphire 800nm optical laser based gaseous plasma wakefield acceleration technique.

However, the mechanism of beam-driven surface modes in bulk crystals and crystal tubes, as opposed to those driven by an x-ray laser, has not yet been modeled and characterized. This is especially important due to the recent opening up of the availability of submicron particle bunches. The beam-driven crystal tube phenomena investigated and the results reported here indicate that the x-ray driven crystal tube wakefields characterized in Refs. 11, 40, and 41 are quite similar to that in the beam-driven crystal tube case. Therefore, our work shows that crystal tube wakefields have both longitudinal and focusing fields similar to the x-ray driven wakefields.[11,40,41] It may be noted that our work on beam-driven wakefields in a crystal tube is distinctive from previously modeled gaseous hollow-plasma wakefields because in gaseous hollow-plasma the wakefields of a relativistic particle beam are proven to have zero focusing fields.[42] The preliminary analysis and computational results presented below demonstrate the experimental realizability of Solid-State Tube Wakefield Accelerator (SOTWA).

In the following sections on modeling of beam-driven wakefields in crystal tubes, we introduce and characterize the beam-driven solid-state tube accelerator using surface wave wakefields in crystals. The model and significance of solid-state collective electron or plasmon oscillation modes is presented in Sec. 2. An analytical model of the tube wall electron oscillations extending into the tube is presented in Sec. 3. Preliminary proof-of-principle particle-in-cell method based computational modeling of beam-driven solid-state tube wakefield accelerator is detailed in Sec. 4. We also study the novel "crunch-in" behavior shown by the wakefields in a tube which includes wakefield amplitudes close to the Tajima-Dawson acceleration gradient for relatively small beam to tube density ratios as well as the existence of transverse fields that extend down to the tube axis.

2. Collective Oscillation in Quantum Mechanical Systems: Oscillation Modes of Electron Gas in Crystal Ionic Lattice

Collective electron oscillations in crystals have for long been established a critical yet physically valid simplification of the many body interaction in solid-state materials. The many body problem of solid-state electrons can be either described using an assembly of Fermions or using collective oscillation theory. The collective oscillation approach was exhaustively modeled in theory[32-35] (phonon, plasmon and polaritons) and experimentally proven to result in observable effects[36] in 1950s.

Solid-state collective oscillations were first investigated with great details in the context of the modeling of the stopping power of an incident electron beam in metals with an inherent crystal structure.[37] The predictions of energy loss of a particle beam incident on a metal were found to be in excellent agreement with the theory of excitation of collective electron oscillations in the crystal, driven as a wake of the incident particles.

The terminology of excitation of collective oscillations in the "wake" of an incident particle was introduced in 1950s. Moreover, to explain quantization in beam energy loss when interacting with a thin foil with thickness of the order of mean free path of the bulk plasma oscillations in crystals, these oscillations where referred to as plasmons. The theoretical plasmon model of the collective oscillations of electrons in crystals showed good agreement with experiments on energy loss of injected beam electrons. In addition to the explanation of the quantized energy loss of the beam, the conditions for the excitation of collective oscillations in the wake of an incident particle were also detailed. The collective oscillations of valence electrons were demonstrated to be quite similar to the plasma oscillations observed in gaseous plasmas.

Bloch[32] was the first to model the excitations of a Fermi gas as collective gas oscillations as opposed to excited states of single particles. Bloch treated the Fermi gas collective oscillations both with and without quantum mechanics. However, when density fluctuations were important to be considered for understanding the phenomena, then quantum aspects of the problem were critical. In Tomonaga's work[33] on collective oscillations it was demonstrated that modeling and understanding the many Fermion interaction in solid-state electron gas in a crystal was greatly simplified by the use of collective modes of many Fermion oscillations. These collective electron oscillations were first investigated in 1D in 1950 by Tomonaga through the use of density fluctuation method (where density if the field variable) as opposed to the conventional quantum mechanical method of computation of expectation values from the wave functions of the system. This was because the equations of motion of collective oscillations are linear in field variable (density fluctuation) as opposed to bilinear field variable terms in the conventional method. Moreover, linearity of the field variables in the equation of motion holds irrespective of the presence or absence of inter-particle forces (direct interaction between single particles).

Subsequently, in 1953 Pines and Bohm[34] used a collective canonical transformation method to analyze the collective many Fermion (electron) oscillations in crystal lattices in metals. They recognized the dominance of the long-range nature of the Coulomb forces which controls the phenomena and produces collective oscillations of clouds of electrons over spatial scale much greater than Debye length. Here the characteristic dimension of an electron cloud is of the order of a Debye length (in quantum mechanical treatment this characteristic dimension is modified).

The collective behavior is therefore critical to explain physical phenomena over micron or nanometer scales. In their work the term "plasmon" was introduced to describe the quantum of elementary electron excitation associated with this high-frequency collective motion in bulk crystals with the dimensions of the order of one plasmon oscillation wavelength. This is a quantum of energy of collective oscillations of valence electrons. The energy of a plasmon was shown to be[34,35]

$$\hbar\omega_{pe} = \hbar\left(\frac{4\pi n_0 e^2}{m_e}\right). \tag{1}$$

When the dimension of the solid-state material is below mean free path of the collective electron oscillations, quantization of electron plasma frequency is observed. Plasmon energy is greater than the energy of any individual conduction band electron. Although, these plasmonic oscillations are the quantum analog of the collective oscillations of plasma electrons in gaseous plasmas their extremely high energies ($\hbar\omega_{pe} \gg k_B T_e$) and small spatial scales necessitate the consideration of the quantum nature of these oscillations. Typical, valence electron density in crystals which lies in the range of $n_0 \simeq 10^{22} - 10^{24}\,\mathrm{cm}^{-3}$ result in plasma energies in crystals of $\hbar\omega_{pe} \simeq 4$ to $30\,\mathrm{eV}$. As a result of this, plasmonic collective oscillation are not thermally excitable and under normal conditions metals do not sustain plasmonic oscillations driven by a valence electron.

The dispersion relation of a plasmon using the Hamiltonian approach of a Fermi electron gas in the presence of an ionic lattice was derived in Refs. 34 and 35. This approach is essential when a quantum-mechanical treatment of the electronic motion is required, as is the case for the electrons in a metal. The particle based or density fluctuation approach taken in this work was argued to be a microscopic approach to the modeling of collective oscillations. In the density fluctuation method, the Coulomb interaction was effectively split up into a long-range and a short-range part. The conditions under which an externally injected electron beam can excite collective electron oscillations in a crystal.

In contrast certain other plasmon models simply used the dielectric constant of a medium to represent its plasma behavior, which is a macroscopic approach. To model the effect of electron-electron interaction on the stopping power of a metal for high-energy charged particles, the electron gas is described as a classical fluid with an artificially introduced coefficient of internal friction (Kronig and Korringa).

As the collective behavior of the electron gas is essential to model phenomena over distances greater than the Debye length (or a critical spatial scale, quantum mechanically), cumulative potential of all the electrons involved in the oscillation is quite large since the long range of the Coulomb interaction permits a very large number of electrons to contribute to the potential at a given point. The higher the density the larger is the number of electron that contribute to the potential and thus higher is the collective field and potential.

Surface wave modes in solid-state plasmas at the interface of crystals with vacuum or metals have also been well modeled.[38] Using both microscopic as well as macroscopic modeling, the "surface plasmon" oscillation frequency of a metal vacuum interface is determined to be $\omega_p^s = \omega_{pe}/\sqrt{1+\epsilon}$, where ϵ is the dielectric constant of the metal. The dispersion characteristics of surface plasmon and phonon modes have been well characterized in a linearized perturbative regime.[39]

3. Solid-state Surface Waves in Crystal Tube: Surface Electron Oscillation Model in Tube Nanostructure

Using the collective electron oscillation models described above and under the condition that the dominant behavior of solid-state media is that of an ideal electron gas, we analytically model surface electron oscillations in a tube structure driven in the wake of an electron beam. These analytically modeled surface electron oscillations also sustain a propagating surface wave which propagates at nearly the same velocity as the drive beam. Moreover, as collision-less behavior dominates it is possible to treat the density fluctuations using a single particle oscillation model.

Because the crystal tube under consideration here conforms with a cylindrical geometry, in our analysis we model the surface electron oscillations in a cylindrical coordinate system. Moreover, as these surface oscillations and the surface wave sustained by these collective oscillations co-propagate with the electron beam, the longitudinal dimension of the cylindrical coordinates will be transformed to a co-moving frame behind the drive beam. An preliminary analysis of a similar nature has been previously attempted.[43]

Classifying the onset of non-linearity and wavelength of the density oscillations both require understanding of the electron dynamics within the plasma. The plasma considered is one of density 0 for $r < r_{\text{tube}}$ and n_0 for $r > r_{\text{tube}}$. An electron or positron driving beam of density n_b and volume V_b – moving with velocity $v_b \hat{\mathbf{z}} = c\beta_b \hat{\mathbf{z}}$ — perturbs electrons within the plasma electrostatically. Subsequently, the plasma electrons oscillate freely in the radial direction as a result of the electric field that has been set up due to the no longer quasineutral plasma.

This electric field is found using Gauss' law

$$\int_S \mathbf{E} \cdot d\mathbf{S} = 4\pi \, Q_{enc} \tag{2}$$

where S is a closed Gaussian surface, $\mathbf{E}$ is the electric field, $d\mathbf{S}$ is the surface area element of S, Q_{enc} is the total charge enclosed within S, and $\epsilon_0 = 1/(4\pi)$ is the permittivity of free space (in cgs units). As the plasma is cylindrically symmetric, a cylinder of radius r and length l is used as the Gaussian surface S. $d\mathbf{S}$ can therefore be simplified to $d\mathbf{S} = r \, d\theta \, dz \hat{\mathbf{r}}$. Assuming the electric field to be purely radial, $\mathbf{E} = E_r \hat{\mathbf{r}}$, the left hand side of Gauss' law is simplified (after integrating) to $2\pi r l E_r$.

The enclosed charge is given by the integral of the ion charge density over the volume enclosed by S (cylinder of radius, r). The ions within the plasma are of density n_0. As the plasma density n_0 is constant, the enclosed charge is given by the net volume of plasma within S multiplied by en_0, $Q_{enc} = en_0\pi(r^2 - r_{\text{tube}}^2)l$. Gauss' law thus gives the following form for the radial electric field set up by the non-quasineutral plasma

$$E_r(r) = 4\pi e n_0 \frac{1}{2r}(r^2 - r_{\text{tube}}^2). \tag{3}$$

As the electric field vanishes for $r = r_{\text{tube}}$, Eq. (3) describes the electric field for an electron situated initially on the channel wall.

The force experienced by a given plasma electron is found by multiplying the electric field by $-e$, the electronic charge. Finally, a transformation to the frame of the driving beam $\xi = \beta_b ct - z$ is made. This is to allow for direct comparisons to be made between the model and Particle-In-Cell simulations (Sec. 4). The equation of motion is

$$m_e \frac{d^2 r}{d\xi^2} + \frac{4\pi n_0 e^2}{c^2 \beta_b^2} \frac{1}{2r}(r^2 - r_{\text{tube}}^2) = 0 \tag{4}$$

where m_e is the electron mass. Defining the plasma frequency $\omega_p = \sqrt{\frac{4\pi n_0 e^2}{m_e}}$ and $\rho = r/r_{\text{tube}}$, Eq. (5) is rewritten as

$$\frac{d^2 \rho}{d\xi^2} + \frac{1}{2\beta_b^2}\left(\frac{\omega_p}{c}\right)^2 \frac{1}{\rho}(\rho^2 - 1) = 0. \tag{5}$$

The above equation is a non-linear second order differential equation and describes the natural oscillations of a plasma electron about the channel wall. The lack of charge within the channel wall gives rise to an asymmetry in these oscillations. Setting $r_{\text{tube}} = 0$ returns the standard simple harmonic oscillations seen in homogeneous plasmas about a cylindrical axis.

3.1. *Weakly driven surface charge dynamics:*
Linear surface electron oscillations

Equation (4) is readily solvable when considering small displacements of the electron from the tube wall. These small displacements are valid for very low driving beam charges or large tube radii, when the electrostatic force of the beam acting on the plasma electrons is small. In this limit, $r \approx r_{\text{tube}}$, and Eq. (5) is linearized using

$$r^2 - r_{\text{tube}}^2 = (r - r_{\text{tube}})(r + r_{\text{tube}}) \approx 2r\,(r - r_{\text{tube}}) \tag{6}$$

where the first term on the right hand side is a second order term and has been removed. The linearised equation of motion is thus

$$\frac{d^2 r}{d\xi^2} = -\frac{(\omega_p/c)^2}{\beta_b^2}(r - r_{\text{tube}}) \tag{7}$$

which has the solution

$$r(\xi) = r_{\text{tube}} + A \sin\left(\frac{\omega_p}{c\beta_b}\xi\right) \tag{8}$$

where A is a constant. For an ultrarelativistic driving beam, $\beta_b = 1$, and so an immediate form for the oscillation wavelength in the linear/weakly excited case is

$$\lambda_{\text{linear}} = 2\pi\frac{c}{\omega_p} \tag{9}$$

which is the well-known result for plasma oscillations in homogeneous plasma.

3.2. *Strongly driven surface charge dynamics:*
Non-linear surface oscillations

Solving Eq. (4) in general requires calculation of the plasma electron's initial effective velocity ρ' for a given radial position ρ. Three basic assumptions are made to simplify the calculation to a good approximation.

The first is that the driving beam is assumed to be a quasi-static point charge of total charge Q_b. This assumption is valid provided the drive beam density changes over multiple electron oscillations and its charge is conserved. Gauss' law states that the electric field intersecting a Gaussian surface S is the same regardless of the shape of the charge distribution within S. Quasistaticity ensures the shape or size of the beam do not change significantly over time such that beam-plasma intersections do not arise.

The second assumption requires that electrons excited by the driving beam are no longer influenced by the driving beam beyond the first collapse to the axis. In other words, the primary electron collapse occurs at $\xi \gg 0$, corresponding an electric potential of approximately zero. This assumption simplifies calculation of the kinetic energy gained by the electron due to the driving beam, as the radial position of the electron at collapse need no longer be determined.

The final assumption is that the kinetic energy gained by the electron is primarily radial kinetic energy. This simplifies determination of the electron velocity at the crystal tube wall (Subsec. 3.2.2).

3.2.1. *Transforming Surface oscillation equation to first order*

As Eq. (4) is an autonomous ODE (i.e. an ODE with no dependence on ξ), the following manipulation can be made:

$$\frac{d}{d\xi}\left[\frac{1}{2}\left(\frac{d\rho}{d\xi}\right)^2\right] = \frac{d\rho}{d\xi}\frac{d^2\rho}{d\xi^2}. \tag{10}$$

Using the chain rule on the left-hand side of (10)

$$\frac{d}{d\xi} = \frac{d\rho}{d\xi}\frac{d}{d\rho}\rho'' = \frac{d}{d\rho}\left(\frac{1}{2}\rho'^2\right)$$

where $\rho' = d\rho/d\xi$. Equation (11) can then be substituted in (4) and integrated, resulting in

$$\frac{1}{2}\rho'^2 + \frac{1}{2\beta_b^2}\left(\frac{\omega_p}{c}\right)^2\left(\frac{1}{2}\rho^2 - \ln\rho\right) = C_1 \tag{11}$$

where C_1 is a constant of integration to be determined.

3.2.2. *Surface oscillation: Initial condition for velocity*

To find C_1, one must know a value of ρ' for a given ρ. At $\xi = 0$, an electron at the surface ($\rho = 1$) sees the repulsive potential (attractive potential) of the electron (positron) beam. As the electron is pulled to the axis, its energy will be converted from potential energy between it and the beam to potential energy from the no longer neutral plasma. As it recoils back towards the crystal tube wall, the electron gains kinetic energy which will become maximized at $\rho = 1$ as, beyond $\rho = 1$, the force will be directed anti-parallel to the electron velocity. Therefore, the kinetic energy of the electron at the crystal tube wall (after the excitation from the beam) is effectively equal to the potential energy it has at $(\rho, \xi) = (1, 0)$ due to the electron or positron beam under the assumptions given at the start of this section.

By letting $\rho'(\rho = 1) = \rho_0'$, i.e. some initial effective velocity to be determined later, one arrives at an equation for C_1 after substitution in (11)

$$C_1 = \rho_0'^2 + \frac{1}{2\beta_b^2}\left(\frac{\omega_p}{c}\right)^2. \tag{12}$$

Determining ρ_0' requires consideration of the energy gained by the electron using the simplifying assumptions made in the introduction of this section. The potential energy of an electron due to an electron or positron beam is

$$U(\rho, \xi) = -\frac{e^2 N_b}{4\pi\epsilon_0} \frac{1}{\sqrt{\rho^2 r_{tube}^2 + \xi^2}}, \tag{13}$$

where $N_b = Q/e$ is the total number of electrons or positrons in the beam. Due to the conservative nature of the potential, the kinetic energy E gained by a surface electron due to the electron or positron beam, initially at $\xi = 0$, is

$$E = U(\rho_1, \xi_1) - U(1, 0) \tag{14}$$

where (ρ_1, ξ_1) defines the position of the particle when its radial velocity is zero. If it is assumed that ξ_1 is large, then $U(\rho_1, \xi_1) \approx 0$ and Eq. (14) reduces to

$$E = \frac{e^2 N_b}{4\pi\epsilon_0 r_{tube}} \approx \frac{1}{2} m_e \dot{r}_0^2 \tag{15}$$

where $\dot{r}_0 = r_{tube} v_b \rho_0'$ is the initial condition velocity, v_b is the beam velocity, and the final term on the right-hand side assumes that the energy gain occurs primarily in the radial direction. Rearranging Eq. (15) yields

$$\dot{r}_0^2 \approx \frac{e^2 n_b}{m_e \epsilon_0} \frac{V_b}{4\pi r_{tube}}$$
$$= \omega_{pb}^2 \frac{V_b}{4\pi r_{tube}} \tag{16}$$

where $N_b = n_b V_b$, V_b is the effective volume of the beam, and n_b is the beam density. For a Gaussian beam distribution of width σ_r and length σ_z, n_b is defined as the electron or positron density at the beam's centre (or the peak density) with $V_b = \sigma_r^2 \sigma_z \sqrt{2\pi}$. Upon substituting $\dot{r}_0 = r_{tube} v_b \rho_0'$ in the above equation and rearranging, an approximate form for ρ_0' is determined as

$$\rho_0' \approx \frac{\omega_{pb}}{c} \frac{1}{\beta_b r_{tube}^{3/2}} \sqrt{\frac{V_b}{2\pi}} \tag{17}$$

and, for a Gaussian beam profile

$$\rho_0' \approx \frac{\omega_{pb}}{c} \frac{1}{\beta_b r_{tube}^{3/2}} \sqrt{\frac{\sigma_r^2 \sigma_z}{\sqrt{2\pi}}}. \tag{18}$$

3.2.3. Non-linearity parameter and radial boundary conditions

Due to the oscillatory nature of the problem, it is clear that there will exist two solutions for ρ where $\rho' = 0$. After substituting expressions for C_1 and ρ_0', Eq. (11) reduces to

$$\rho^2 - 2\ln\rho = 1 + 2\frac{n_b}{n_0} \frac{V_b}{2\pi r_{tube}^3}. \tag{19}$$

Defining $\alpha = 1 + 2\frac{n_b}{n_0}\frac{V_b}{2\pi r_{\text{tube}}^3}$, Eq. (19) describes a transcendental equation with two solutions:

$$\rho_+ = \sqrt{-W_{-1}(-e^{-\alpha})}, \tag{20}$$

$$\rho_- = \sqrt{-W_0(-e^{-\alpha})}, \tag{21}$$

$$\alpha = 1 + 2\frac{n_b}{n_0}\frac{V_b}{2\pi r_{\text{tube}}^3} \tag{22}$$

$$\left(= 1 + 2\frac{n_b}{n_0}\frac{\sigma_r^2\sigma_z}{\sqrt{2\pi}\,r_{\text{tube}}^3}, \text{ Gaussian Profile}\right), \tag{23}$$

where $W_{-1,0}(x)$ are the decreasing and increasing branches of the lambert W function respectively. Each solution respectively describes the amplitude of the crests and troughs of the plasma density oscillations. Looking at the extreme cases for α,

$$\rho_+ \to \begin{cases} 1 & \alpha \to 1 \\ \infty & \alpha \to \infty \end{cases}, \tag{24a}$$

$$\rho_- \to \begin{cases} 1 & \alpha \to 1 \\ 0 & \alpha \to \infty \end{cases}, \tag{24b}$$

which suggest that, for large plasma densities and tube wall radii or physically small, low density beams, $|\rho_+ - 1| \approx |\rho_- - 1|$ yielding a linear wave. In the opposite case, the wave amplitudes are different and thus the wave is non-linear. It is therefore deduced that α must describe the strength of non-linearity of the wave, and that increasing n_b, σ_r, σ_z, or decreasing n_0 or r_{tube} results in increased non-linearity.

Increasing the beam charge will correspond to a stronger driving potential experienced by the plasma electrons. As a consequence, electrons have more energy to collapse closer to the axis. This is similar for the crystal tube radius; electrons will initially be closer to the driving beam and thus experience a stronger potential. Conversely, decreasing the plasma density for a fixed beam density will reduce the number of plasma ions which weakens the restoring force allowing the tube wall electrons to collapse closer to the axis.

3.2.4. Wavelength of surface density oscillation: Analytical model

Equation (11) can be rearranged in terms of ρ', leading to an integral solution $\xi(\rho)$ with no closed form expression:

$$\xi(\rho) - \xi_0 = \frac{c\beta_b}{\omega_p}\int_1^\rho \frac{dx}{\sqrt{\frac{1}{2}\alpha - \left(\frac{1}{2}x^2 - \ln x\right)}} \tag{25}$$

where ξ_0 is a constant of integration and x is a dummy variable.

Equation (25) is restricted to a range spanning half the wavelength of the density oscillation, and a domain of (ρ_-, ρ_+). The wavelength is thus

$$\lambda = 2\big[\xi(\rho_+(\alpha)) - \xi(\rho_-(\alpha))\big] \tag{26a}$$

$$= 2\frac{c\beta_b}{\omega_p} \int_{\rho_-(\alpha)}^{\rho_+(\alpha)} \frac{dx}{\sqrt{\frac{1}{2}\alpha - \left(\frac{1}{2}x^2 - \ln x\right)}} = 2\frac{c}{\omega_p} I(\alpha) \tag{26b}$$

where in the last expression (26b), $\beta_b = 1$. The integral $I(\alpha)$ converges to π as $\alpha \to 1$, which is consistent with the linear solution. However, the wavelength of nonlinear surface oscillations is greater than that in the linear regime by the factor $2 \times I(\alpha)$,

$$\lambda_{\text{crunch-in}} = 2 \times I(\alpha) \, 2\pi c \omega_p^{-1}. \tag{27}$$

The ω_p dependence in (26b) is as per expectations that the density oscillation wavelength is strongly affected by the plasma density in the tube walls.

The parameter α also affects the wavelength in the model, suggesting a dependence of the wavelength on changing n_b, σ_r, σ_z and r_{tube}. Thus α the nonlinearity factor provides a correction to the oscillation wavelength under linear approximation. The factor $I(\alpha)$ does not include the effect of relativistic enlargement of the wavelength of radial oscillations which is a well-known additional factor.

This model therefore suggests the wavelength of oscillation may be tuneable by adjusting the plasma density n_0, keeping α constant. Conversely, the model suggests the possibility of directly controlling the strength of non-linearity while maintaining a constant wavelength, simply by adjusting multiple parameters at once.

The dependence on the plasma density, n_0, outside of the integral agrees with simulation data that adjusting the plasma frequency results in strong changes in wavelength. In addition, the solution suggests that the wavelength is tunable by adjusting n_0 while α remains constant. Conversely, the model suggests the possibility of directly controlling the strength of non-linearity while maintaining a constant wavelength, simply by adjusting multiple parameters at once.

4. Proof-of-Principle Simulations Results: Solid-State Electron Oscillations with Particle-In-Cell Simulations

Multi-dimensional Particle-In-Cell simulations using the EPOCH code[44] have been carried out to model collective electron oscillation phenomena in solid-state or crystal plasma. The use of a PIC code for analyzing collective electron oscillations at densities, $n_0 > 10^{22} \, \text{cm}^{-3}$ is justified under the conditions where the phenomena is collision-less as was shown above to be the case under strong excitation of valence electrons in crystals. Moreover, as the length of the driver particle beam is chosen to be of the order of the wavelength of collective electron oscillations, it is possible to sustain plasmonic oscillations of electrons without triggering phonons and other mixed modes, etc. Furthermore, crystal tube structures are known to naturally have mean free path lengths of several hundreds of nanometers.[45]

In this work we model the interaction of a crystal tube with intense sub-micron scale electron beam, especially its bunch length being sub-micron scale. We restrict the choice of maximum crystal tube internal diameter to 1 micron. Currently, crystal tubes are grown by folding multiple layers of mono-atomic sheets into a cylinder and in the process closing a sheet upon itself. The precise nano-engineering process of growing hundreds of nanometer tube radius is yet to be fully characterized to determine the electron density profile spanning the cross-section from the edge of the tube wall to the axis of the tube.

Graphene based carbon nanotubes (CNT) are in recent years shown to be relatively straightforward to manufacture in the sense that sophisticated machinery is generally not required.[45] These crystal tubes have valence electron densities in the range of 10^{22-24} cm^{-3} and a mean free path of about ~ 1 micron. These tube characteristics are ideally suited for supporting collective electron oscillations. Commercially, silica (SiO_2) or molten glass based tubes with internal diameters ranging from 200–1000 nm are sold variously as nano-capillary,[46] etc. The gradient of density at the interface of the tube wall and hole region has been characterized using scanning electron microscope (SEM) and found to be less than its nanometer scale resolution limit. Similarly, the uniformity of the diameter is found to be quite consistent over several meters of tube length. However, the extent of surface deformities and imperfections of the exact tubes grown from atomic monolayer deposition is not precisely modeled or characterized here and will form a part of the future work.

Our computational modeling effort characterizes an experimentally realizable interaction scenario due to the recently reported availability of sub-micron scale bunch lengths at Stanford Linear Accelerator Center[10] and possibly at other accelerator facilities in the near future.

This computational modeling effort currently serves as a proof-of-principle and is an attempt to demonstrate the possibilities that can be opened up by the solid-state tube wakefield acceleration technique. In our modeling we utilize the fact that beams of several hundred nanometer bunch lengths are accessible. For experimental relevance the modeling effort is carried out under the following constraints:

(i) In Subsec. 4.1 – an experimentally available beam density of $n_b = 1.0 \times 10^{21}$ cm^{-3} is considered. The beam waist-size is $\sigma_r = 500$ nm which is also chosen to be the tube radius $r_{tube} = 500$ nm. The PIC simulations indicate that a tube wall electron density of $n_{tube} = 2.0 \times 10^{22}$ cm^{-3} is suitable. This tube wall density may need customized nanoengineering.

(ii) In Subsec. 4.2 – under an experimental constraint that the beam waist-size exceeds the tube diameter we model a beam-tube interaction scenario such that the beam particles radially in the Gaussian wings of the tube interact with the tube wall, $\sigma_r > r_{tube}$ while the most intense part of the beam propagates within the tube.

(iii) In Subsec. 4.3 – we assume that a beam density of $n_b = 1.0 - 5.0 \times 10^{22}$ cm^{-3}

is experimentally accessible. In this case the suitable tube wall densities of $n_{\text{tube}} = 1.0 - 5.0 \times 10^{23} \, \text{cm}^{-3}$ are known to be available commercially off-the-shelf.

4.1. *Beam waist comparable with tube radius,* $r_{\text{tube}} \gtrsim \sigma_r$

Using PIC simulations we model and make preliminary investigations of beam-driven solid-state tube acceleration in a parameter regime where the waist-size of the beam injected into a crystal tube is comparable to the crystal tube radius.

The electron density in the tube walls is chosen to be $n_0 = 2.0 \times 10^{22} \, \text{cm}^{-3}$ with a fixed ion background. In the simulation results presented below a 2D cartesian grid is chosen such that it resolves the reduced plasmonic wavelength of $\lambda_{\text{pe}}/(2\pi) = 38 \, \text{nm}$ with 15 cells in the longitudinal and 15 cells in the transverse direction. Thus each grid cell in these simulations is about 2.5 nm × 2.5 nm (the Debye length, conservatively assuming a few eV thermal energy is $\lambda_D \simeq 1\text{Å}$). The cartesian box co-propagates with the electron beam. The box dimensions span 7 μm in longitudinal direction and at least 7 μm in the transverse (it is wider in transverse to incorporate wider beams). The tube electrons are modeled with 10 particles per cell of the cartesian grid. Absorbing boundary conditions are used for both fields and particles.

The electron beam has a $\gamma_b = 10,000$ (roughly 5.1 GeV) with a Gaussian bunch profile of a fixed bunch length with $\sigma_z = 400 \, \text{nm}$. Typical beam density of $1.0 \times 10^{21} \, \text{cm}^{-3}$ is considered in this experimentally relevant modeling effort. The beam is initialized with 16 particle per cell. In order to analyze the interaction, the waist-size of the beam, σ_r and the tube radius r_{tube} are varied. The moving simulation box tracks the particle beam. The particle beam is initialized in vacuum and propagates into the crystal tube before the simulations box begins to move.

PIC simulation snapshots in Fig. 1 correspond with solid-state tube accelerator interaction parameters of crystal tube radius, r_{tube} of 500 nm and beam waist-size, $\sigma_r = 500 \, \text{nm}$ and bunch length, $\sigma_z = 400 \, \text{nm}$. From this snapshot we observe that a surface wave is sustained by the oscillations of the electrons across the interface of the tube wall with density, $n_{\text{tube}} = 2.0 \times 10^{22} \, \text{cm}^{-3}$. In Fig. 1(a) the tube wall density snapshot in real-space shows *three* distinct spatial oscillations of a surface plasmon sustained by radial electron oscillations across the surface. These snapshots are at a simulation time of 250 fs which corresponds to a beam-tube interaction length of around 72 μm (the interaction has a delayed start as the beam is initialized in vacuum and pushed to propagate into the plasma).

Beam density profiles from PIC simulation for the drive beam with $\sigma_z = 400 \, \text{nm}$ at a density of $n_b = 1.0 \times 10^{21} \, \text{cm}^{-3}$ show that the beam electrons experience the transverse or focusing fields of the "crunch-in" wakefields of the surface wave and exhibit betatron oscillations. This effect of beam density modulation can be prominently observed in the beam density snapshots presented in later sections in Fig. 5(b) and Fig. 6(d). These coherent density modulations of the drive beam were first modeled in an innovative plasma beam dump proposal.[47] This coherent drive beam density modulation has been observed and labelled as scalloping in

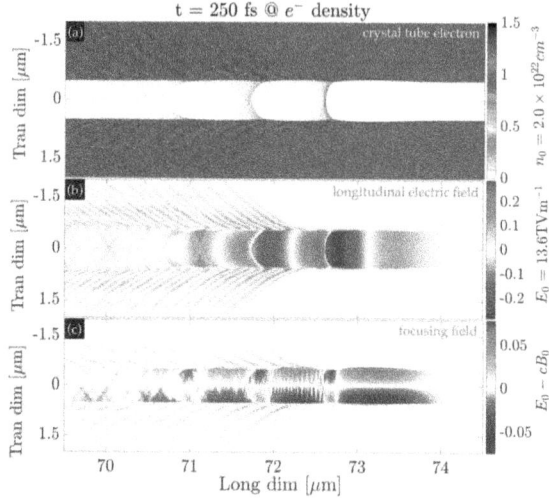

Fig. 1. 2.5D PIC simulation snapshot at around $72\,\mu$m of beam-tube interaction showing solid-state tube accelerator with crystal tube radius of 500nm and beam waist-size, $\sigma_r = 500$ nm and bunch length, $\sigma_z = 400$ nm. The beam density is $n_b = 1.0 \times 10^{21}$ cm^{-3} whereas the channel wall density of the tube is $n_{\text{tube}} = 2.0 \times 10^{22}$ cm^{-3}. Solid-state tube wakefield accelerator dynamics extracted from a 2.5D PIC simulation at around $72\,\mu$m of beam-tube interaction showing the tube wall electron density (in a, normalized to $n_{\text{tube}} = 2.0 \times 10^{22}$ cm^{-3}), longitudinal electric field of the surface wave (in b, normalized to $E_0 = 13.6$ TVm^{-1}) and the focusing field (in c, $E_0 - cB_0$).

some recent works. In the near-term, beam-tube interaction can be experimentally studied by observing the small spatial-scale drive beam density modulations.

In Fig. 1, the longitudinal (in b) and focusing forces (in c) are shown along with the density wave in real-space. The fields are normalized to the Tajima-Dawson acceleration gradient ($E_{wb} = E_0 = m_e c \omega_{pe} e^{-1}$ or the cold-plasma wavebreaking limit). The Tajima-Dawson limit for the tube density of $n_{\text{tube}} = 2.0 \times 10^{22}$ cm^{-3} is $E_0 = 13.6$ TVm^{-1}.

The simulation results lead to several interesting possibilities. It is observed that although the beam to tube density ratio n_b/n_{tube} is only 0.05, the longitudinal wakefields approach $\langle E_{\text{acc}} \rangle \simeq 0.25 E_0$ which is an acceleration gradient of around $\langle E_{\text{acc}} \rangle \simeq 2.5$ TVm^{-1}. Moreover, the "crunch-in" behavior results in the excitation of focusing fields of the order of $0.1 E_0$ which is a focusing gradient of several 100 GVm^{-1}.

Figure 2 describes the longitudinal phase-space of beam-tube interaction at around $72\,\mu m$ in the "crunch-in" surface wakefields regime. From the longitudinal momentum against transverse momentum is shown in Fig. 2(a) and transverse real space dimension in Fig. 2(b) it can be observed that only those beam particles that are within the crystal tube and that experience both the longitudinal as well as the transverse "crunch-in" wakefields undergo acceleration. Longitudinal phase-space plotted against longitudinal real space dimension in Fig. 2(c) and the corresponding on-axis field, electron density lineout in Fig. 2(d) demonstrate that the longitudinal dimension of the wakefield is such that the beam particles in the

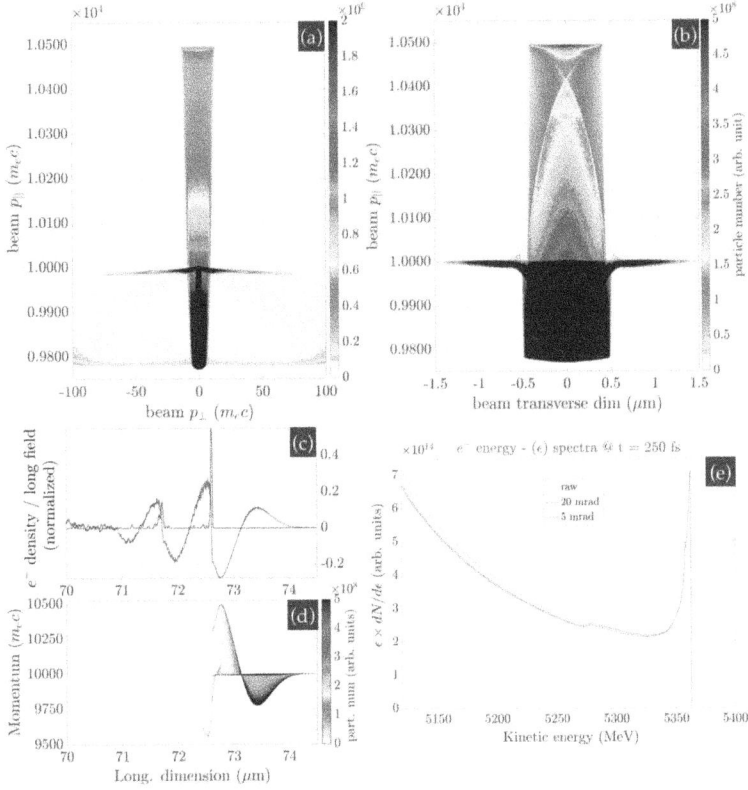

Fig. 2. Longitudinal momentum phase-space against transverse momentum (in a) and transverse real space dimension (in b) from 2.5D PIC simulation snapshot after around $72\,\mu m$ of beam-tube interaction with parameters same as Fig. 1. Longitudinal phase-space against longitudinal real space dimension (in c) and corresponding on-axis field, electron density lineout (in d) from 2.5D PIC simulation snapshot are also shown around $72\,\mu m$ of beam-tube interaction.

tail of the beam undergo acceleration. This opens up the possibility of using a single hundreds of nanometer scale bunch to observe a few $\mathrm{TVm^{-1}}$ acceleration gradients.

The energy spectra in Fig. 2(e) shows some of the particles in the tail of the beam being accelerated from the initial beam energy centered around 5110 MeV to 5360 MeV, a gain of about 250 MeV in $72\,\mu m$. This gives an average acceleration gradient over $72\,\mu m$ of around $\langle E_{\mathrm{acc}}\rangle \simeq 3.0\,\mathrm{TVm^{-1}}$. It is also quite evident from the snapshots in (a) and (b) that only those beam electrons that are within the tube get accelerated in the surface wave, whereas the electrons in the Gaussian wings of the beam undergo minimal energy change.

The possibilities of accessing high average acceleration gradients of the order of several $\mathrm{TVm^{-1}}$ are unprecedented and being based upon a modeling effort where realistic parameters are utilized call out for an experimental verification campaign.

In Fig. 3, a comparison of bulk plasma ($n_0 = n_{\mathrm{tube}} = 2\times 10^{22}\,\mathrm{cm^{-3}}$) and crystal tube wakefields is presented by plotting side-by-side the electron density (in a,d), longitudinal electric field (in b,e) and the focusing field (in c,f) profiles. It is quite evident from the comparison of the snapshots in (a) and (d) that whereas the surface wave wakefield amplitude for $n_b/n_0 = 0.05$ is significantly high and in the nonlinear regime for a tube radius, $r_{\text{tube}} = 500\,\text{nm}$, the wakefields driven in homogeneous plasma are almost non-existent due to the low beam to plasma density ratio.

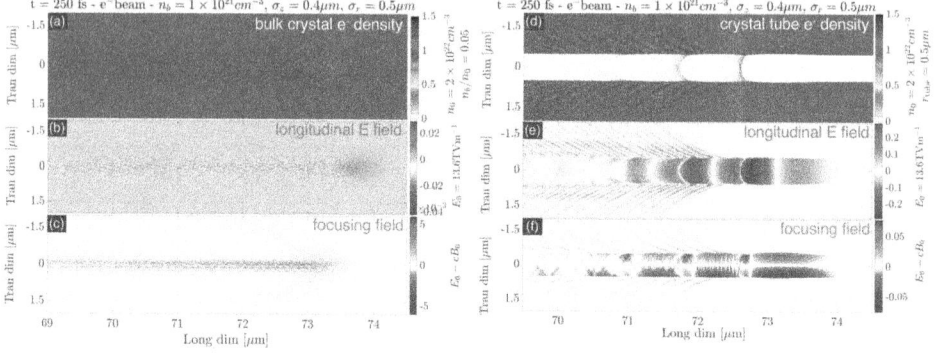

Fig. 3. Comparison of homogeneous plasma wakefield (in a,b,c) with the crystal tube wakefield (in d,e,f; repeated from Fig. 1) from 2.5D PIC simulation snapshot after around $72\,\mu\text{m}$ of beam-tube interaction with parameters same as Fig. 1. The comparison of (a) and (d) shows that whereas the surface wave wakefield for $n_b/n_0 = 0.05$ is in the nonlinear regime, the wakefields driven in homogeneous plasma are almost non-existent due to the low beam to plasma density ratio.

In order to characterize the effect of the ratio of drive beam density to tube wall density, n_b/n_{tube}, in Fig. 4 we compare the PIC simulation snapshots of tube electron density for different n_b/n_{tube} ratio at a beam-tube interaction length of $72\,\mu\text{m}$. These beam densities are: (a) $n_b = 0.5 \times 10^{21}\,\text{cm}^{-3} = 0.025\,n_{\text{tube}}$, (b) $n_b = 1.0 \times 10^{21}\,\text{cm}^{-3} = 0.05\,n_{\text{tube}}$, (c) $n_b = 2.0 \times 10^{21}\,\text{cm}^{-3} = 0.1\,n_{\text{tube}}$, and (d) $n_b = 4.0 \times 10^{21}\,\text{cm}^{-3} = 0.2\,n_{\text{tube}}$.

The corresponding peak longitudinal on-axis fields or acceleration gradient over varying beam density as extracted from PIC simulations are summarized as follows:

(a) $\langle E_{\text{acc}} \rangle \simeq 0.1 E_0$ for $n_b = 0.5 \times 10^{21}\,\text{cm}^{-3} = 0.025\,n_{\text{tube}}$

(b) $\langle E_{\text{acc}} \rangle \simeq 0.25 E_0$ for $n_b = 1.0 \times 10^{21}\,\text{cm}^{-3} = 0.05\,n_{\text{tube}}$

(c) $\langle E_{\text{acc}} \rangle \simeq 0.65 E_0$ for $n_b = 2.0 \times 10^{21}\,\text{cm}^{-3} = 0.1\,n_{\text{tube}}$

(d) $\langle E_{\text{acc}} \rangle \simeq 1.5 E_0$ for $n_b = 4.0 \times 10^{21}\,\text{cm}^{-3} = 0.2\,n_{\text{tube}}$

It is quite evident that as the drive beam density is increased in "crunch-in" regime of solid-state tube, the surface electron trajectories become increasing nonlinear. The nonlinear surface wave results in wakefields that are not only higher (of the order of the Tajima-Dawson acceleration gradient limit) but also result in the excitation of stronger focusing fields within the tube.

4.2. Beam waist size larger than the tube diameter, $\sigma_r > r_{\text{tube}}$

An experimentally accessible parameter regime in the short-term where the beam waist-size is a few times larger than the crystal tube radius is investigated below using preliminary PIC simulations. In these simulations it is assumed that the peak

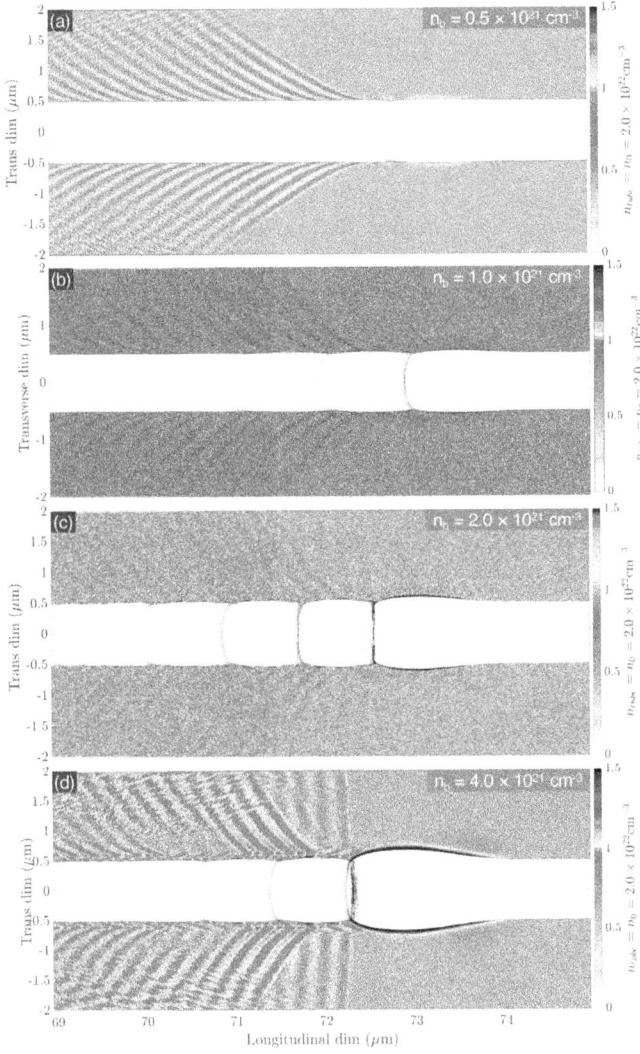

Fig. 4. Comparison of the "crunch-in" surface wave modes in crystal nanotube for different drive beam densities from 2.5D PIC simulation snapshot after around $72\,\mu$m of beam-tube interaction in a crystal tube with wall density, $n_{\text{tube}} = 2.0 \times 10^{22}$ cm^{-3}. The drive beam densities n_b are: (a) 0.025 n_{tube}, (b) 0.05 n_{tube}, (c) 0.1 n_{tube}, and (d) 0.2 n_{tube}.

of the beam density coincides with the axis of the crystal tube such that the most intense part of the externally focussed beam travels in the low density region of the tube. The PIC simulation setup and beam density are as described in Subsec. 4.1.

From the PIC simulation results that are summarized below in Fig. 5, it is quite clear that surface wave wakefields in the "crunch-in" regime are sustained within the tube even when $\sigma_r > r_{\text{tube}}$. From the results in this section we observe that when the beam density is retained same, spatial profiles of the wakefields and the acceleration gradient of the order of 2.5 TVm^{-1} sustained in the case of $\sigma_r > r_{\text{tube}}$

being studied here are nearly equal to the case where $\sigma_r \lesssim r_{\text{tube}}$.

This excitation of near Tajima-Dawson acceleration gradient ($0.1E_{wb}$) limit in a crystal tube is quite interesting because the peak drive beam density of $1.0 \times 10^{21}\,\text{cm}^{-3}$ is much smaller than the tube wall density of $2.0 \times 10^{22}\,\text{cm}^{-3}$. The simulations show an energy gain of about $50\,\text{MeV}$ in around $23.5\,\mu\text{m}$ which is an average acceleration gradient of $\langle E_{\text{acc}} \rangle > 2.0\,\text{TVm}^{-1}$.

Moreover, the "crunch-in" regime wakefields observed in this work show that strong coherent focusing fields are also excited within the tube of the order of several $100\,\text{GVm}^{-1}$. It is evident from the beam density snapshots in Fig. 5 that only those beam particles that are within the tube and which as a result experience the $\langle E_{\text{acc}} \rangle \sim \text{TVm}^{-1}$-scale fields of the surface plasmon wave undergo significant density perturbation.

From comparison of the density and wakefield characteristics of crystal tube wakefields in Fig. 1 for the case of $\sigma_r \sim r_{\text{tube}}$ and the same (not shown) for the case of $\sigma_r \gg r_{\text{tube}}$, it is observed that the wakefield characteristics, amplitude and spatial profile, are quite similar. As the drive electron beam density in the case of Fig.1 and $\sigma_r \gg r_{\text{tube}}$ case are equal, $n_b = 1.0 \times 10^{21}\,\text{cm}^{-3}$, it is possible to postulate that drive beams of a given density are equally effective at the excitation of crystal tube wakefields irrespective of their transverse properties (given that the peak of the beam spatial distribution is aligned with the axis of the tube).

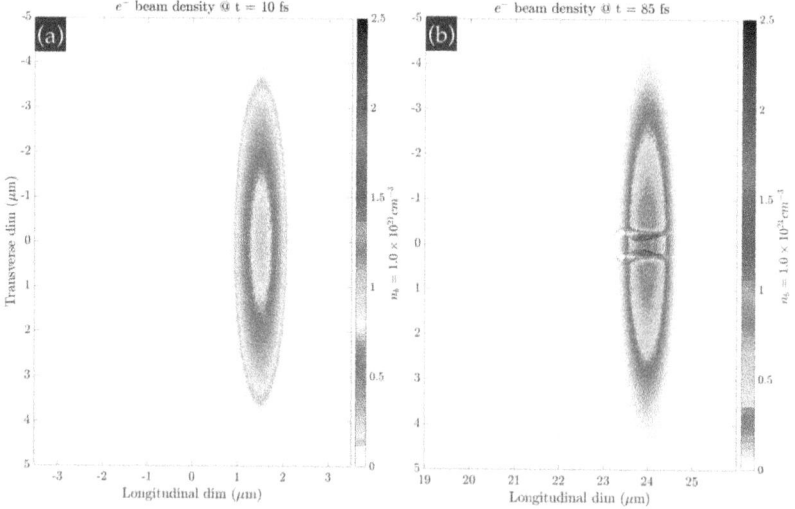

Fig. 5.   2.5D PIC simulation snapshots comparing the electron beam density at initialization and after around $23.5\,\mu m$ of beam-tube interaction.

4.3. *Scaling to off-the-shelf tube wall densities:* $n_{\text{tube}} \sim 10^{23}\,\text{cm}^{-3}$

Crystal nanotubes[45] that are currently available off-the-shelf have mass densities in the range of $1.3-2.0$ g-cm^{-3}. For purely Carbon atom based nanomaterial this mass density translates to ionic and electron densities in highly ionized states of between $n_{\text{tube}} \sim 10^{23-24}$ cm^{-3}. In this section we present our examination of the scaling of the "crunch-in" modes in crystal tubes when the crystal wall densities are around $n_{\text{tube}} \sim 10^{23-24}$ cm^{-3}. The PIC simulation setup and beam density are the same as described in Subsec. 4.1. These simulations show that if certain beam densities may be experimentally within reach of existing electron beam facilities, then it may be possible to excite beam-driven solid-state tube surface wave wakefields using off-the-shelf nanotubes.

In the previous sections, Subsecs. 4.1 and 4.2, we have presented proof-of-principle PIC simulation results under the constraint that the beam densities are limited to around $n_b \leq 2 \times 10^{21}$ cm^{-3} whereas the beam bunch length is characterized by $\sigma_z \simeq 400$ nm. Under this constraint on the beam density, the "crunch-in" regime was observed to be accessible only for tube wall densities, $n_{\text{tube}} \sim 10 \times n_b$.

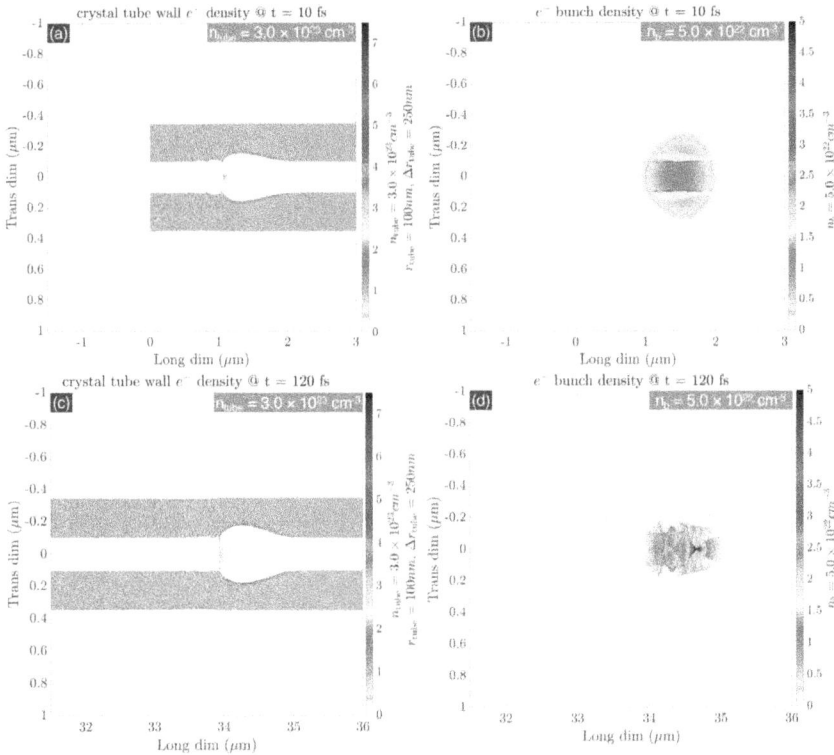

Fig. 6. Density snapshots for off-the-shelf solid-state tube parameters from 2.5D PIC simulations showing tube electron density (in a,b) with fixed background ions and drive beam electron density (in b,d) after around 36 μm of beam-tube interaction. From (a,c) it follows that "crunch-in" mode is excited and that the beam evolves as it experiences the transverse fields of this mode.

In this section, we assume that beam densities as high as $n_b \sim 5.0 \times 10^{22}$ cm^{-3} are experimentally accessible using currently available accelerator facilities. With these range of beam densities, using PIC simulations snapshots presented below,

we observe that coherent "crunch-in" wakefields supported by collective oscillations of crystal electrons are accessible using nanotube structures that are available
off-the-shelf. Therefore, if beam densities of the order of $n_b \sim 1.0 \times 10^{23} \mathrm{cm}^{-3}$
are experimentally accessible, then proof of concept experimental verifications of
SOTWA mechanism can be carried out in the near term.

In the simulations presented in this section the electron density in the tube
walls is chosen to be $n_0 = 3.0 \times 10^{22} \, \mathrm{cm}^{-3}$ with a fixed ion background. A 2D
cartesian grid is chosen such that it resolves the reduced plasmonic wavelength of
$\lambda_{\mathrm{pe}}/(2\pi) = 10 \, \mathrm{nm}$ with 20 cells in the longitudinal and 20 cells in the transverse
direction. Thus each grid cell in these simulations is about $500 \, \text{Å} \times 500 \, \text{Å}$ (the Debye
length is $\lambda_{\mathrm{D}} \leq 1 \, \text{Å}$).

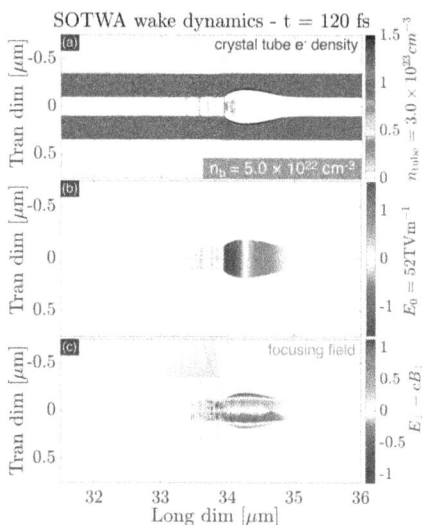

Fig. 7. 2.5D PIC simulation snapshot of the tube wall electron density (in a), longitudinal field
(in b) and focusing field (in c) at around $36 \, \mu m$ of beam-tube interaction with exactly the same
beam and tube parameters as in Fig.6. From (b,c) both the longitudinal as well as transverse
wakefields are of the order of the Tajima-Dawson acceleration gradient limit, $52 \, \mathrm{TVm}^{-1}$.

The crystal tube of radius, $r_{\mathrm{tube}} = 100 \, \mathrm{nm}$ is here is modeled to have a finite
thickness of $250 \, \mathrm{nm}$, with the outer radial extent of the tube thus terminating at
$350 \, \mathrm{nm}$ from the axis of the tube. The cartesian box co-propagates with the electron
beam. The box dimensions span $5 \, \mu m$ in longitudinal direction and at least $3 \, \mu m$
in the transverse. The tube electrons are modeled with 4 particles per cell.

The electron beam has $\gamma_b = 10,000$ (roughly 5.1 GeV) with a Gaussian bunch
profile of a fixed bunch length with $\sigma_z = 400 \, \mathrm{nm}$. In the simulation snapshots presented below the beam density is $5.0 \times 10^{22} \, \mathrm{cm}^{-3}$ which is experimentally relevant.
The beam is initialized with 9 particles per cell. A comparison of the PIC simulation
results in Subsecs. 4.1 and 4.2 provides enough confidence that the critical parame-
ter in beam tube interaction is the beam density, n_b with the σ_r to r_{tube} ratio being
relatively insignificant. In consideration of this we use a beam with $\sigma_r = 250 \, \mathrm{nm}$,

with a good and previously justified approximation that a beam of higher waist-size (for example, $\sigma_r \sim 2.5\,\mu m$) but the same density will have the same characteristics of beam tube interactions and excite considerably similar wakefields.

From these simulation snapshots summarized in Figs. 6 and 7 it is possible to conclude that if beam densities as high as $n_b \sim 5.0 \times 10^{22}\,cm^{-3}$ are experimentally accessible at current accelerator facilities, then it may be possible to excite strong "crunch-in" wakefields in off-the-shelf crystal tubes of nominal tube dimensions. In our simulations, the tube has a radius (r_{tube}) of 100nm and a wall thickness (Δr_{tube}) of 250nm. The beam density is initialized to $n_b = 5.0 \times 10^{22}\,cm^{-3}$ and tube wall density is initialized to $n_{tube} = 3.0 \times 10^{23}\,cm^{-3}$, with the $n_b/n_{tube} = 0.17$. The beam properties are: $\gamma_b = 10,000$, $\sigma_z = 400\,nm$ and $\sigma_r = 250\,nm$.

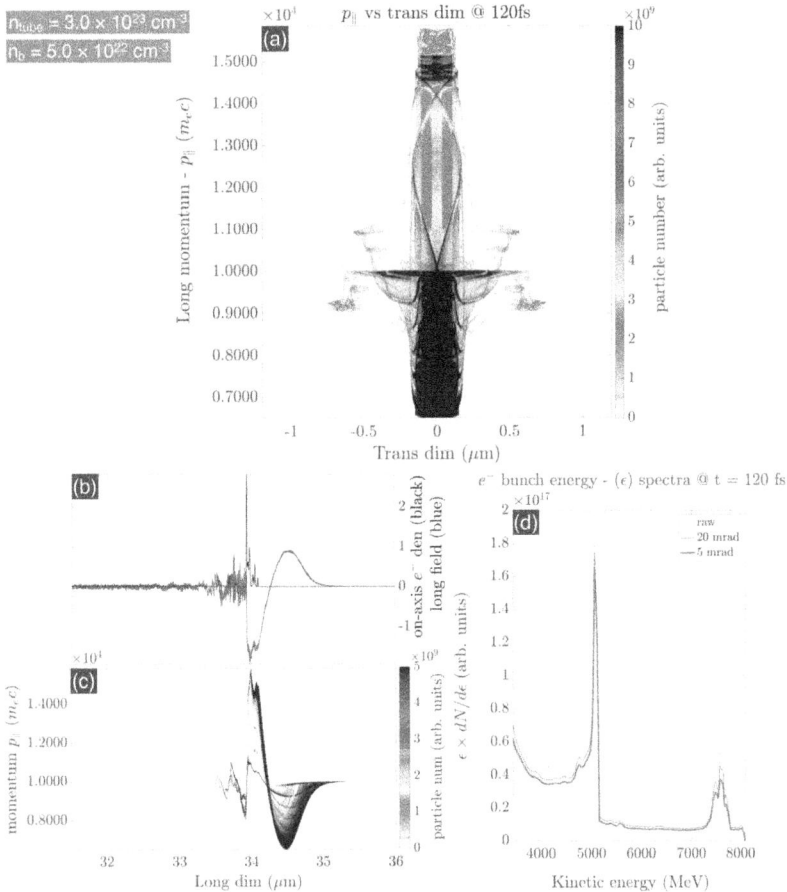

Fig. 8. Longitudinal momentum phase-space against transverse real space dimension (in a), the same against longitudinal real space dimension (in c) and the corresponding on-axis field, electron density lineout (in b) from 2.5D PIC simulation snapshot after $36\,\mu m$ of beam-tube interaction.

From Fig. 8 we can infer that a few $10\,TVm^{-1}$ acceleration gradient may be experimentally realizable using current accelerator facilities. The accelerated energy spectra shown in Fig. 8(d) shows the acceleration of a small fraction of the drive beam from the initial beam energy centered around 5110 MeV to 7570 MeV, a gain of about 2.46 GeV in 36 μm gives an average gradient of around $\langle E_{\text{acc}} \rangle \simeq 70.0$ TVm^{-1}. It is quite evident from the transverse real-space vs longitudinal momentum snapshot in (a) that only those beam electrons that are within the tube ($r_{\text{tube}} = 100$ nm) get accelerated in the wakefield. This increase in the acceleration gradient simply follows the electron density scaling of the Tajima-Dawson acceleration gradient.

It is quite attractive to have the possibility to accelerate a part of the 5GeV particle beam by around 2.5 GeV in sub millimeter-scale crystal tubes while sustaining unprecedented many tens of TVm^{-1} acceleration gradients.

5. Discussion and Future Work

In this work we have presented a preliminary analytical and computational model of beam-driven solid-state acceleration mechanism in crystal tubes. The solid-state tube wakefield acceleration or SOTWA mechanism presented here utilizes collective electron oscillation modes on and across the surface of a crystal tube. These plasmonic oscillations sustain propagating surface waves driven as the wakefield of a charged particle beam of submicron bunch length (and, ideally submicron waistsize). A tube shaped nanostructured crystal is not only found to offer the possibility to minimize the direct high-intensity interaction of the beam with bulk crystal but also the possibility of excitation of significantly higher wakefield amplitude compared to the direct interaction of same density particle beam with bulk crystal.

The experimentally available submicron scale bunch length (for instance, $\sigma_z = 400$ nm [10]) is shown to have the potential for resonant excitation of collective electron oscillations in crystal tube. The resonant excitation of a surface mode in a crystal tube driven by the beam at a given density is shown to be experimentally realizable within a range of tube wall densities. A preliminary analytical model of the crystal tube wall surface electron oscillations has been presented based upon the seminal works on modeling many body crystal phenomena as collective modes of collisionless Fermi electron gas. Our model currently assumes minimal ion motion over the relevant attosecond timescales (from our preliminary mobile ion simulations) but crystal lattice ion motion effects will be a major part of the future work.

In the computational models presented in Sec. 4 using beam densities of the order of $n_b \sim 10^{21}$ cm^{-3}, as some of the electrons in the tail of the drive bunch experience strong wakefields of the tube surface wave, they are shown to rapidly gain energy. The average acceleration gradients experienced by the tail particles are shown to be of the order of several TVm^{-1} as per the expectations of the Tajima-Dawson acceleration gradient limit for crystals. The particles in the tail of the beam gain several hundred MeVs in a few hundred microns under the influence of surface wakefields (Subsecs. 4.1 and 4.2). The possibility of accessing average acceleration gradients that are at least *two* orders of magnitude higher than the gaseous plasma wakefield acceleration techniques will pave the way forward in accelerator research.

It is further demonstrated that if experimentally accessible beam densities may be of the order of $n_b \sim 5.0 \times 10^{22}$ cm^{-3} with other beam properties being the same, then off-the-shelf crystal tubes of a few hundred nanometer diameter can be utilized (Subsec. 4.3). With the densities of the these off-the-shelf tubes being of the order of 10^{23} cm^{-3}, the accessible Tajima-Dawson acceleration gradients are of the order of 10 TVm^{-1}. Our simulations suggest the possibility of SOTWA fields being at least *three* orders of magnitude higher than gaseous plasma acceleration technology.

The possibility that an increase in the drive beam density allows access to a nonlinear surface wave "crunch-in" regime has been demonstrated. In the "crunch-in" regime both strong transverse fields of the order of many 100 GVm^{-1} as well as longitudinal wakefields of the order of many TVm^{-1} are excited. Thus, this regime using a crystal tube is useful to control the accelerated bunch transverse properties while the accelerated particles do not directly experience high ion density in their propagation path resulting in the minimization of associated instabilities.

Controlled crystal tube photon source: Moreover, the strong transverse fields of the crystal tube wake make controlled and tunable generation of gamma-ray photons as an electron or positron beam particle trajectories undergo oscillations during their interaction with "crunch-in" transverse fields of the order of many 100GVm^{-1}. The use of specifically structured crystal tube such as with a super-lattice, allows significantly higher control of the gamma-ray flux as opposed to the uncontrolled filamentation driven interaction in a metal.[8]

Nano-modulation of drive beam: In the very near-term, beam-tube interaction can be experimentally investigated by the observation of coherent density modulations of the drive beam[47] which may be related to the effect of beam scalloping observed in some gaseous plasma studies elec-Beam-Wakefield-Expt. Beam-tube interaction can be experimentally diagnosed by observing the small spatial-scale beam density modulations after the interaction.

In future work, we will extend the analytical and computational modeling of the SOTWA mechanism presented here. Moreover, we will determine the optimal conditions for the excitation of strong many TVm^{-1} acceleration (and in-tube focusing) gradients under various tube and beam parameters by modeling the plasmonic surface wave in crystal tube. It is also critical to understand non-ideal conditions of the interaction, such as misalignment of the beam and tube axis, effect of limited thickness of tube wall and electron density profile of tube, secondary high-field ionization of the channel walls, ion motion processes, modification of tube density profile due to ablation of the crystal tube to beam irradiation etc. The extent and time-scales of damage caused to the crystal tube structure by the drive beam, the possibility of reuse in consideration of effects such as atomic stabilization as well as the effect of these non-ideal structural properties on collective electron oscillations will also be carefully modeled.

Laser Wakefield Accelerator injector for SOTWA: Our future work will also study the external injection of the inherently micron-scale electron and positron beams[48] that are accelerated using laser-driven wakefields in gaseous plasmas and are thus likely to be more accessible. Furthermore, it is well known that the radiation from muons interacting with the high transverse or focusing field of the tube and undergoing oscillations is significantly smaller than electrons or positrons, ($\propto (m_e/m_\mu)^4$) we will also model the injection and acceleration of muons. Laser acceleration of muons,[49] also put forth and investigated as part of this XTALS 2019 workshop, is modeled to be able to produce ultra-short micron-scale muon beams that are suitable for injection into crystal tube wakefields.

Through the proposed extensive modeling effort, our work will seek the parameter regime and feasible diagnostics for demonstration of an experimental prototype of possibly many TVm^{-1} average acceleration gradient of the SOTWA mechanism.

Acknowledgments

A. A. S. was supported by the College of Engineering and Applied Science, University of Colorado, Denver. V. D. S. was supported by Fermi National Accelerator Laboratory, which is operated by the Fermi Research Alliance, LLC under Contract No. DE-AC02-07CH11359 with the United States Department of Energy. We appreciate valuable discussions with S. Chattopadhyay, U. Winenands, G. Stupakov and V. Lebedev. This work used the Extreme Science and Engineering Discovery Environment (XSEDE), which is supported by National Science Foundation grant number ACI-1548562.[50] This work utilized the RMACC Summit supercomputer through the XSEDE program, which is supported by the National Science Foundation (awards ACI-1532235 and ACI-1532236), the University of Colorado Boulder, and Colorado State University.[51]

References

1. R. Hofstadter, An atomic accelerator, Stanford University HEPL, Report No. 560 (1968)
2. T. Tajima and M. Cavenago, Crystal x-ray accelerator, *Phys. Rev. Lett.* **59**, 1440 (1987), doi:10.1103/PhysRevLett.59.1440
3. P. Chen and R. J. Noble, A solid state accelerator, *AIP Conf. Proc.* **156**, 222 (1987); P. Chen and R. J. Noble, Crystal channel collider: Ultra-high energy and luminosity in the next century, *AIP Conf. Proc.* **398**, 273 (1997), [SLAC-PUB-7402 (1998)].
4. T. Tajima and J. M. Dawson, Laser electron accelerator, *Phys. Rev. Lett.* **43**, 267–270 (1979), doi:10.1103/PhysRevLett.43.267
5. P. Chen *et al.*, Acceleration of electrons by the interaction of a bunched electron beam with a plasma, *Phys. Rev. Lett.* **54**(7), 693–696 (1985), doi:10.1103/PhysRevLett.54.693
6. N. M. Naumova *et al.*, Relativistic generation of isolated attosecond pulses in a λ^3 focal volume, *Phys. Rev. Lett.* **92**, 063902 (2004), doi:10.1103/PhysRevLett.92.063902
7. A. Zholents, M. Zolotorev and W. Wan, Generation of attosecond electron bunches, in *Proc. of the 2001 Particle Accelerator Conference, Chicago, IL, USA*; N. Naumova *et al.*, Attosecond electron bunches, *Phys. Rev. Lett.* **93**, 195003 (2004).
8. A. Benedetti, M. Tamburini and C. H. Keitel, Giant collimated gamma-ray flashes, *Nat. Photonics* **12**, 319–323 (2018).

9. D. Bohm and D. Pines, A collective description of electron interactions: III. Coulomb interactions in a degenerate electron gas, *Phys. Rev.* **92**, 609 (1953).

10. V. Yakimenko, Ultimate beams at FACET-II, in *Proc. of the Workshop on Acceleration in Crystals and Nanostructures (XTALS 2019), 24–25 June 2019, Fermilab*; V. Yakimenko *et al.*, Prospect of studying nonperturbative QED with beam-beam collisions, *Phys. Rev. Lett.* **122**, 190404 (2019).

11. X. Zhang *et al.*, Particle-in-cell simulation of x-ray wakefield acceleration and betatron radiation in nanotubes, *Phys. Rev. Accel. Beams* **19**, 101004 (2016).

12. D. Strickland and G. Mourou, Compression of amplified chirped optical pulses, *Opt. Commun.* **55**, 447–449 (1985).

13. A. W. Chao, Gymnastics in phase space, SLAC-PUB-14832.

14. E. Adli *et al.*, Acceleration of electrons in the plasma wakefield of a proton bunch, *Nature* **561**, 363–367 (2018), doi:10.1038/s41586-018-0485-4

15. S. P. D. Mangles *et al.*, Monoenergetic beams of relativistic electrons from intense laser-plasma interactions, *Nature* **431**, 535 (2004), doi:10.1038/nature02939; C. G. R. Geddes *et al.*, High-quality electron beams from a laser wakefield accelerator using plasma-channel guiding, *Nature* **431**, 538 (2004), doi:10.1038/nature02900; J. Faure *et al.*, A laser-plasma accelerator producing monoenergetic electron beams, *Nature* **431**, 541 (2004), doi:10.1038/nature02963.

16. M. Litos *et al.*, High-efficiency acceleration of an electron beam in a plasma wakefield accelerator, *Nature* **92**, 515, (2014), doi:10.1038/nature13882.

17. J. B. Rosenzweig *et al.*, Effects of ion motion in intense beam-driven plasma wakefield accelerators, *Phys. Rev. Lett.* **95**, 195002 (2005), doi:10.1103/PhysRevLett.95.195002

18. C. B. Schroeder *et al.*, Trapping, dark current, and wave breaking in nonlinear plasma waves, *Physics of Plasmas* **13**, 033103 (2006).

19. R. Gholizadeh *et al.*, Preservation of beam emittance in the presence of ion motion in future high-energy plasma-wakefield-based colliders, *Phys. Rev. Lett.* **104**, 155001 (2010), doi:10.1103/PhysRevLett.104.155001.

20. A. A. Sahai, Excitation of a nonlinear plasma ion wake by intense energy sources with applications to the crunch-in regime, *Phys. Rev. Accel. Beams* **20**, 081004 (2017), doi:10.1103/PhysRevAccelBeams.20.081004.

21. T. Tajima, Laser accelerators for ultra-high energies, in *Proc. 12th Int. Conf. on High-Energy Accelerators (HEACC 1983)*, Fermilab, Batavia, August 11–16, 1983, pp. 470–472, C830811.

22. D. C. Barnes, T. Kurki-Suonio and T. Tajima, Laser self-trapping for the plasma fiber accelerator, *IEEE Trans. Plasma Sci.* **15**(2), 154–160 (1987).

23. A. J. Gonsalves *et al.*, Petawatt laser guiding and electron beam acceleration to 8 GeV in a laser-heated capillary discharge waveguide, *Phys. Rev. Lett.* **122**, 084801 (2019).

24. A. W. Trivelpiece and R. W. Gould, Space charge waves in cylindrical plasma columns, *Journal of Applied Physics* **30**, 1784 (1959), doi:10.1063/1.1735056.

25. T. C. Chiou and T. Katsouleas, High beam quality and efficiency in plasma-based accelerators, *Phys. Rev. Lett.* **81**, 3411 (1998) doi:10.1103/PhysRevLett.81.3411.

26. S. Gessner *et al.*, Demonstration of a positron beam-driven hollow channel plasma wakefield accelerator, *Nature Communications* **7**, 11785 (2016), doi:10.1038/ncomms11785.

27. C. A. Lindstrom *et al.*, Measurement of transverse wakefields induced by a misaligned positron bunch in a hollow channel plasma accelerator, *Phys. Rev. Lett.* **120**, 124802 (2018), doi:10.1103/PhysRevLett.120.124802.

28. A. Pukhov and J. P. Farmer, Stable particle acceleration in coaxial plasma channels, *Phys. Rev. Lett.* **121**, 264801 (2018).

29. A. A. Sahai, On certain non-linear and relativistic effects in plasma-based particle acceleration, Chapter 8, Ph.D. thesis, Duke University, 2015, 3719664.

30. W. D. Kimura *et al.*, Hollow plasma channel for positron plasma wakefield acceleration, *Phys. Rev. ST Accel. Beams* **14**, 041301 (2011).

31. L. Yi *et al.*, Positron acceleration in a hollow plasma channel up to TeV regime, *Sci. Rep.* **4**, 4171 (2015).

32. F. Bloch, Bemerkung zur Elektronentheorie des Ferromagnetismus und der elektrischen Leitfahigkeit, *Zeitschrift fur Physik* **57**(7–8), 545–555 (1929);
    F. Bloch, Bremsvermogen von Atomen mit mehreren Elektronen, *Zeitschrift fur Physik* **81**(5–6), 363–376 (1933).

33. S.-i. Tomonaga, Remarks on Bloch's method of sound waves applied to many-fermion problems, *Prog. Theor. Phys.* **5**, 544–569 (1950).

34. D. Bohm and D. Pines, A collective description of electron interactions: III. Coulomb interactions in a degenerate electron gas, *Phys. Rev.* **92**, 609 (1953).

35. D. Pines, A collective description of electron interactions: iv. electron interaction in metals, *Phys. Rev.* **92**, 626 (1953).

36. R. H. Ritchie, Plasma losses by fast electrons in thin films, *Phys. Rev.* **106**, 874 (1957);
    R. H. Ritchie and H. B. Eldridge, Optical emission from irradiated foils. I, *Phys. Rev.* **106**, 874 (1957).

37. D. Pines, Collective energy losses in solids, *Rev. Mod. Phys.* **28**, 184 (1956).

38. E. A. Stern and R. A. Ferell, Surface plasma oscillations of a degenerate electron gas, *Phys. Rev.* **120**, 130 (1960).

39. T. Tajima and S. Ushioda, Surface polaritons in LO-phonon-plasmon coupled systems in semiconductors, *Phys. Rev. B* **18**, 1892 (1978).

40. S. Hakimi *et al.*, Wakefield in solid state plasma with the ionic lattice force, *Phys. Plasmas* **25**, 023112 (2018).

41. S. Hakimi *et al.*, X-ray laser wakefield acceleration in a nanotube, in *Proc. of the Workshop on Acceleration in Crystals and Nanostructures (XTALS 2019), 24–25 June 2019, Fermilab.*

42. T. C. Chiou and T. C. Katsouleas, High beam quality and efficiency in plasma-based accelerators, *Phys. Rev. Lett.* **81**, 3411 (1998); S. Corde *et al.*, Multi-gigaelectronvolt acceleration of positrons in a self-loaded plasma wakefield, *Nature* **524**, 442 (2015).

43. A. A. Sahai, Crunch-in regime - Non-linearly driven hollow-channel plasma, arXiv:1610.03289.

44. T. D. Arber *et al.*, Contemporary particle-in-cell approach to laser-plasma modelling, *Plasma Phys. Control. Fusion* **57**, 113001 (2015).

45. S. Ijima, Helical microtubules of graphitic carbon, *Nature* **354**, 56–58 (1991); P. M. Ajayan and S. Iijima, Capillarity-induced filling of carbon nanotubes, *Nature* **361**, 333–334 (1993).

46. Molex nano-capillary Polymicro tubing, part number 1068150033, 1068150035, 1068150037.

47. H.-C. Wu *et al.*, Collective deceleration: Toward a compact beam dump, *Phys. Rev. ST Accel. Beams* **13**, 101303 (2010).

48. A. A. Sahai *et al.*, Quasimonoenergetic laser plasma positron accelerator using particle-shower plasma-wave interactions, *Phys. Rev. Accel. Beams* **21**, 081301 (2018).

49. A. A. Sahai, T. Tajima and V. Shiltsev, Schemes of laser muon acceleration: Ultrashort, micron-scale beams, in *Proc. of the Workshop on Acceleration in Crystals and Nanostructures (XTALS 2019), 24–25 June 2019, Fermilab.*

50. J. Towns *et. al.*, XSEDE: Accelerating scientific discovery, *Comput. Sci. Eng.* **16**, 62–74 (2014).

51. J. Anderson *et al.*, Deploying RMACC Summit: An HPC resource for the Rocky mountain region, *Proceedings of PEARC17*, New Orleans, LA, USA, July 9–13, (2017).

Electron Acceleration at ELI-Beamlines: Towards High-Energy and High-Repetition Rate Accelerators

C. M. Lazzarini,[+] L. V. Goncalves, G. M. Grittani, S. Lorenz, M. Nevrkla,

P. Valenta, T. Levato, S. V. Bulanov and G. Korn

Institute of Physics of the ASCR, ELI Beamlines,
Na Slovance 2, 18221 Prague, Czech Republic
[+] CarloMaria.Lazzarini@eli-beams.eu

The high energy electron experimental platform* at ELI-Beamlines will give to the users high energy tunable electron beams with low energy spread and divergence, by employing laser-wakefield-acceleration scheme (LWFA) driven by PW-class laser system working at 10 Hz. The platform will offer great flexibility over electron beam parameter space and is foreseen to exploit different targets, acceleration and laser-guiding advanced schemes. In this paper we summarize about more compact accelerators that can be envisioned by the use of really short (near single-cycle) fem-mJ-level laser pulses interacting with nanoparticle and solid targets, as well as with specific near-critical density targets.

Keywords: LWFA; laser-driven electron acceleration; ultrashort lasers; high-density targets; nano-structures; ELI-Beamlines.

1. Introduction

Laser-plasma acceleration[1] is a technique capable of producing very high-energy particles bunch (GeV-level) with short time duration (down to 1 fs) and in relatively short distance compared to conventional accelerator systems,[2–4] thanks to high accelerating gradients well beyond 100 MV/mm.[5] This opened the way to generate ultra-relativistic particle beams for different applications in smaller size research infrastructure such as ELI-Beamlines.[6,7] Moreover, due to its versatility, Laser Wakefield Acceleration (LWFA) is an attractive technique for many applications because of the possibility to tune within the same configuration the electron beam parameters (energy, charge, and emittance) throughout different orders of magnitude. Nowadays, laser-accelerated electron beams are produced with table-top laser systems from TW to PW peak power thus accessing a broad energy range up to the multi-GeV level.[8–10] The use of an ultrashort laser pulse driver, made available by the pioneering work on Chirped-Pulse Amplification (CPA) by Strickland and Mourou,[11] enables at the same time the possibility to pump and probe a sample with secondary laser or particle beams, thus offering ultrafast temporal resolution in the 10-fs range. These features are extremely interesting for several potential multidisciplinary applications.[12] However, the not yet optimized electron beam quality in terms of pointing stability, average energy and energy spread, com-

---

*Originally developed as H.E.L.L., within the *Particle acceleration by Laser* program (RP3).

pared to conventional accelerators, impose severe limitations to the usability of such beams. At the same time, such challenging requirements have increased the number of test experiments aiming to improve the beam reliability for applications. It has been shown that different injection mechanisms can help to stabilize the average electron energy and to reduce the energy spread to less than 1%,[13] the electron beam pointing however is showing a stability 2 to 3 orders of magnitude worse than top-class laser systems (few μrad). These recent achievements for electrons parameters are paving the way for further improvement to extend the acceleration with, for example, multiple stage accelerators[14,15]; anyway, at the same time, they already appear at a useful level for specific applications,[15] as emerged in the detailed user workshop. In trend with this workshop topic, to accelerate beams in solid-state plasmas that promise very high accelerating gradient toward the final goal to reach TeV/m to build future particle colliders much more compact. A step in this direction can be, for example, to use compact size laser system to drive relativistic particles, or continuous focusing of single-cycle laser pulses (eventually X-ray lasers to further reduce the wavelength and so the accelerator plasma-cavity volume) inside solid nanostructured targets as crystals, carbon nanotubes (CNT),[16] or in between metasurfaces to exploit the metasurface laser accelerator (MLA) scheme in the NIR spectrum.[17] Usually compact lasers can work at higher repetition rate compared to big PW-class laser system but, even if more stable and reliable, bring very low energy for every laser shot, resulting in an intensity not enough to generate good quality beam, for example through the so-called blowout regime.[18,19] For this reason solutions to this problem are:

(i) drastically reduced the pulse duration to single-cycle by pulse post-compression (for example in gas-filled fibers as shown in Refs. 20 and 21 or by the new thin film compression approach proposed by G. Mourou et al.[22]),

(ii) tight focusing by small f# off-axis parabola (OAP) together with a deformable mirror (DM) and automatic wavefront curvature correction software,

(iii) by exploiting higher density targets like solid-plasma or near-critical plasma.

By using combinations of said points it has been shown the possibility to produce MeV-pC electron beam with mJ-class laser systems, which can operate at high repetition rate (kHz).[23–26] Two different regimes have been used to produce such beams, one based on an improved laser driver, compressed near single-cycle (3.4 fs)[24]; the second consists in increasing the plasma density close to critical value ($n_c = 1.7 \times 10^{21}$ cm^{-3} for 800 nm driving laser) to enable relativistic self-focusing for longer (> 30 fs) laser pulses.[26] In this work it is considered and shown the possibilities to implement some of these schemes in the platform at ELI-Beamlines; as well the first demonstrations, both experimentally and theoretically, that could pave the way toward the goal of realizing a compact high-repetition rate (few Hz) ultra-relativistic (MeV to GeV) particle accelerator with pointing, divergence and energy spread good enough to use electrons as secondary source for further experiments. The paper is structured in the following way. In the second section is given an overview of the ELI-Beamlines research center and the available lasers for particle acceleration. Sections 3 and 4 give a general overview of the LWFA idea and typical setup of a laser-plasma acelerator, together a brief discussion on laser pulse optimization. In Sec. 5 we summarized few aspects of the high energy electron platform and related diagnostics. Section 6 is discussing recent trends towards the realization of kHz electron acceleration. Finally, in the last two sections are presented recent theoretical results obtained by considering what happens when a single-cycle laser pulse travels (a) inside a single nanoparticle (NP), by resolving the E-field dynamics in the fs-timescale, and (b) inside an over-dense plasma, to study wake-field modulations either analytically and numerically.

2. Electron Acceleration Possibilities at ELI-Beamlines

The ELI-Beamlines project[†] main mission is to deliver laser-produced secondary sources to the scientic community, both for applied and fundamental research in different elds of plasma and ultrashort laser-matter interaction physics. For this reason, different user-oriented beamlines are being implemented with the goal of delivering high repetition-rate highly synchronized laser-driven sources of particle and radiation with unique features, especially when compared to conventional facilities. Practical applications are medical and bio-molecular research (i.e. radiotherapy, pulse radiolysis), new material research, and the development of new technologies including high rep-rate DPSSL pumped PW-systems. The building has been constructed in 2015 and, since then, four lasers and many different beamlines have been or are under installation and commissioning. What is unique of ELI-Beamlines, other than having unique laser as the 10 Hz PW, it is the possibility to combine in different experimental rooms multiple lasers with dedicated beam transport (as shown in Fig. 1), to be used synchronized in complex experiments. This is possible thanks to the fact that the lasers have a common timing system for referencing and there is a centralized beam distribution control system.

Different unique lasers (parameters in Table 1) are available at ELI-Beamlines, and for experimental platforms like Plasma Physics (E3), Ion Acceleration (E4) and Electron Acceleration (E5) is foreseen the possibility use in combination different laser beams with dedicated beam transport and diagnostics. The first laser to be ready is the L1 Allegra laser system,[27] designed and built by the ELI-Beamlines laser team. This is based on picosecond Optical Parametric Chirped-Pulse Amplification (OPCPA) pumped Yb:YAG thin-disk lasers and provides < 20 fs compressed pulse at a wavelength $830 < \lambda < 860$ nm, with an energy up 100 mJ (upgradable to 200 mJ in the future). The unique feature of this system is the high-repetition rate of 1 kHz. This is used for all the experiments that do not need high energy or intensity but rather good stability and repetition rate to increase the data statistics

---

[†]https://www.eli-beams.eu/

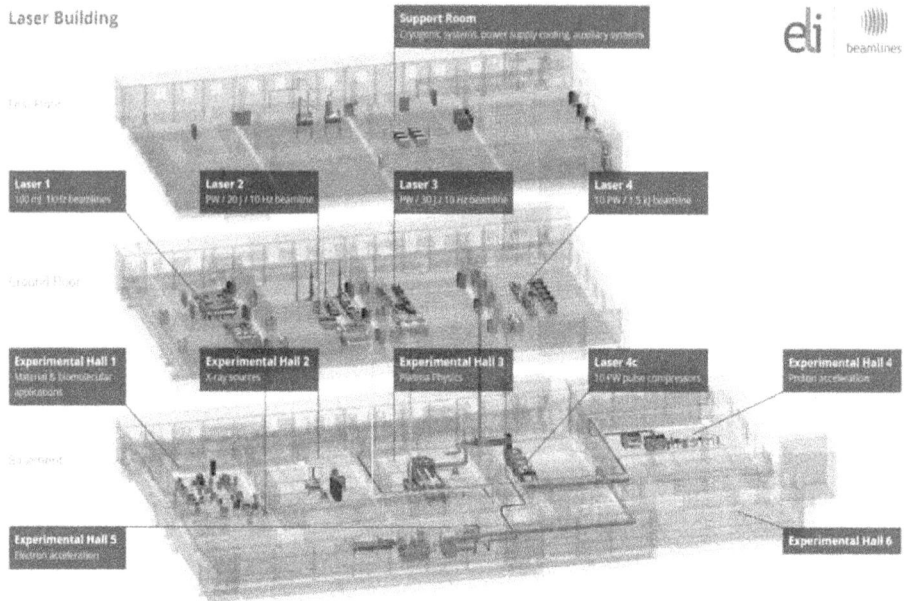

Fig. 1. Experimental building scheme at ELI-Beamlines: 4 main lasers are located in separate rooms at the ground floor, and are transported underground to different experimental halls. In violet is L1 laser 100 mJ, 1 kHz; in green the PW-class laser working at 10 Hz and in red the 55 cm diameter 10 PW, 150 fs laser. The experimental hall (E5) is dedicated for high-energy electron acceleration can allow future-proof scheme of acceleration with long focusing optics, extendable to 100 m, as originally planned into the former *Particle Acceleration by Laser* program (RP3).

and the signal-to-noise ratio. This is mainly used to generate secondary sources as High-Harmonic Generation (HHG) in the range 10–120 eV and Plasma X-ray source (PSX) station for the generation of incoherent ultrashort, high brightness, hard X-ray. Applications in the Atomic, Molecular and Optical fields are coherent diffractive imaging, sub-ps VUV ellipsometer and time-resolved scattering, diffraction and spectroscopy imaging. The main working laser is the L3 HAPLS (High-Repetition-Rate Advanced Petawatt Laser System) laser system based on Ti:Sa crystal with record breaking peak power pumping diode arrays, developed by LLNL, US.[‡] This provides PW pulses with energy of at least 30 J and duration < 30 fs at 820 nm central wavelength, at a repetition of 10 Hz. It also provides very high contrast and a high degree of real-time control and timing synchronization. Last more powerful laser is the L4 Aton laser system developed by a consortium of ELI-Beamlines, National Energetics and EKSPLA. It will provide up to 2 kJ pulses with duration < 150 fs, therefore a peak power of 10 PW and the highest possible intensity on focus at ELI-Beamlines, with a repetition rate of 1 shot/minute due to the Nd:glass

---

[‡]More info at: https://www.llnl.gov/news/lawrence-livermore-developed-petawatt-laser-system-installed-eli-beamlines.

slabs amplification technology. This laser will be intrinsically synchronized with a 0.5 to 5 ns (chirped) pulse laser with programmable temporal shape, just by by-passing the final compressor in the same distribution chain.

Table 1. Lasers available at ELI-Beamlines.[6]

Laser name	Pulse energy [J]	Pulse duration [fs]	Wavelength [nm]	Repetition Rate [Hz]	Peak Power [PW]
L1 - Allegra	0,1 (up. to 0,2)	30	830-860	1000	0.005 (0.01)
L2 - Amos	10–20	15	810	10	1
L3 - HAPLS	30	30	820	10	1
L4 - Atos	1500	150	810	< 0.02	10

3. LWFA Experimental Diagnostics

The idea of laser wakefield acceleration (LWFA) come out originally as a theoretical possibility in 1979,[1] and the first proofs showing high quality relativistic beams accelerated by laser in short distance were given only more than 16 years later by different groups.[2-4,28] It consists in a short duration laser pulse from CPA system that drives plasma waves in a medium by the ponderomotive driving force, proportional to the variation of the laser intensity:

$$F_{NL} = -\nabla I_{\text{laser}}. \tag{1}$$

These waves, referred to as the *wakefield*, are basically plasma oscillations that can produce very high E-field by electrons-ions displacement. The longitudinal electromagnetic plasma wave (EPW) are responsible for the acceleration of electrons, where the final gain is proportional to the E-field and the displacement (e.g. the acceleration length). The biggest advantage of this acceleration scheme, compared to the conventional linear accelerators based on radio-frequency (RF) cavities, is the extreme reduction in dimensions. Indeed, the limiting factor for RF accelerators is the electric breakdown of the material that limits the accelerating gradients to around 100 MeV/m, this does not happen in an ionized plasma. The E-field responsible for the particle energy gain is dependent both on the laser intensity (that means are needed really short pulse duration lasers) and on the plasma electron density n_e, and can be higher than the so-called cold non-relativistc wave-breaking field[29]:

$$E_0 \left[\frac{V}{m}\right] = \frac{c m_e \omega_{pl}}{e} \approx 96\sqrt{n_e\,[\text{cm}^{-3}]}, \tag{2}$$

$$\omega_{pl} = \sqrt{\frac{4\pi n_e e^2}{m_e}}, \tag{3}$$

where m_e and e are the electron rest mass and charge, c the speed of light, and ω_{pl} the plasma frequency. This implies that working with few tens of fs laser at a plasma density $n_e \approx 10^{18}\,\mathrm{cm}^3$ is possible to get accelerating field $> 100\,\mathrm{GV/m}$. When the plasma perturbation length is equal to the plasma wavelength

$$\lambda_{pl} = \frac{2\pi}{\omega_{pl}} = c\tau_{\mathrm{laser}} \qquad (4)$$

with τ_{laser} the pulse duration, resonant high amplitude wake-fields are produced and the electrons, if injected in phase inside these waves, can be accelerated to very high energy. By using PW-class laser it has been recently shown a record electron energy gain up to 8 GeV[10] by extending the acceleration length over 20 cm inside a channel formed in a plasma discharge capillary. Electrons can be injected in the accelerating part of the wake-fields by different methods,[5] as for example by self-injection, ionization-injection or even external injection when a pre-formed beam is sent with proper timing after the main laser driving the wake. To get good quality electron beam is critical to tune the laser-plasma interaction to assure good excitation of waves and timed injection of electrons, dependent on the plasma density (e.g. type of gas, backing pressure, height of laser focal spot compared to gas target). To do so usually the gas-target, that can be a sub- or super-sonic nozzle, a gas-cell or a small-diameter capillary, is moved in 3D to find the best position compared to the focal spot of the laser, given by the off-axis parabola (OAP) optimization at a previous stage. During this process the laser-plasma interaction is monitored by different diagnostics techniques, the most basic setup consist in a side-view probing beam to get lateral shadowgraphy (or interferometry) images from which is possible to know height and longitudinal position of the focus, a high-magnification top-view color-CCD to view the Thomson scattered light from plasma (visible to IR by laser) to infer the laser propagation into plasma, and eventually a laser spatial profile and spectrum monitoring before and after the plasma propagation. The electron beam spectrum is typically recorded by a deflecting B-field and an imaging Lanex screen, imaged off-axis by a CCD. From this signal is also possible to estimate the beam divergence, charge and pointing at every shot.

4. Laser Pulse Optimization

In order to boost the acceleration gradient inside the plasma cavity it is important to have a high intensity of the laser concentrated inside a very small focal spot (as can be understood by Eq. (1)). Indeed to drive highly relativistic electron motion the normalized (Gaussian) laser vector potential amplitude, defined as:

$$a_0 = \left(\frac{2e^2\lambda^2 I}{\pi m_e^2 c^5} \right)^{\frac{1}{2}} \simeq 8.6 \times 10^{-10} \lambda\,[\mu\mathrm{m}]\,\sqrt{I\,[\mathrm{W/cm}^2]} \qquad (5)$$

must be $a_0 \geq 1$, that means an intensity on spot $I > 10^{18}\,\mathrm{W/cm}^2$ for a central laser wavelength at 1 μm. The focal spot can be dependent experiment by experiment on the different acceleration techniques, as for example if the laser pulse must be coupled into a guiding channel with acceptance aperture (radius) similar to the laser spot. However, other than generating a small laser waist in focus w_0 (eventually close to the optical diffraction limit $w_{limit} = 1.22\lambda f_\#$), it is crucial both (i) to maximize the energy encircle in the spot size $2w_0$ and (ii) to reduce to the minimum the pulse duration in time. To accomplish the first goal (i) usually a deformable mirror (DM) is used as one of the last optics in the path before the final focusing element. This is composed by a very thin membrane behind which different actuators are moved to deform the surface in a way to correct for optical aberrations and guarantee flat wavefront curvature, measured by a wavefront sensor device (WFS) optically conjugted with the DM surface. An additional step would be to re-optimize the wavefront at the real focus of the OAP, to further correct aberrations introduced by non-perfect alingment or coating imperfections of the OAP itself. For the second goal (ii) it is important to fine-tuning the laser final compressor (in particular the relative distance of the two gratings) measuring as reference the pulse duration on target. This can be done either in-air or in-vacuum by considering the additional B-integral due to the vacuum-air optical window. To measure really compressed laser pulse in the fs-domain it is needed a self-reference technique. Commercial devices are available for this scope as the Wizzler, based on single-shot Spectral Interferometry[30] to reconstruct the spectral phase and intensity, and the D-Scan,[31] based on ultrafast dispersion-scan setup.

5. High-Energy Platform Scheme

Since the capability to produce secondary beams using ultra-high-power laser systems is still in a fast development phase, the high energy electron platform available at ELI-Beamlines (developed within the Particle Acceleration by Laser program RP3 and reviewed in detail in T. Levato *et al.*[32]) is geometrically organized to be exible and modular to offer the possibility to arrange advanced optical setups for future development of the source and for the preparation of complex fundamental physics experiments that require a high-power super-stable counter-propagating laser.[33,34] All the optics have been designed to work with the (250 mm clear aperture) PW-laser and having optimized coating for said laser. The first laser-driver vacuum chamber consists of all the optics for focusing down the large diameter to a spot size of $\approx 15~\mu m$ (33 μm) in the case with an OAP of focal length $f = 4.4$ m (10 m), plus all the online in-air diagnostics (pulse energy, contrast, pointing, near-field distribution) and the in-vacuum ones (pulse duration and wavefront curvature measurements). The two OAP allow to work in two different regimes of LWFA, respectively the self-injection regime[18,35,36] that requires a high intensity, and the guided regime[2] by increased focal length and, therefore, increased accelerating region inside the channel. The DM is needed to compensate diffraction and phase-front aberrations effects (that limit the minimum focal spot size and so the maximum intensity reachable) due to multiple reflections on optics and long

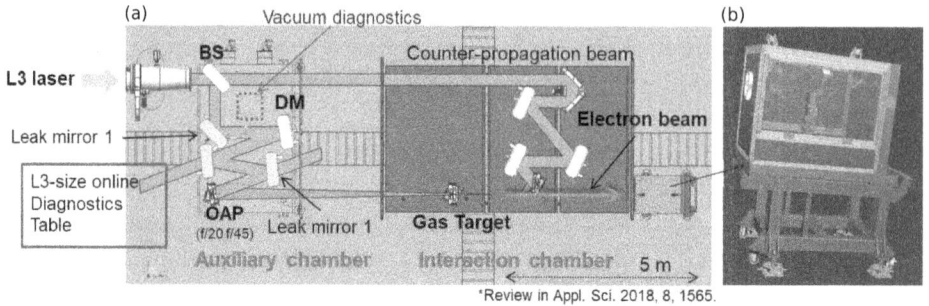

Fig. 2. (Courtesy of T. Levato) The High-energy Electron by Laser Light (H.E.L.L.) experimental platform: optical scheme (a) of the two chambers composing the beamline and (b) the User station equipped with an high-precision positioning system. The beamline is designed to host the PW-class laser (indicated here with orange arrow), offering a high-degree of experimental flexibility thanks to the possibility to move the interaction (big) chamber with respect to the (first) laser focusing one. To optimize the interaction the platform will provide in-air and in-vacuum shot-to-shot diagnostics for the laser and its propagation in the plasma. Moreover, is considered the possibility to have a PW-class counter-propagating laser beam to generate high-energy γ-ray by laser-particle collider.[33] Adapted reprint from Ref. 32.

propagation. For lower energy applications (100-MeV-range), users can benefit of a dedicated user-station[§] (shown in Figs. 2(a) and 2(b)) designed to work at 10 Hz and to host up to 5 kg target (in air), offering both a high-precision positioning system and the possibility to monitor online the electron beam characteristics before and after the interaction with the target.

6. Towards High-Repetition Rate Accelerators

In the view to use nanostructures either for direct particles acceleration or for laser pulse guiding, it is critical to have a LWFA system with very good pointing stability (ideally tens of μrad) and reliability over time, working at high-repetition rate (tens to hundreds of kHz). The first requirement will assure a good focusability of the laser pulse to be guided or matched to nanostructures as carbon nano-tubes (CNT) or plasma channels that can have a matching opening aperture from tens to sub-μm. The second point will assure the enhancement of the particle beam current (needed in many applications), a better signal-to-noise ratio, and the possibility to introduce a live feedback control and beam optimization.[37] Most Laser Plasma Accelerators (LPA) rely on 100 TW-class or PW-class laser to generate GeV-range high-charge electron bunches, however these systems usually work at low repetition rate (≤ 1 Hz) or at best at 10 Hz (L3-laser project). Increasing the repetition rates bring a series of advantages as: more stable acceleration process because kHz laser reach thermal steady operation state, it is possible to average over many shots increas-

---

[§]Developed in the *Particle Acceleration by Laser* group (RP3) together with *Elettra-Sincrotrone Trieste*.

ing both the statistics and the signal-to-noise ratio, it is possible to introduce very fast active or passive feedback control and optimization, and finally can increase by few orders of magnitude the current brought within the fs electron bunch.[26,38] Nowadays few-mJ range compact lasers working higher than kHz are commercially available. One of the technical challenges to work with such systems to drive LWFA is the fact that, being available only low energy per pulse, it is needed to work in tight focusing regimes shooting very close to super-sonic (high-Z) gas targets (that means very high-plasma density). In this way is possible to have shorter plasma wavelength to be in resonance with the laser pulse in the so-called bubble regime, that leads to the generation of high-quality relativistic electrons with narrow energy spread and small divergence.[18,19] The direct consequence of this is the fact that by scaling down the spot waist, also the Rayleigh length z_R, the accelerating cavity radius R and the dephasing length $L_{deph} \propto \frac{\omega_0^2}{\omega_p^2} R$ (limiting the acceleration process) are reduced. This is clearly visible by scaling laws with requirements for LWFA in bubble regime highlighted in the Table 2. This makes the optimal condition more difficult to find and it is usually required a very careful fine-scanning of the focal spot compared to the gas target.

Table 2. Scaling laws for LWFA in the blowout regime,[37] considering a driving pulse of 800 nm wavelength.

Laser class	a_0	E_L	τ_L	w_0	z_R	n_e	L_{deph}	ΔE
0.5 PW	4.8	30 J	60 fs	26 μm	2.6 mm	$6.6 \times 10^{17} cm^{-3}$	4.5 cm	4.2 GeV
30 TW	3.5	1 J	25 fs	10 μm	400 μm	$4.2 \times 10^{18} cm^{-3}$	2.8 mm	500 MeV
1 TW	2	3 mJ	5 fs	2.1 μm	18 μm	$10^{20} cm^{-3}$	25 μm	10 MeV

Another technical experimental issue working at high-repetition rate is the synchronization of the gas valve opening time with the laser and the gas release into the chamber, a too slow evacuation could result in an increase of the vacuum level over the working threshold of 10^{-3} mbar, potential harmful for the optics especially if the target chamber is in close proximity with the laser compressor gratings. To mitigate these problems it is possible to perform hydrodynamics simulations, as shown in Ref. 39, in order to get the gas density profile and to find the best interaction position for of the focal spot compared to the target.

7. Theoretical Considerations

In the view to use nanostructure with solid-density plasmas in which to drive LWFA by reduced wavelength (eventually from X-ray lasers), in our group at ELI-Beamlines we are focusing the attention on two points we consider relevant and not fully explored from a theoretical point of view. The first consists of analysis of the interaction and response on the fs-timescale of a single nanoparticle (NP) to a ultra-fast broadband driving pulse. The second point to mention here is the study of the propagation of a single-cycle laser pulse in a near-critical density plasma, useful to analyze eventual laser self-modulations when working at high-repetition rate at densities $> 10^{20}$ cm^{-3}.

7.1. NP response to broadband pulse on the fs-timescale

Optical properties of plasmonic nanoparticles, such as their large optical cross sections and the enhancement of the optical near eld in subwavelength regions, are well known in the literature[40] and the rst theories of light interacting with small objects (called later on plasmonics) based on classical electrodynamics go back more than a century.[41-43] However in the recent years the development of advanced nano-fabrication techniques, high-sensitivity single-particle optical characterization techniques,fast numerical modeling tools, and the availability of compact size of fs-laser pulse have led to an increasing interest of the scientic community in the so-called ultrafast nanoplasmonics world due to huge number of applications.[44,45] In the latter, the key point is the control of the strong-eld enhancement localized in space on the nanometer scale and in time on the fs and sub-fs scale, either in the linear or nonlinear regime. For many plasmon-based applications in the visible, because of the very narrow resonance, silver and gold are the usual choices however, plasmonic excitation can be also obtained in the far-infrared zone using doped semiconductors and in the ultraviolet zone using aluminium or silicon. Most of these studies on plasmonics and magnetic light are limited to the monochromatic, continuous-wave regime. The work by C. M. Lazzarini et al.,[46] developed into the *Particle acceleration by Laser* research program (RP3), presents a further step into the research field unifying the areas of nanoplasmonics and ultrafast high-power lasers. The key point here, still not clear even in the linear case, was the time-response of the near-field excited inside a single NP placed in the focus of a broadband Fourier-limited Gaussian laser pulse in the long wavelength limit, in the cases its duration is longer, comparable or smaller than the NP resonance lifetime (as shown in Fig. 3 for a sub-wavelength Ag NP). The enhancement and decay have been resolved on a fs-timescale both analytically and numerically for a 40 nm-diameter Ag NP (e.g. plasmonic resonance in metal) and for a bigger 460 nm-diameter Si one (magnetic resonance in high-index dielectric). Similar delayed near-field enhancement and decays have been observed for the metal and dielectric NPs, by considering respectively a modified Drude-Sommerfeld model[47] in the dipole-limit and the exact Mie theory[48] needed for bigger particles.

We started by solving the Helmholtz equation in the paraxial limit (neglecting spatio-temproal couplings):

$$\nabla^2 \mathbf{E} + k^2 \mathbf{E} = 0 \,, \tag{6}$$

$$\mathbf{E}(\mathbf{r}, \tau) = \mathbf{E}_{\text{space}}(\mathbf{r}) E_{\text{time}}(\tau) \tag{7}$$

with $k = |\mathbf{k}| = \omega/c$. Solutions of above equations are a Gaussian (mode TEM$_{00}$)

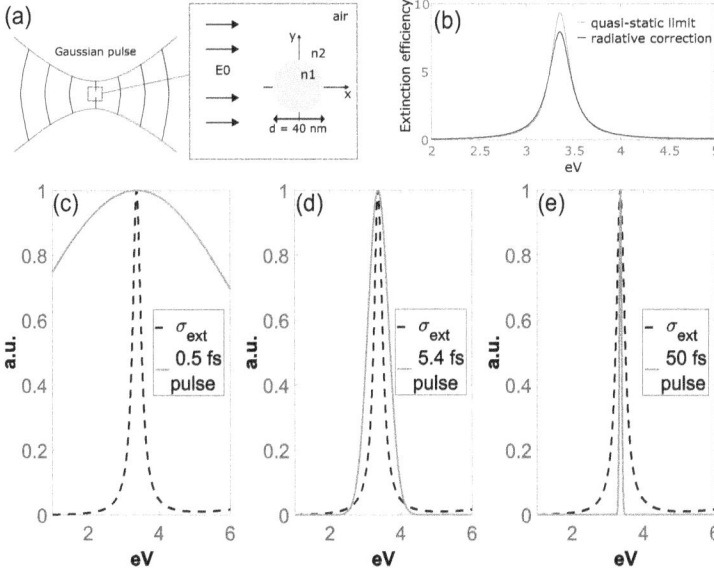

Fig. 3. (a) Scheme of the idea from C. M. Lazzarini *et al.*[46]: a 40 nm diameter Ag nanoparticle placed in the beam waist of an incident Fourier-limited Gaussian pulse in the long-wavelength limit ($\lambda \gg d$) under the paraxial approximation. In the beam waist the electric eld is polarized along the x-axis. (b) Plane wave extinction efciency for the NP computed considering radiative correction effect. (c), (d), (e) Normalized spectra of the laser pulse (E_{inc} in red) centered at $\lambda_{\text{res}} = 369$ nm having a bandwidth bigger, similar, or smaller compared to the plasmon resonance one, considering a pulse duration respectively of 0.5 fs, 5.4 fs, and 50 fs.

propagating along the x-axis and a Gaussian function in time:

$$\mathbf{E}_{\text{space}}(\mathbf{r}) = E_0 e^{ikz} \hat{x} \tag{8}$$

$$E_{\text{time}}(\tau) = e^{-\alpha_p \tau^2} e^{-i\omega_0 \tau} \tag{9}$$

where ω_0 is the central pulse frequency and α_p is related to the pulse duration by $\alpha_p = 4\ln(2)/\tau_p^2$. The pulse duration range here considered (0.5 to 250 fs) covers from the state-of-the-art laser pulses for pump-probe experiments to study collective charge excitation and dephasing (110 fs), the non-thermal electron distribution effects, up to the the e-e (or e-h in a semiconductor) scattering regime (100 fs range).

To calculate the broadband response, the incoming pulse temporal part has been Fourier-expanded and for each frequency the incident eld on the small sphere becomes

$$\mathbf{E}_{\text{space}}(\mathbf{r}) = E_0 e^{ikz} \hat{x} \tag{10}$$

and the solution can be calculated solving the Laplace equation

$$\nabla^2 \Phi = 0, \tag{11}$$

Near-field enhancement

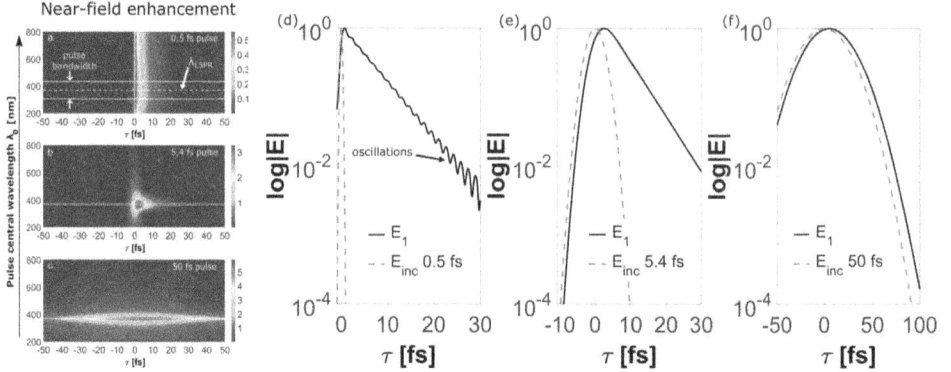

Fig. 4.   Left: Color maps showing the normalized E-eld enhancement inside the 40 nm Ag sphere on an ultrafast time scale. The incoming pulse is a Gaussian Fourier-limited pulse centered at λ_0 with a pulse duration of (a) 0.5 fs, (b) 5.4 fs, and (c) 50 fs. The dashed lines represent the plasmon resonance wavelength $\lambda_{\mathrm{LSPR}} = 369$ nm, while the solid white lines show the pulse bandwidths. Right: Semi-log plots of the respective induced E-field for a laser pulse centered at λ_{LSPR} (solid black lines) compared to the incoming one (dashed red) having a pulse duration of (d) 0.5 fs, (e) 5,4 fs, and (f) 50 fs. Data and reprint from Ref. 46.

$$\mathbf{E} = -\nabla\Phi \tag{12}$$

giving a near-field for single frequency:

$$\mathbf{E}_1(\omega) = E_0 \sqrt{\frac{\pi}{\alpha}} e^{-\frac{(\omega-\omega_0)^2}{4\alpha_p}} \frac{3\epsilon_2}{\epsilon_\infty + 2\epsilon_2 - \frac{\omega_p^2}{\omega(\omega+i\gamma)}} \hat{x} \,. \tag{13}$$

In this we consider air as background ($\epsilon_2 = 1$), a dielectric function at innite frequency $\epsilon_\infty = 5$, a plasma frequency $\hbar\omega_p = 8.9$ eV and a relaxation (electron scattering) rate $\gamma = 17^{-1}$ fs. By inverse Fourier transforming this equation is possible to get the NP response in time, results are shown in Fig. 4 for three different cases. A rich physics has been observed for a pulse duration shorter than the NP resonance lifetime (of 5.4 fs for Ag). That is a delayed retarded Lorentz mode decay with parameter $\Gamma/2 = 0.186$ fs^{-1} and we observed that by increasing the pulse duration the response delay saturate at the exact value of the plasma resonance $\tau_{\mathrm{delay}} \simeq \tau_{\mathrm{res}} = 5.4$ fs. A similar scenario happen for bigger Si particle, where the resonances mode are magnetic resonances centered in the NIR spectrum ($\lambda > 1100$ nm). Extension works on these findings will look at the non-linear NP and multi-particle interference response, the far-field dynamics resolved in time, and could lead to new mechanisms for time-shaping ultra-fast laser pulses or be used for laser-plasma diagnostics and direct particle acceleration in nano-structured targets.

7.2. Single-cycle laser propagation in near-critical plasma

Since as explained above LWFA at high repetition rate is possible with mJ-class lasers but requires very high-density plasma targets and really short pulse durations, the dispersion and the carrier envelope phase (CEP) effects of few-cycles pulses must be taken into consideration when considerating the generated wake waves and their effect back on the driving pulse. Indeed, usually in this scenario the laser pulse driving the wakefield is longer in space than the plasma wavelength, meaning that self-modulation effects can occur to eventually split the pulse into different waves. In this last chapter is presented an analysis[49] showing Particle-In-Cell (PIC) simulations to study the propagation of the pulse in near-critical plasma (where $n_{cr} = 1.8 \times 10^{21}$ cm^{-3} for a 800 nm laser). Starting from the dispersion relation for a small amplitude E.M. wave in a collisionless plasma:

$$\omega^2 = k^2 c^2 + \omega_p^2 \tag{14}$$

with $\omega_p = \sqrt{4\pi n_e e^2/m_e}$ the Langmuir frequency, c the speed of light in vacuum, e and m_e the electron charge and mass and n_e the electron density, the phase and group velocity of laser pulse are:

$$v_{ph} = \frac{\omega}{k} = \frac{c\omega}{\sqrt{\omega^2 - \omega_p^2}}, \tag{15}$$

$$v_{gr} = \frac{\partial \omega}{\partial k} = \frac{c\sqrt{\omega^2 - \omega_p^2}}{\omega}. \tag{16}$$

The simulated laser pulse, characterized by angular frequency $\omega_0 = 2\pi c/\lambda_0$, where λ_0 is the vacuum center laser wavelength, is Gaussian in both spatial and temporal proles. The pulse contains one optical cycle only, thus its duration is $\tau_{\text{FWHM}} = 2\pi/\omega_0$ (2.66 fs for 800 nm laser). The laser is focus to $w_0 = 4\lambda_0$, with a corresponding normalized amplitude at focus of $a_0 = 2$, at the beginning of a cold, collisionless, uniform hydrogen plasma with electron density $n_e = 0.1 * n_c$ through a smooth density ramp in the front vacuum interface to mitigate the wave-breaking that could occur from a too sharp plasma edge. The resolution of the 2D Cartesian grid is 100 cells per λ in both directions. The electromagnetic elds are calculated using the standard second-order Yee solver[50] by applying absorbing boundary conditions for both elds and particles. Results are presented in Fig. 5, where is clear that the laser entering the density plateau undergoes self-focusing exciting high-amplitude waves and, consequently, a substantial red-shift happens in the front part of the pulse. Due to the dispersion, the lower frequencies propagate through plasma slower than the higher frequencies, that by consequence remains in the front part of the pulse. This effects can ends in a separation of the pulse in two parts (as evidenced in Fig. 5) traveling at different velocities. During this event the group velocity v_g of the laser keeps decreasing and, therefore, the excitation of the wake waves becomes less efficient since their phase velocity is proportional v_g. The on-axis electron density n_e composing the decelerating waves is also showing periodic localized modulations in time with a period that is reducing proportionally to the wake wave phase velocity. Starting by considering a single-cycle driver wave-packet

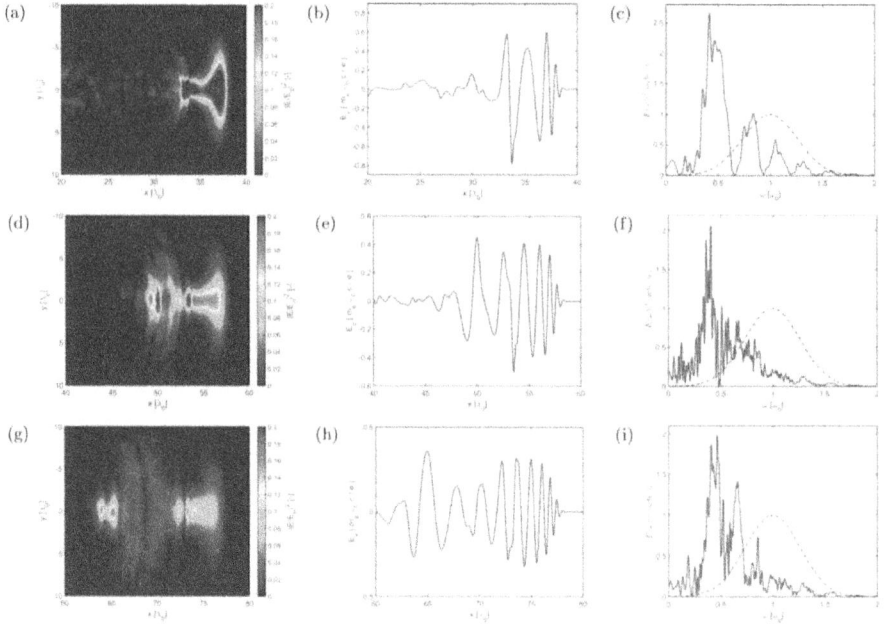

Fig. 5. Data from P. Valenta *et al.*[49] Laser field normalized (left column), transverse E-field on the y = 0 axis (center) and the corresponding on-axis spectrum (right) for three different time moments after the laser entered the uniform plasma: (a), (b), (c) $t_1 = 106$ fs, (d), (e), (f) $t_1 = 160$ fs, and (g), (h), (i) $t_1 = 212$ fs.

propagating in a collisionless plasma, and resolving the generated wake equations below the wavebreaking limit ($\gamma < 1$), it is possible to calculate the dispersion time

$$t_{\text{disp}} = \frac{\pi k_0 c}{\chi}, \tag{17}$$

where the dispersion effects result in a space-time modulation with a modulation period determined by the parameter

$$\chi = \frac{v_{gr}}{c} - \frac{c}{v_{gr}}. \tag{18}$$

By such calculations and simulations can be inferred that laser pulse dispersion of a small amplitude single-cycle laser can be responsible for strong wake-field modulations, presenting a period related to the laser v_g. These modulations are in turn responsible for the observed modulations in the plasma density of the waves behind the pulse and can limit the propagation of the driver pulse and so set an additional limit to the acceleration length. Further studies are needed to better understand how these modulations could be useful in the acceleration process itself or in the particle injection in separated wave packets.

8. Conclusions

All the laser sources and experimental possibilities available at ELI-Beamlines, together with the challenges to be solved framed in the general goal to better understand the laser-driven acceleration process in near-solid-state plasma targets and potentially reach accelerating gradients as high as TeV/m, have been summarized. Firstly, it has been presented a general overview of our User Research facility, the available lasers for particle-acceleration and, in more detail, the high electron platform scheme[32] with related challenges to achieve very energetic and good quality particle beams. Secondly, it has been discussed the importance of high-repetition rate accelerators and relevant challenges to be solved in an experiment. Finally two theoretical results have been discussed concerning the description of a nano-particle time-resolved response to very short laser pulses and the propagation of a single-cycle pulse in a near-critical density plasma. These could result potentially useful for improved diagnostics to monitor and to improve the propagation of ultra-short pulses in sub-μm nano-structures.

Acknowledgments

Major part of the research work presented has been carried out within the Particle Acceleration by Laser Research Program (RP3) at ELI-Beamlines. Supported by the project Advanced research using high intensity laser produced photons and particles ($CZ.02.1.01/0.0/0.0/16\_019/0000789$) from European Regional Development Fund (ADONIS). The results of the Project LQ 1606 were obtained with the financial support of the Ministry of Education, Youth and Sports as part of targeted support from the National Programme of Sustainability II.

References

1. T. Tajima and J. M. Dawson, Laser electron accelerator, *Phys. Rev. Lett.* **43**, p. 267 (1979).
2. C. G. R. Geddes, C. Toth, J. van Tilborg, E. Esarey, C. B. Schroeder, D. Bruhwiler, C. Nieter, J. Cary and W. P. Leemans, High-quality electron beams from a laser wake-eld accelerator using plasma-channel guiding, *Nat.* **431**, p. 538 (2004).
3. S. P. D. Mangles, C. D. Murphy, Z. Najmudin, A. G. R. Thomas, J. L. Collier, A. E. Dangor, E. J. Divall, P. S. Foster, J. G. Gallacher, C. J. Hooker, D. A. Jaroszynski, A. J. Langley, W. B. Mori, P. A. Norreys, F. S. Tsung, R. Viskup, B. R. Walton and K. Krushelnick, Monoenergetic beams of relativistic electrons from intense laserplasma interactions, *Nat.* **431**, p. 535 (2004).
4. J. Faure, Y. Glinec, A. Pukhov, S. Kiselev, S. Gordienko, E. Lefebvre, J.-P. Rousseau, F. Burgy and V. Malka, A laser-plasma accelerator producing mono-energetic electron beams, *Nat.* **431**, p. 541 (2004).
5. E. Esarey, C. B. Schroeder and W. P. Leemans, Physics of laser-driven plasma-based electron accelerators, *Rev. Mod. Phys.* **81**, p. 1229 (2009).
6. G. A. Mourou, G. Korn, W. Sandner and J. L. Collier, *ELI-Extreme Light Infrastructure Whitebook, Science and Technology with Ultra-Intense Lasers* (THOSS Media GmbHy, Germany, 2011).

7. B. Rus, P. Bakule, D. Kramer, J. Naylon, J. Thoma, M. Fibrich, J. T. Green, J. C. Lagron, R. Antipenkov, J. Bartoncek, F. Batysta, R. Base, R. Boge, S. B. J. Cupal, M. A. Drouin, M. Durk, B. H. T. Havlcek, P. Homer, A. Honsa, M. Horcek, P. Hrbek, J. Hubacek, Z. Hubka, G. Kalinchenko, K. Kasl, L. Indra, P. Korous, M. Koelja, L. Koubkov, M. Laub, T. Mazanec, A. Meadows, J. Novk, D. Peceli, J. Polan, D. Snopek, V. Sobr, P. Trojek, B. Tykalewicz, P. Velpula, E. Verhagen, S. Vyhldka, J. Weiss, C. Haefner, A. Bayramian, S. Betts, A. Erlandson, J. Jarboe, G. Johnson, J. Horner, D. Kim, E. Koh, C. Marshall, D. Mason, E. Sistrunk, D. Smith, T. Spinka, J. Stanley, C. Stolz, T. Suratwala, S. Telford, T. Ditmire, E. Gaul, M. Donovan, C. Frederickson, G. Friedman, D. Hammond, D. Hidinger, G. Chriaux, A. Jochmann, M. Kepler, C. Malato, M. Martinez, T. Metzger, M. Schultze, P. Mason, K. Ertel, A. Lintern, C. Edwards, C. Hernandez-Gomez and J. Collier, Eli-beamlines: Progress in development of next generation short-pulse laser systems, *Proceedings of the SPIE 10241* (2017).
8. H. T. Kim, K. H. Pae, H. J. Cha, I. J. Kim, T. J. Yu, J. H. Sung, S. K. Lee, T. M. Jeong and J. Lee, Enhancement of electron energy to the multi-gev regime by a dual-stage laser-wakefield accelerator pumped by petawatt laser pulses, *Phys. Rev. Lett.* **111**, p. 165002 (2013).
9. W. P. Leemans, A. J. GGonsalves, H. S. Mao, K. Nakamura, C. Benedetti, C. B. Schroeder, C. Toth, J. Daniels, D. E. Mittelberger, S. S. Bulanov, J. L. Vay, C. G. R. Geddes and E. Esarey, Multi-gev electron beams from capillary-discharge-guided sub-petawatt laser pulses in the self-trapping regime, *Phys. Rev. Lett.* **113**, p. 245002 (2014).
10. A. J. Gonsalves, K. Nakamura, J. Daniels, C. Benedetti, C. Pieronek, T. C. de Raadt, S. Steinke, J. Bin, S. S. Bulanov, J. van Tilborg, C. Geddes, C. B. Schroeder, C. Toth, E. Esarey, K. Swanson, L. Fan-Chiang, G. Bagdasarov, N. Bobrova, V. Gasilov, G. Korn, P. Sasorov and W. P. Leemans, Petawatt laser guiding and electron beam acceleration to 8 gev in a laser-heated capillary discharge waveguide, *Phys. Rev. Lett.* **122**, p. 084801 (2019).
11. D. Strickland and G. Mourou, Compression of amplified chirped optical pulses, *Opt. Commun.* **56**, p. 219 (1985).
12. *HELL Detailed User Workshop (2014).* *Available online: https://www.eli-beams.eu/en/media-en/events/hell-dur-2014-2/.*
13. W. T. Wang, W. T. Li, J. S. Liu, Z. J. Zhang, R. Qi, C. H. Yu, J. Q. Liu, M. Fang, Z. Y. Qin, C. Wang, Y. Xu, F. X. Wu, Y. X. Leng, R. X. Li and Z. Z. Xu, High-brightness high-energy electron beams from a laser wakefield accelerator via energy chirp control, *Phys. Rev. Lett.* **117**, 124801 (2016).
14. S. V. Bulanov, T. Esirkepov, Y. Hayashi, H. Kiriyama, J. Koga, H. Kotaki and M. Kando, On some theoretical problems of laser wake-field accelerators, *J. of Plasma Phys.* **82**, p. 905820308 (2016).
15. S. Steinke, J. van Tilborg, C. Benedetti, C. G. R. Geddes, C. B. Schroeder, J. Daniels, K. K. Swanson, A. J. Gonsalves, K. Nakamura, N. H. Matlis, B. H. Shaw, E. Esarey and W. P. Leemans, Multistage coupling of independent laser-plasma accelerators, *Nat.* **530**, p. 190 (2016).
16. T. Tajima, Laser acceleration in novel media, *The European Physical Journal Special Topics* **223**, p. 1037 (2014).
17. D. Bar-Lev, R. J. England, K. P. Wootton, W. Liu, A. Gover, R. Byer, K. J. Leedle, D. Black and J. Scheuer, Design of a plasmonic metasurface laser accelerator with a tapered phase velocity for subrelativistic particles, *Phys. Rev. Acc. And Beams* **22**, p. 021303 (2019).

18. A. Pukhov and J. M. ter Vehn., Laser wake field acceleration: the highly non-linear broken-wave regime, *Appl. Phys. B* **74**, p. 355 (2002).
19. W. Lu, C. Huang, M. Zhou, W. B. Mori and T. Katsouleas, Nonlinear theory for relativistic plasma wakefields in the blowout regime, *Phys. Rev. Lett.* **96**, p. 165002 (2006).
20. F. Bohle, M. Kretschmar, A. Jullien, M. Kovacs, M. Miranda, R. Romero, H. Crespo, U. Morgner, P. Simon and R. Lopez-Martens, Compression of cep-stable multi-mj laser pulses down to 4 fs in long hollow fibers, *Laser Phys. Lett.* **11**, p. 095401 (2014).
21. A. Jullien, A. Ricci, F. Bhle, J.-P. Rousseau, S. Grabielle, N. Forget, H. Jacqmin, B. Mercier and R. Lopez-Martens, Carrier-envelope-phase stable, high-contrast, double chirped-pulse-amplification laser sytem, *Opt. Lett..* **2**, p. 3774 (2014).
22. G. Mourou, S. Mironov, E. Khazanov and A. Sergeev, Single cycle thin film compressor opening the door to zeptosecond-exawatt physics, *The European Physical Journal Special Topics* **223**, 1181 (2014).
23. B. Baurepaire, A. Vernier, M. Bocoum, F. Bhle, A. Jullien, J.-P. Rousseau, T. Lefrou, D. Douillet, G. Iaquaniello, R. Lopez-Martens, A. Lifschitz and J. Faure, Effect of the laser wave front in a laser-plasma accelerator, *Phys. Rev. X* **5**, p. 031012 (2015).
24. D. Guenot, D. Gustas, A. Vernier, B. Beaurepaire, F. Bhle, M. Bocoum, M. Lozano, A. Jullien, R. Lopez-Martens, A. Lifschitz and J. Faure, Relativistic electron beams driven by khz single-cycle light pulses, *Nat. Phys.* **11**, p. 293 (2015).
25. A. J. Goers, G. A. Hine, L. Feder, B. Miao, F. Salehi, J. Wahlstrand and H. M. Milchberg, Multi-mev electron acceleration by subterawatt laser pulses, *Phys. Rev. Lett.* **115**, p. 194802 (2015).
26. F. Salehi, A. J. Goers, G. A. Hine, L. Feder, D. Kuk, B. Miao, D. Woodbury, K. Y. Kim and H. M. Milchberg., Mev electron acceleration at 1 khz with < 10 mj laser pulses, *Opt. Lett.* **42**, p. 215 (2017).
27. F. Batysta, R. Antipenkov, J. T. Green, J. A. Naylon, J. Novk, T. Mazanec, P. Hbek, C. Zervos, P. Bakule and B. Rus, Pulse synchronization system for picosecond pulse-pumped opcpa with femtosecond-level relative timing jitter, *Opt. Express* **22**, p. 30281 (2014).
28. A. Modena, Z. Najmudin, A. E. Dangor, C. E. Clayton, K. A. Marsh, C. Joshi, V. Malka, C. B. Darrow, C. Danson, D. Neely and F. N. Walsh, Electron acceleration from the breaking of relativistic plasma wave, *Nat.* **377**, p. 606 (1995).
29. J. M. Dawson, Nonlinear electron oscillations in a cold plasma, *Phys. Rev.* **113**, p. 383 (1959).
30. T. Oksenhendler, Self-referenced spectral interferometry, *Appl. Phys. B* **99**, p. 7 (2010).
31. A. Miranda, P. Rudawski, C. Guo, F. Silva, C. L. Arnold, T. Binhammer, H. Crespo and A. LHuillier, Ultrashort laser pulse characterization from dispersion scans: a comparison with spider *OSA Technical Digest paper*
32. T. Levato, S. Bonora, G. M. Grittani, C. M. Lazzarini, M. F. Nawaz, M. Nevrkla, L. Villanova, R. Ziano, S. Bassanese, N. Bobrova, K. Casarin, E. Chacon-Golcher, Y. Gu, D. Khikhlukha, D. Kramer, M. Lonza, D. Margarone, V. Olovcov, M. Rosinski, B. Rus, P. Sasorov, R. Versaci, A. Zara-Szydowska, S. V. Bulanov and G. Korn, Hell: High-energy electrons by laser light, a user-oriented experimental platform at eli beamlines, *Appl. Sci.* **8**, p. 1565 (2018).
33. S. V. Bulanov, T. Z. Esirkepov, Y. Hayashi, M. Kando, H. Kiriyama, J. K. Koga, K. Kondo, H. Kotaki, A. S. Pirozhkov, S. S. Bulanov, A. G. Zhidkov, P. Chen, D. Neely, Y. Kato, N. B. Narozhny and G. Korn, On the design of experiments for the study of extreme eld limits in the interaction of laser with ultrarelativistic electron

beam, *Nucl. Instrum. MethodsPhys. Res. Sect. A* **660**, p. 31 (2011).

34. A. Di Piazza, C. Mller, K. Z. Hatsagortsyan and C. H. Keitel, Extremely high-intensity laser interactions with fundamental quantum systems, *Rev. Mod. Phys.* **84**, p. 1177 (2012).

35. S. V. Bulanov, I. Inovenkov, V. I. Kirsanov, N. Naumova and A. S. Sakharov, Ultrafast depletion of ultra-short and relativistically strong laser pulses in an underdense plasma, *Phys. Fluids B* **4**, p. 1935 (1992).

36. S. F. Martins, R. A. Fonseca, W. Lu, W. B. Mori and L. O. Silva, Exploring laser-wakeeld-accelerator regimes for near-term lasers using particle-in-cell simulation in lorentz-boosted frames, *Nat. Phys. B* **6**, p. 311 (2010).

37. J. Faure, D. Gustas, D. Gunot, A. Vernier, F. Bhle, M. Ouill, S. Haessler, R. Lopez-Martens and A. Lifschitz, A review of recent progress on laser-plasma acceleration at khz repetition rate, *Phys. and Cotr. Fus.* **61**, p. 014012 (2018).

38. D. Gustas, D. Gunot, A. Vernier, S. Dutt, F. Bhle, R. Lopez-Martens, A. Lifschitz and J.Faure, High-charge relativistic electron bunches from a khz laser-plasma accelerator, *Phys. Rev. Acc. Beams* **21**, p. 013401 (2018).

39. S. Lorenz, G. M. Grittani, E. Chacon-Golcher, C. M. Lazzarini, J. Limpouch, M. F. Nawaz, M. Nevrkla, L. Vilanova and T. Levato, Characterization of supersonic and subsonic gas targets for laser wafefield electron acceleration experiments, *Matter and Radiation at Extremes* **4**, p. 015401 (2019).

40. C. F. Bohren and D. Huffman, *Absorption and Scattering of Light by Small Particles* (John Wiley and Sons, New York, 1993).

41. F. Lord Rayleigh, On the scattering of light by small particles, *Philos. Mag.* **41**, p. 274 (1871).

42. F. Lord Rayleigh, On the transmission of light through an atmosphere containing small particles in suspension, and on the origin of the blue of the sky, *Philos. Mag.* **47**, p. 375 (1899).

43. G. Mie, Beitrage zur optik truber medien, speziell kolloidaler metallosungen, *Ann. Phys* **330**, p. 377 (1908).

44. V. Giannini, A. I. FernndezDomnguez, Y. Sonnefraud, T. Roschuk, R. FernndezGarca and S. A. Maier, Controlling light localization and light-matter interactions with nano-plasmonics, *Small* **6**, p. 2498 (2010).

45. S. A. Maier, *Plasmonics: Fundamentals and Applications* (Springer, New York, 2007).

46. C. M. Lazzarini, L. Tadzio, J. M. Fitzgerald, J. A. Snchez-Gil and V. Giannini, Linear ultrafast dynamics of plasmon and magnetic resonances in nanoparticles, *Phys. Rev. B* **96**, p. 235407 (2017).

47. H. Yang, J. D'Archangel, M. L. Sundheimer, E. Tucker, G. D. Boreman and M. B. Raschke, Optical dielectric function of silver, *Appl. Phys. B* **91**, p. 235137 (2015).

48. M. Quinten, *Optical Properties of Nanoparticle Systems: Mie and Beyond* (Wiley-VCH Verlag GmbH and Co, Weinheim, 2011, 2011).

49. P. Valenta, O. Klimo, G. M. Grittani, T. Z. Esirkepov and S. V. B. G. Korn, Wakefield excited by ultrashort laser pulses in near-critical density plassmas, *arXiv* **1905.02043v1** (2019).

50. K. S. Yee, Numerical solution of initial boundary value problems involving maxwells equations in isotropic media, *IEEE Trans. Ant. Prop.* **14**, p. 302 (1966).

X-ray Laser Wakefield Acceleration in a Nanotube

Sahel Hakimi,* Xiaomei Zhang, Calvin Lau,

Peter Taborek, Franklin Dollar and Toshiki Tajima

Department of Physics and Astronomy,
University of California, Irvine, CA 92697, USA
www.physics.uci.edu
**sahelh@uci.edu*

Plasma-based accelerator technology enables compact particle accelerators. In Laser Wakefield Acceleration, with an ultrafast high-intensity optical laser driver, energy gain of electrons is greater if the electron density is reduced. This is because the energy gain of electrons is proportional to the ratio of laser's critical density to electron density. However, an alternative path for higher energy electrons is increasing the critical density via going to shorter wavelengths. With the advent of Thin Film Compression, we now see a path to a single cycle coherent X-ray beam. Using this X-ray pulse allows us to increase the plasma density to solid density nanotube (carbon or porous alumina) regime and still be under-dense for a Laser Wakefield Acceleration technique. We will discuss some implications of this below.

Keywords: LWFA in solid; nanotube; X-ray laser; TFC.

1. Motivation

The proposed technique of Thin Film Compression (TFC)[1] combined with the Relativistic Compression (RC) technique[2] could generate a coherent single cycle X-ray pulse. TFC technique could be used to compress a commonly existing short (< 100 [fs]) optical pulse to near a single cycle pulse. The spectrum of the pulse is broadened via Self Phase Modulation (SPM) as the pulse travels in a thin film. This pulse can then be compressed down to a shorter duration due to its now broader spectrum with $> 50\%$ efficiency.[1] If this few cycle optical pulse is focused onto an overcritical solid target, it pushes the surface electrons inward setting up a large electric field. As these electrons bounce back toward their parent ions, the reflected pulse is relativistically compressed to X-ray wavelengths. Therefore, the optical pulse from a TFC stage could be up-converted to an X-ray pulse via interacting with an overcritical target by RC at $\sim 10\%$ efficiency.[2] This X-ray pulse could potentially have Joule level energy and duration of a few attosecond or even zeptosecond translating to 1–10 keV energy photons.[1] This is potentially a way to reach the ExaWatt regime in the future. If we are able to focus this pulse to its much smaller diffraction limited spot size (on the order of nm), we could achieve unprecedented intensities[3] in the Schwinger regime of 10^{29} [W/cm^2]. This evolution is shown in Fig. 1 by a dashed black line. Thus the combination of TFC and RC

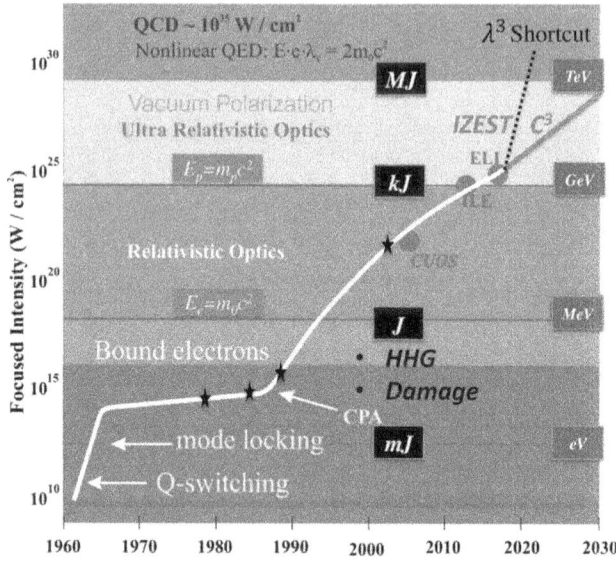

Figure 1. Evolution of Focused peak intensity is shown as a a function of time. The (λ^3) shortcut to the Nonlinear QED regime via pulse compression is also shown by a black dashed line. Image credit: Ref. 4.

techniques beautifully paves the path to a coherent single cycle X-ray pulse as shown in Fig. 2.

There is interest in using such a pulse for the purpose of accelerating electrons to ultra-high energies in micron to millimeter distances for the future TeV/cm accelerators. In Laser Wakefield Acceleration (LWFA), electron energy gain is approximately given by $\epsilon = 2a_0^2 m_e c^2 (n_c/n_e)$, where a_0 is the normalized vector potential of the laser and it relates to the intensity according to $a_0 = 0.85\sqrt{I}\lambda$, where intensity is in units of $[10^{18} \text{ W/cm}^2]$ and wavelength is in $[\mu]$. $n_c = m_e \omega_l^2 / 4\pi e^2 \propto 1/\lambda_l^2$ is the critical density of the laser and is inversely proportional to the laser wavelength. An X-ray pulse has a critical density 6-8 orders of magnitude higher compared to currently existing optical lasers. This matches the much higher electron density provided by solid materials and amounts to an energy gain increase by 3 orders of magnitudes compared to LWFA with optical lasers and gaseous materials as summarized in Table 1. These X-ray photons have energies on the order of 1–10 keV. To these photons, weakly-bound electrons with smaller binding energies would be considered effectively free although they are bound to an atom in reality. These electrons may be regarded as plasma-like in the timescale of the X-ray pulse, while more strongly bound electrons remain intact. Thus, solid material may be considered as a metallic plasma to this X-ray pulse. Further, material is not ionized during the interaction as the timescale for the above threshold ionization is long

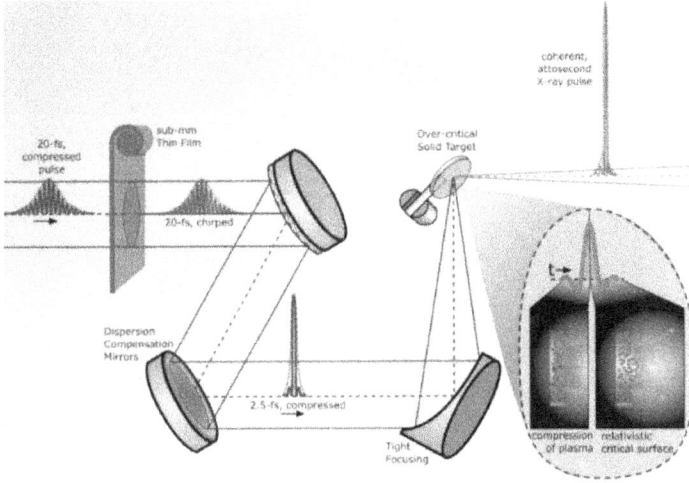

Figure 2.   Demonstration of the two-stage single cycle coherent X-ray production scheme. TFC setup is shown on the left side where a 20 [fs] short pulse travels through a thin piece of material and is chirped and spectrally broadened as a result. Dispersion compensation mirrors are used to undo the linear chirp and compress the beam to it's now shorter duration due to broader spectra. This results in a single cycle optical pulse. The scheme for Relativistic Compression is shown on the right where this compressed optical pulse is focused onto a solid target. Surface electrons are pushed in and bounce back toward the positive charges. This relativistically compresses the optical pulse to an X-ray pulse. Image Credit: Ref. 5.

Table 1.   Comparison of acceleration gradient with optical and X-ray laser.

	Optical (1000 [nm])	X-ray (1 [nm])
n_c	10^{21} $[cm]^{-3}$	10^{29} $[cm]^{-3}$
n_e	10^{18} $[cm]^{-3}$	10^{23} $[cm]^{-3}$
n_c/n_e	10^3	10^6
ϵ	1 GeV	1 TeV

compared to the few attosecond timescale of the passage of the X-ray pulse. Electrons oscillate a relatively small distance, on the order of an angstrom, about their original location to produce a wakefield. If the amplitude of wakefield reaches the wave-breaking limits, some electrons will get ripped from the atoms and will be accelerated. Electrons can also be injected into this structure.

In order to minimize collisions between the accelerated electrons and electrons within the solid, it is suggested to use fabricated nanotubes.[7] Nanoporous alumina structures can be fabricated on quartz crystals.[6] The pore density and diameter of this highly regular pore structure can be controlled during the fabrication to match the laser parameters and produce an optimized wakefield structure. Figure 3 shows an example of this regular honeycomb channel structure and its dimensions.

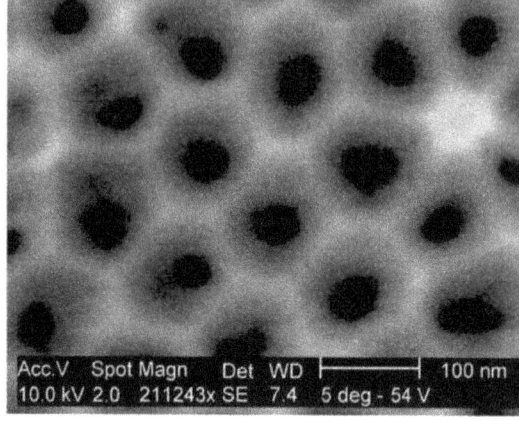

Figure 3.   SEM image of the top surface of a porous alumina fabricated on a QCM. Image Credit: Ref. 6.

A point to note is that an X-ray laser pulse can couple with ionic motions in solid material through optical phonon modes. Tajima and Ushioda have worked out the dispersion relation for polaritons in phonon-plasmon coupled systems.[8] The equations for this derivation are the continuity and momentum equations for both species and Poisson's equation:

$$\frac{\partial n_e}{\partial t} + \nabla \cdot (n_e(v_D + v_e)) = 0, \frac{\partial n_i}{\partial t} + \nabla \cdot (n_i v_i) = 0, \tag{1}$$

$$m_e n_e \left(\frac{\partial v_e}{\partial t} + (v_D + v_e) \cdot \nabla v_e \right) = n_e q_e E - \nabla P_e, \tag{2}$$

$$m_i n_i \left(\frac{\partial v_i}{\partial t} + v_i \cdot \nabla v_i \right) = n_i q_i E - K_i (x_i - x_{i0}), \tag{3}$$

$$\nabla \cdot E = 4\pi e(n_i - n_e), \tag{4}$$

$$\xi_e = \frac{eE}{m_e \omega^2 - m_e k_x v_D \omega - \frac{k_x^2 \gamma T_e \omega}{\omega - k_x v_D}}, \tag{5}$$

$$\xi_i = \frac{-eE/m_i}{\omega^2 - \frac{K_i}{m_i}}, \tag{6}$$

where v_D, v_e and v_i are the drift velocity of electrons, thermal velocity of electrons and ions respectively. E and P_e represent the electric field and electrons pressure, K_i is the effective spring constant for the lattice force of the neighboring ions and $\xi_\alpha (\alpha = e, i)$ are the displacements of charged particles. The dispersion relation where $\epsilon(k, \omega)$ is the relative permittivity is then given by:

$$\epsilon(k, \omega) = 1 - \frac{\omega_{pi}^2}{\omega^2 - \omega_{TO}^2} - \frac{\omega_{pe}^2}{\omega^2 - k_x^2 v_e^2}. \tag{7}$$

Therefore, this model is capable of including the important effects of ionic motions such as the polaritons and collective modes at solid densities by including the transverse optical phonon frequency, $\omega_{TO} = \sqrt{K_i/m_i}$, in the dispersion relation.[8]

2. Verification by Simulations

X-ray LWFA in solid density nanotube has previously been investigated by Zhang et al.[9] Two-dimensional (2D) EPOCH[10] Particle in Cell (PIC) code was used to simulate this interaction and study the appropriate scaling laws. These simulations compared LWFA driven by an X-ray pulse, $\lambda_l = 1$ [nm], in a nanotube with wall density of $n_e = 5 \times 10^{24}$ [cm^{-3}]; with LWFA driven by an optical pulse, $\lambda_l = 1$ [μm], in a nanotube with wall density of $n_e = 5 \times 10^{18}$ [cm^{-3}]. The later case is not a physical nanotube since it has a lower density than solid material but for the purpose of a fair comparison, it was modeled as a nanotube with a central vacuum channel. In these simulations $a_0 = 10$ was kept constant. This translates to an intensity of 1.4×10^{20} [W/cm^2] and 1.4×10^{26} [W/cm^2] for the optical and X-ray pulse respectively. Diameter of these nanotubes were chosen to be $5 \times \lambda_l$ corresponding to 5 [μm] and 5 [nm] for the optical and X-ray case. These simulations confirmed an accelerating gradient of TeV/cm when wakefield is driven by an X-ray pulse in solid density compared to GeV/cm when it is driven by an optical pulse in gaseous material as shown in Fig. 4.

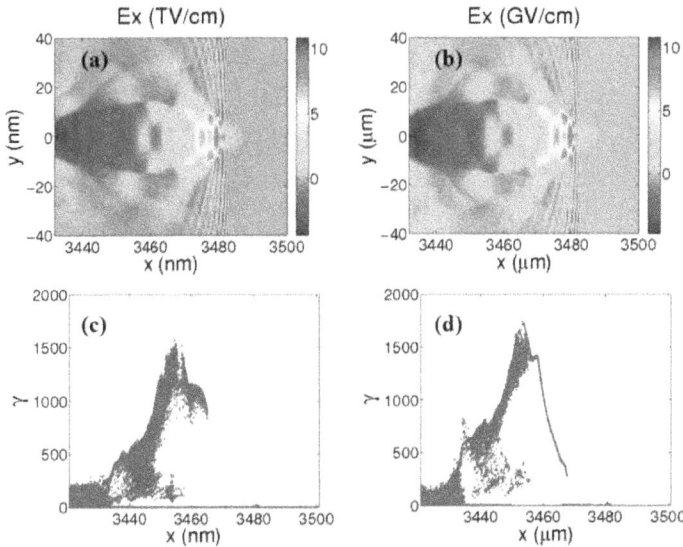

Figure 4. Comparison between the x-ray regime and optical regime. Distributions of (a) and (b) the longitudinal wakefield and (c) and (d) electron longitudinal momentum γv_x induced by (a) and (c) the x-ray laser pulse and (b) and (d) optical laser pulse in a tube when $a_0 = 10$. Image Credit: Ref. 9.

176

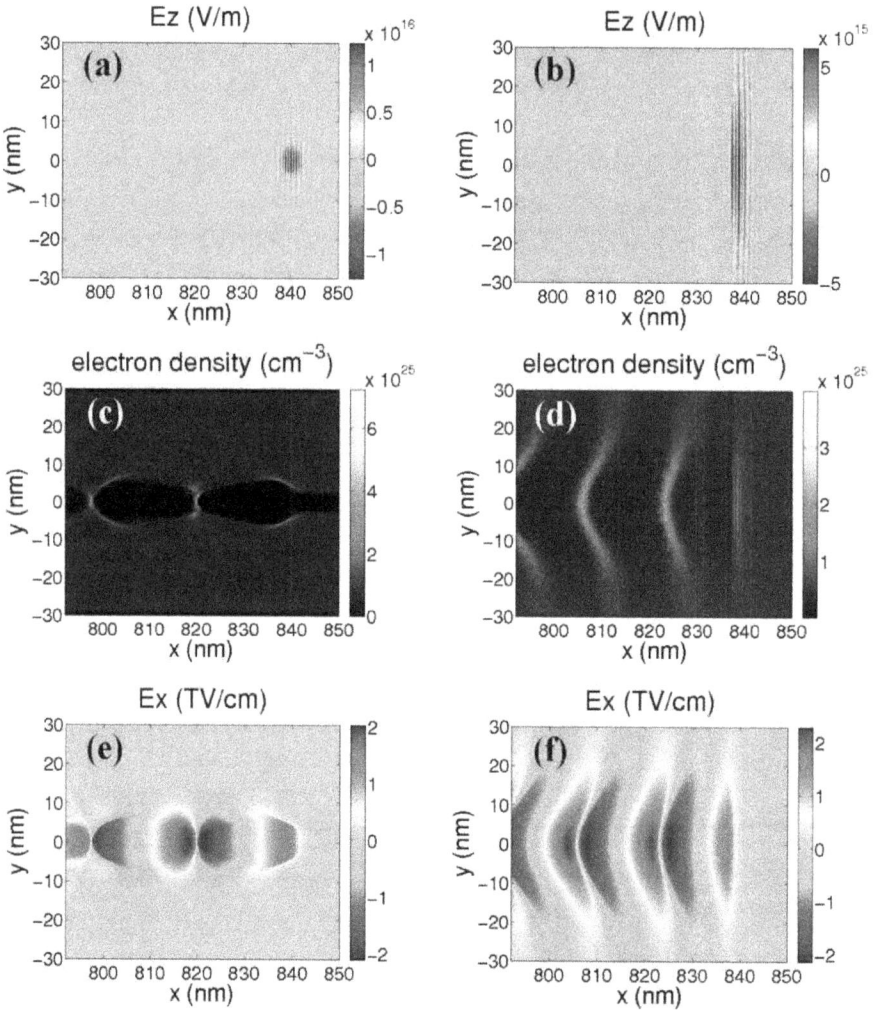

Figure 5. Comparison of X-ray laser in a hollow tube with a uniform system. Distributions of
(a)(b) the laser field, (c)(d) electron density, and (e)(f) wakefield in terms of a (a,c,e) tube and
(b,d,f) uniform density driven by the X-ray pulse. Image Credit: Ref. 9

An additional set of simulations compared the X-ray, $\lambda_l = 1$ [nm], driven LWFA
in a uniform solid density with a nanotube case modeled with a central vacuum
channel and solid density walls.[9] In this case, $n_e = 5 \times 10^{24}$ [cm^{-3}] and $a_0 = 4$
corresponding to 2.2×10^{25} [W/cm^2]. Diameter of the nanotube was $5 \times \lambda_l = 5$
[nm]. The simulation results are shown in Fig. 5 where the nanotube is seen to have
a higher quality wake formation and the additional advantage of guiding the laser.
Laser was guided for $4500\lambda_l$ in the nanotube case versus $2000\lambda_l$ in the uniform
density case.[9] The effects of the nanotube size were also studied[9] and are shown
in Fig. 6(a). Here, σ_{tube} refers to the radius of the nanotube and σ_L is the laser

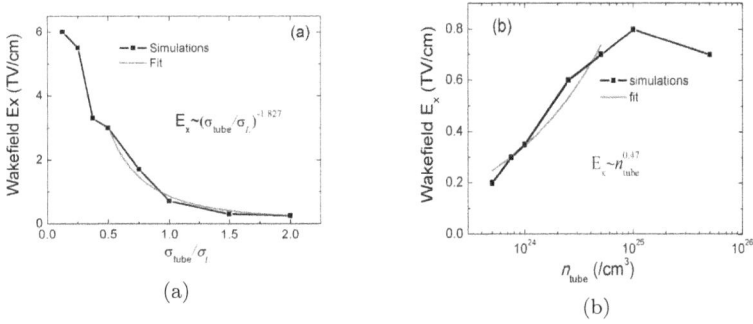

Figure 6.   Wakefield scalings in the X-ray regime with (a) tube radius, (b) tube wall density when the tube radius is fixed $\sigma_{tube}/\sigma_L = 1$. Image Credit: Ref. 9.

pulse width. When the ratio of $\sigma_{\text{tube}}/\sigma_L$ is small, wakefield evolution is nonlinear and approximated as wakefield in a plasma column. If the tube radius is matched to the pulse width as is the case in these simulations,[9] wakefield evolution is still qualitatively similar to a uniform plasma case but it is more organized. As this ratio increases, the effective density and thus the wakefield strength is reduced. If it is increased even more, laser pulse interacts less with the walls and eventually approximates a laser pulse travelling in vacuum. Figure 6(b) shows the variation of wall density when $\sigma_{\text{tube}}/\sigma_L = 1$. Wakefield strength is increased as the wall density is increased as expected. However, exciting a wakefield becomes more difficult as the density is increased further since laser and plasma parameters no longer match.

In these simulations, the solid state effects have not been included. Simulations have since been performed with the addition of these polariton effects through modification of EPOCH source code and including the ionic lattice force in the solid medium.[11] This was modeled by including the transverse optical phonons, $\omega_{TO} = \sqrt{K_i/m_i}$, and addition of the spring-like force of lattice. In this case an X-ray pulse, $\lambda = 10$ [nm], with $a_0 = 3$ is traveling through a uniform plasma with a density of $n_e = 10^{23}$ [cm^{-3}]. Figure 7 shows the usual evolution of the wakefield when $\omega_{TO} = 0$, at approximately $t \sim 1, 2$ and 3 [fs]. Electrons are seen to be gaining energy at a rate of ~ 80 [MeV] /800 [nm] $= 0.5$ [TeV/cm] as predicted by wakefield theory and in agreement with Zhang et. al. results.

Effects of lattice force and variation of its strength was studied for three cases when transverse optical phonon frequency is greater than, equal to and less than the plasma frequency. The result of each case is summarized in Fig. 8 at $t \sim 2$ [fs] showing the fundamentals of the LWFA, wake formation and acceleration gradient, are unaffected. A noticeable difference in these cases compared to the base case, Fig. 7(b), is the presence of an ion mode related to the lattice frequency. Thus, we conclude that the fundamentals of wakefield is not affected by this lattice addition although ionic phonon modes are now present.

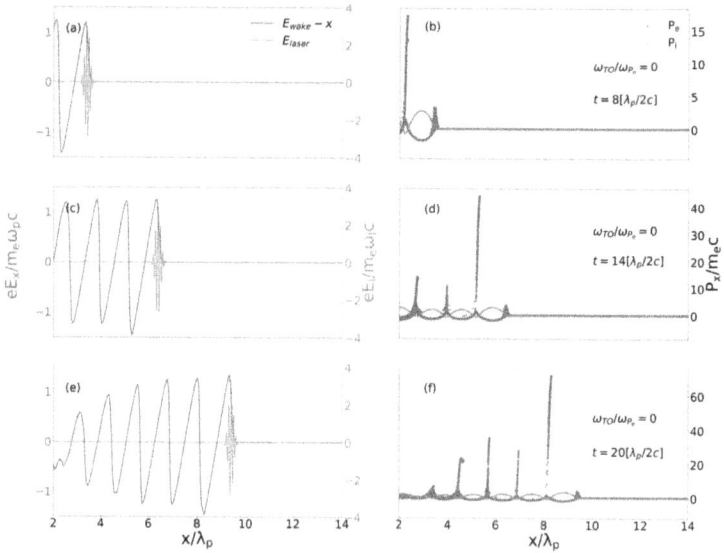

Figure 7. LWFA evolution with $\omega_{TO} = 0$ is shown at $t/[\lambda_{pe}/2c] = 8$ (panels a and b), $t/[\lambda_{pe}/2c] = 14$ (panels c and d) and $t/[\lambda_{pe}/2c] = 20$ (panels e and f). On the left, electric field of the laser pulse, E_y, is shown in red (right axis) and longitudinal electric field, E_x, is shown in blue (left axis) normalized by $E_L^{cr} = m_e\omega_{pe}c/e$. On the right, the phase space is shown for each species.

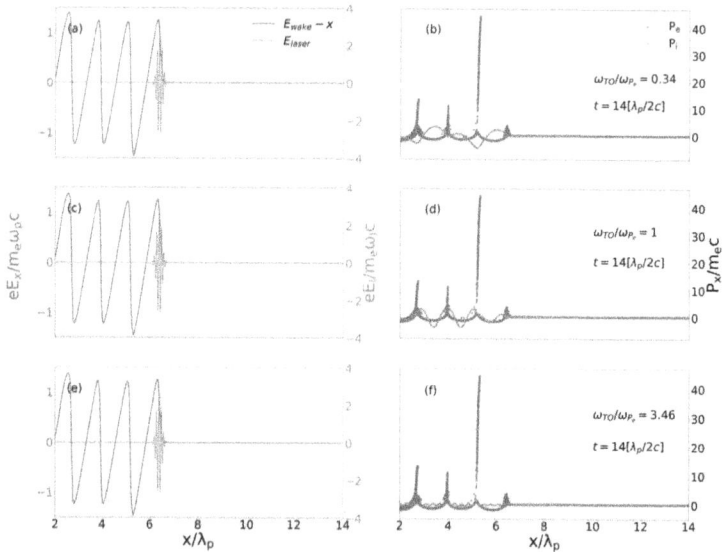

Figure 8. Comparison of different lattice force strength is shown at $t/[\lambda_{pe}/2c] = 14$ for $\omega_{TO}/\omega_{pe} = 0.34$ (panels a and b), $\omega_{TO}/\omega_{pe} = 1$ (panels c and d) and $\omega_{TO}/\omega_{pe} = 3.46$ (panels e and f).

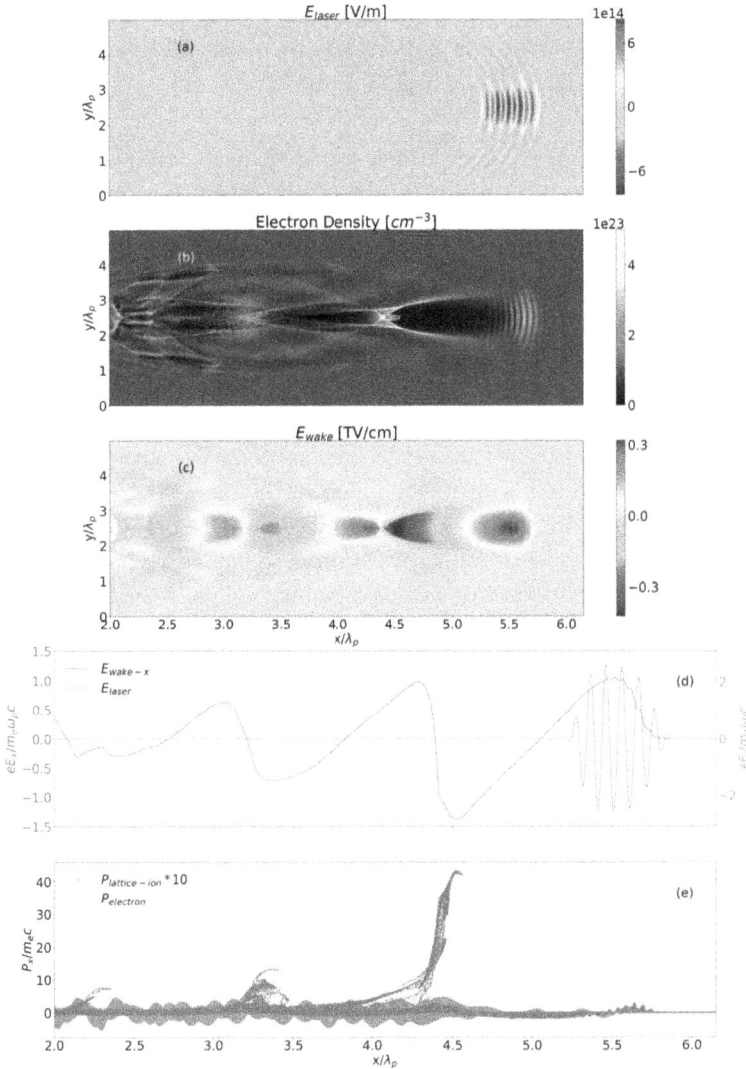

Figure 9.   2D LWFA simulation with lattice force turned on in the x direction, the direction of laser propagation. Here $\lambda_l = 10$ [nm] and $a_0 = 3$, $n_e = 10^{23}$ cm^{-3}, $\omega_{TO}/\omega_{pe} = 3.46$. Electric field of the driving laser is shown in (a), electron density distribution is shown in (b) and wakefield is shown in (c). Lineouts of the laser and wake fields are shown in (d) and phase space for both species is shown in (e).

Further modification to the 2D EPOCH code confirms these phonon modes do not have a deleterious effect on electron acceleration as shown in Fig. 9 for $t \sim 2$ [fs]. In this simulation an X-ray pulse, $\lambda = 10$ [nm], with $a_0 = 3$ is traveling through a uniform plasma with a density of $n_e = 10^{23}$ [cm^{-3}]. The ratio of $\omega_{TO}/\omega_{pe} = 3.46$ and lattice force is only acting in the x-direction (direction of propagation of laser).

Figure 9(d,e) can be compared to Fig. 8 (e,f) to see that in both 1D and 2D cases with the addition of lattice force in x-direction, we observe TV/cm wake formation and an acceleration gradient of ~ 0.5 [TeV/cm].

3. Applications

Acceleration in a crystal channel was studied previously,[12] although at that time it was purely a theoretical consideration. In the light of recent realization of TFC and RC, we are reconsidering this novel regime with a technology-based new entry. The generation of a coherent X-ray pulse,[2] which is well suited for this acceleration scheme, is now enabled via TFC and RC. In fact experimental pursuit of this route to X-ray laser pulse has already begun.[13,14] Further, we can visualize solid state fabrication technology of these nanostructures as well. We note that beam-driven solid-state tube acceleration is currently being studied as well.[15]

The X-ray driven LWFA in solid nanotubes described here is a scheme for an ultra-compact accelerator with an acceleration gradient nearly 6–7 orders of magnitude larger compared to traditional accelerators. This is an improvement of 3 orders of magnitude compared to the current LWFA with gaseous targets, a route toward a "TeV on a chip". Multiple stages of this TeV/cm chip can result in an even larger energy gain in a tabletop accelerator. This could be an alternative and compact route to a TeV linear collider in the future. It is worthy to consider a conceptual scheme of collider path based on this concept.

It is well known that accelerated electrons in a wakefield structure oscillate in the transverse direction and produce X-ray and gamma radiation.[16–18] While electron acceleration is scaled from optical to X-ray case, the radiation generated in these different regimes are quite different as QED effects need to be considered. Photon emission is scaled with the real electric field and can be 2–3 orders of magnitude higher in the X-ray case. Previous simulations[9] show photon emission of hundreds of keV to MeV for the optical case and hundreds of MeV in the X-ray case as shown in Fig. 10. Nanotubes with periodic characterizations (in the direction of propagation of the laser) could also serve as nano-undulators to control the produced radiation. For example, we can envision a specific superlattice period for a particular electron energy to produce a certain energy γ-photons as designed. Such a design may be employed to test the hypothesis of 17 MeV observed γ-photon to be part of the "fifth force" investigation.[19,20] These nanostructures with additional periodicity may be possible by methods such as superlattice of nanomaterial. It should be noted that present day semiconductor technology enables the fabrication of these structures layer-by-layer as specified by design. In this case, the nanotube is doing both the electron acceleration and undulation and there is no need for a superconducting magnet or its cooling systems. Therefore, X-ray driven LWFA could be a new X-ray or gamma radiation source, far more compact in size and greater in photon energy compared to Free Electron Lasers (FELs).

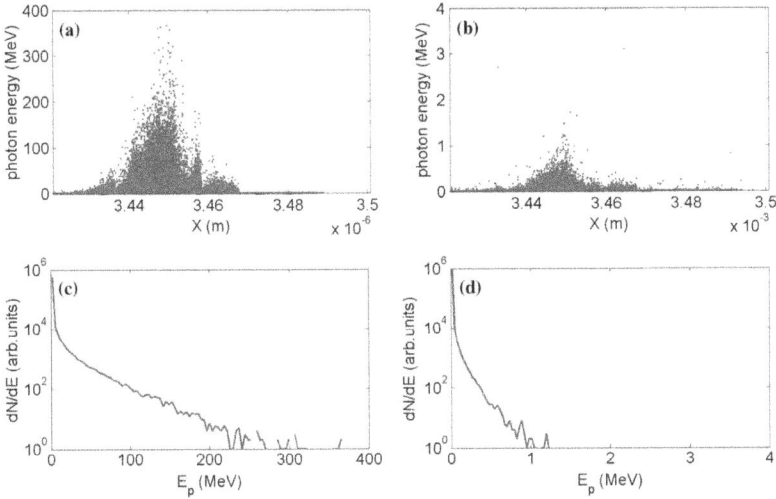

Figure 10. The energy spectrum and spatial distribution of photon emitted from the wakefield driven by x-ray laser and optical one. (a) and (b) Photon energy distributions and (c) and (d) photon energy spectrum in the (a) and (c) x-ray driven case and (b) and (d) 1 eV optical laser driven case in a tube. Image Credit: Ref. 9.

We are fascinated by a broad range of future applications beyond the high energy physics collider and high energy X-ray FELs on a compact chip or chips. These include: (i) the "non-luminosity paradigm" not resorting to the colliding beam to detect high energy physics phenomena; (ii) a candidate for muon source and acceleration;[21] (iii) compactification of the LWFA cancer radiotherapy using nanostructures as the accelerating medium.[22] For example, one of the "non-luminosity paradigm"[23,24] has been proposed to test the texture of the vacuum in ultra high energy γ-photon. In that paper it was assumed to employ the largest energy laser available (Mega Joule lasers) for electrons to be accelerated by LWFA toward PeV. High energy (PeV) γ-photons from these electrons may be used to probe a possible deviation of photons in its ultrahigh energy limit due to the superstring theoretic prediction of the vacuum texture being "stringy". Now that the present X-ray acceleration in nanostructures allows "TeV on a chip" conditions, we imagine the proposed scheme may become more affordable.

4. Conclusion

TFC technique alongside RC, if realized, make a single cycle X-ray pulse a possibility. This pulse can match a much higher solid density for the LWFA technique. The energy gain in this regime could be 3 orders of magnitude higher than LWFA with optical lasers and gaseous materials, thus this is an attractive route to "TeV on a chip" accelerator. Simulations performed in this regime verify a TeV/cm acceleration gradient compared to GeV/cm in the optical regime. Further, effects of ionic motions were investigated, explicitly studying the possibility of the lattice force coupling with the wake formation, and it was observed that ionic motion do not interact with the wake formation. This scheme has a wealth of applications in addition to producing high energy electrons from a super compact accelerator such as muon accelerators[21] and endoscopic oncology.[22] Photon emission in this regime is in a much higher energy range and could amount to a compact and specifically designed radiation sources as well.

Acknowledgments

The authors would like to thank Dr. Gerard Mourou, Dr. Jonathan Wheeler and Dr. Jean-Christophe Chanteloup from Ecole Polytechnique, Dr. Vladimir Shiltsev from Fermilab and Dr. Deano Farinella from UCI. This work was originated from a student term project of the Plasma Physics course (PHYSICS 239B) in the 2016 fall quarter at UC Irvine. This work was partially supported by Rostoker fund.

Bibliography

1. G. Mourou, S. Mironov, E. Khazanov and A. Sergeev, Single cycle thin film compressor opening the door to zeptosecond-exawatt physics, *The European Physical Journal Special Topics* **223**, 1181 (May 2014), arXiv: 1402.5676.
2. N. M. Naumova, J. A. Nees, I. V. Sokolov, B. Hou and G. A. Mourou, Relativistic generation of isolated attosecond pulses in a λ^3 focal volume, *Physical Review Letters* **92**, p. 063902 (February 2004).
3. G. Mourou, J. A. Wheeler and T. Tajima, Extreme light - An intense pursuit of fundamental high energy physics, *Europhysics News* **46**, 31 (September 2015).
4. T. Tajima and G. Mourou, Zettawatt-exawatt lasers and their applications in ultrastrong-field physics, *Physical Review Special Topics - Accelerators and Beams* **5**, p. 031301 (March 2002).
5. J. Wheeler, G. Mourou and T. Tajima, Laser technology for advanced acceleration: accelerating beyond TeV, *Reviews of Accelerator Science and Technology* (February 2017).
6. R. J. Lazarowich, P. Taborek, B.-Y. Yoo and N. V. Myung, Fabrication of porous alumina on quartz crystal microbalances, *Journal of Applied Physics* **101**, p. 104909 (May 2007).
7. T. Tajima, Laser acceleration in novel media, *The European Physical Journal Special Topics* **223**, 1037 (May 2014).
8. T. Tajima and S. Ushioda, Surface polaritons in LO-phonon-plasmon coupled systems in semiconductors, *Physical Review B* **18**, 1892 (August 1978).
9. X. Zhang, T. Tajima, D. Farinella, Y. Shin, G. Mourou, J. Wheeler, P. Taborek, P. Chen, F. Dollar and B. Shen, Particle-in-cell simulation of x-ray wakefield acceleration and betatron radiation in nanotubes, *Physical Review Accelerators and Beams* **19**, p. 101004 (October 2016).
10. T. D. Arber, K. Bennett, C. S. Brady, A. Lawrence-Douglas, M. G. Ramsay, N. J. Sircombe, P. Gillies, R. G. Evans, H. Schmitz, A. R. Bell and C. P. Ridgers, Con-

temporary particle-in-cell approach to laser-plasma modelling, *Plasma Physics and Controlled Fusion* **57**, p. 113001 (September 2015).

11. S. Hakimi, T. Nguyen, D. Farinella, C. K. Lau, H.-Y. Wang, P. Taborek, F. Dollar and T. Tajima, Wakefield in solid state plasma with the ionic lattice force, *Physics of Plasmas* **25**, p. 023112 (February 2018).

12. B. S. Newberger and T. Tajima, High-energy beam transport in crystal channels, *Physical Review A* **40**, 6897 (December 1989).

13. D. M. Farinella, J. Wheeler, A. E. Hussein, J. Nees, M. Stanfield, N. Beier, Y. Ma, G. Cojocaru, R. Ungureanu, M. Pittman, J. Demailly, E. Baynard, R. Fabbri, M. Masruri, R. Secareanu, A. Naziru, R. Dabu, A. Maksimchuk, K. Krushelnick, D. Ros, G. Mourou, T. Tajima and F. Dollar, Focusability of laser pulses at petawatt transport intensities in thin-film compression, *Journal of the Optical Society of America B* **36**, p. A28 (February 2019).

14. M. Masruri, J. Wheeler, I. Dancus, R. Fabbri, A. Nazru, R. Secareanu, D. Ursescu, G. Cojocaru, R. Ungureanu, D. Farinella, M. Pittman, S. Mironov, S. Balascuta, D. Doria, D. Ros and R. Dabu, Optical thin film compression for laser induced plasma diagnostics, in *Conference on Lasers and Electro-Optics* (OSA, San Jose, California, 2019).

15. A. A. Sahai and T. Tajima, Solid-state tube accelerator using surface wave wakefields in crystals.

16. F. Albert and A. G. R. Thomas, Applications of laser wakefield accelerator-based light sources, *Plasma Physics and Controlled Fusion* **58**, p. 103001 (September 2016).

17. G. A. Mourou, T. Tajima and S. V. Bulanov, Optics in the relativistic regime, *Reviews of Modern Physics* **78**, 309 (April 2006).

18. M. Marklund and P. K. Shukla, Nonlinear collective effects in photon-photon and photon-plasma interactions, *Reviews of Modern Physics* **78**, 591 (May 2006).

19. A. J. Krasznahorkay *et al.*, Observation of anomalous internal pair creation in ^{8}Be: A possible indication of a light, neutral boson, *Phys. Rev. Lett.* **116**, 042501 (2016).

20. J. L. Feng, B. Fornal, I. Galon, S. Gardner, J. Smolinsky, T. M. Tait and P. Tanedo, Protophobic fifth-force interpretation of the observed anomaly in ^{8}Be nuclear transitions, *Physical Review Letters* **117**, p. 071803 (August 2016).

21. A. A. Sahai, T. Tajima and V. Shiltsev, Schemes of laser muon acceleration: Ultrashort, micron-scale beams.

22. S. B. Nicks, T. Tajima, D. Roa, A. Necas and G. A. Mourou, Laser-wakefield application to endoscopic oncology.

23. T. Tajima, M. Kando and M. Teshima, Feeling the texture of vacuum: Laser acceleration toward PeV, *Progress of Theoretical Physics* **125**, 617 (2011).

24. G. Amelino-Camelia, J. Ellis, N. E. Mavromatos, D. V. Nanopoulos and S. Sarkar, Tests of quantum gravity from observations γ-ray bursts, *Nature* **393**, 763 (June 1998).

Ultrahigh Brightness Attosecond Electron Beams from Intense X-ray Laser Driven Plasma Photocathode

Ronghao Hu

College of Physics, Sichuan University, Chengdu, Sichuan 610065, China

Zheng Gong

State Key Laboratory of Nuclear Physics and Technology, and Key Laboratory of HEDP of the Ministry of Education, CAPT, Peking University, Beijing 100871, China
Center for High Energy Density Science, The University of Texas at Austin, Austin, Texas 78712, USA

Jinqing Yu

School of Physics and Electronics, Hunan University, Changsha 410082, China
State Key Laboratory of Nuclear Physics and Technology, and Key Laboratory of HEDP of the Ministry of Education, CAPT, Peking University, Beijing 100871, China

Yinren Shou

State Key Laboratory of Nuclear Physics and Technology, and Key Laboratory of HEDP of the Ministry of Education, CAPT, Peking University, Beijing 100871, China

Meng Lv

College of Physics, Sichuan University, Chengdu, Sichuan 610065, China
lvmengphys@scu.edu.cn

Zhengming Sheng

SUPA, Department of Physics, University of Strathclyde, Glasgow G4 0NG, UK
Key Laboratory for Laser Plasmas (MoE), and
Collaborative Innovation Center of IFSA (CICIFSA), and
School of Physics and Astronomy, Shanghai Jiao Tong University, Shanghai 200240, China
Cockcroft Institute, Sci-Tech Daresbury, Warrington WA4 4AD, UK

Toshiki Tajima

Department of Physics and Astronomy, University of California, Irvine, California 92610, USA

Xueqing Yan

State Key Laboratory of Nuclear Physics and Technology, and Key Laboratory of HEDP of the Ministry of Education, CAPT, Peking University, Beijing 100871, China
Collaborative Innovation Center of Extreme Optics, Shanxi University, Taiyuan, Shanxi 030006, China
x.yan@pku.edu.cn

The emerging intense attosecond X-ray lasers can extend the Laser Wakefield Acceleration mechanism to higher plasma densities in which the acceleration gradients are greatly enhanced. Here we present simulation results of high quality electron acceleration driven by intense attosecond X-ray laser pulses in liquid methane. Ultrahigh brightness electron beams can be generated with 5-dimensional beam brightness over 10^{20} A $\cdot$ m^{-2} $\cdot$ rad^{-2}. The pulse duration of the electron bunch can be shorter than 20 as. Such unique electron sources can benefit research areas requiring crucial spatial and temporal resolutions.

1. Introduction

Free Electron Lasers (FEL) have opened new scientific frontiers for chemistry; biology; soft matter; atomic, molecular and optical physics; condensed matter physics; plasma and warm dense matter physics.[1] With the advance of FEL physics and engineering, more powerful and shorter FEL pulses will offer new opportunities to investigate sciences that have never been explored before. The emerging terawatt attosecond X-ray laser pulses[2–6] can extend the optical Laser Wakefield Acceleration (LWFA)[7] to liquid or solid density plasmas in which the acceleration gradients are greatly. enhanced[8] LWFA in gas density plasmas has attracted many research interests in the past few decades.[9,10] The maximum output energy of LWFA in gas density plasmas has been enhanced to 8 GeV.[11] Laser accelerated electron beams have been used to probe electromagnetic fields with micrometer spatial resolution and femtosecond temporal resolution.[12,13] Duration, emittance and brightness of electron beams are essential parameters in such applications. The temporal resolution is determined by the duration of the electron beams, which is typically a few femtoseconds in gas density plasmas.[14,15] The spatial resolution is determined by the emittance of the electron beams, which is about a few microns in gas density plasmas.[16,17] The signal-to-noise ratio depends on the beam brightness, which is about 10^{15}–10^{16} A$\cdot$m$^{-2}\cdot$rad^{-2} in gas density plasmas.[18,19]

In principle, decreasing the laser wavelength can improve the beam qualities. The critical plasma density for reflection depends inversely to the square of the laser wavelength:

$$n_c(\lambda_L) = \frac{4\pi^2 \varepsilon_0 m_e c^2}{e^2 \lambda_L^2} \approx (1.1 \times 10^{27} \text{ [cm}^{-3}])/(\lambda_L \text{ [nm]})^2, \tag{1}$$

where λ_L is laser wavelength, ε_0 is vacuum permittivity, m_e is the rest mass of electron, c is light velocity in vacuum, e is elementary charge. Normalized plasma density n_p/n_c and normalized amplitude of vector potential $a_0 = eA/m_e c$ are the two key parameters in laser plasma interaction.[10] A is the amplitude of vector potential. a_0 is related with laser intensity I_0 and wavelength λ_L by $a_0 \simeq 8.55 \times 10^{-13} (\lambda_L \text{ [nm]})(I_0 \text{ [W/cm}^2])^{\frac{1}{2}}$. When radiation reactions, collisions and atomic processes like ionization and recombination are neglectable, laser plasma interactions are self-similar if n_p/n_c and a_0 are the same and all characteristic lengths, such as laser pulse waist and longitudinal length, scale as the laser wavelength, $L \propto \lambda_L$. The plasma density increases as laser wavelength decreases, $n_p \propto n_c \propto \lambda_L^{-2}$.

For FELs with wavelengths of a few nanometers, liquid or solid density (10^{23}–10^{24} cm^{-3}) plasmas can be witnessed as underdense plasmas. As the plasma frequency ω_p is proportional to the square root of plasma density n_p, $\omega_p = \sqrt{n_p e^2 / \varepsilon_0 m_e}$, increasing plasma density can increase the plasma frequency and reduce the plasma oscillation period. The duration of the bunch τ is proportional to the initial bunch length, which is proportional to laser wavelength, $\tau \propto \lambda_L$. Shorter electron bunch can be generated with smaller laser wavelength. Optical lasers with wavelengths above 100 nm are difficult to be focused down to a spot size below 1 μm due to the diffraction limit. [20] FELs with wavelengths of a few nanometers can be efficiently focused down to sub-100 nm foci with optics like Kirkpatrick-Baez Mirror pairs. [21,22] Emittance is proportional to phase space area occupied by the electron beam, $\varepsilon_n \propto y_m p_{ym}$, where y_m is the maximum transverse position, and p_{ym} is the maximum transverse momentum. y_m is proportional to the initial transverse position of the electron, $y_m \propto \lambda_L$. The transverse focusing force is proportional to the plasma density and transverse position, $F_\perp \propto n_p L$. [10,23] The maximum transverse momentum is proportional to the product of focusing force and initial transverse position, $p_{ym} \propto F_\perp L \propto n_p L^2 \propto \lambda_L^0$. Then emittance is proportional to the laser wavelength, $\varepsilon_n \propto \lambda_L$. The injected charge decreases as laser wavelength decreases, $Q \propto n_p L^3 \propto \lambda_L$. The peak current I is not related with laser wavelength as $I \propto Q/\tau \propto \lambda_L^0$. The 5D brightness B_{5D} is inversely proportional to the square of laser wavelength as $B_{5D} \propto I/\varepsilon_n^2 \propto \lambda_L^{-2}$. With reduced laser wavelength, electron beams with shorter duration, smaller emittance and much higher brightness can be generated.

Here we report Particle-in-Cell (PIC) simulation results of ultrahigh brightness attosecond electrons beams generated by intense attosecond FEL pulse interacting with liquid methane targets. The simulation setup is illustrated in Fig. 1. The wavelength of the FEL pulse is 2.5 nm, corresponding to a photon energy of 495.93 eV, which is above the K-shell binding energy of carbon atoms. The pulse contains about 9.38 mJ energy with a pulse duration of about 100 as. The peak power is 88.08 TW with a gaussian temporal profile. With focusing optics like Kirkpatrick-Baez Mirror pairs, the FEL pulse is focused down to a 100 nm spot, reaching an intensity of 5.61×10^{23} W/cm^2. For the given FEL parameters, a_0 is about 1.6. The cryogenic liquid methane has a density of about 0.4 g/cm^3. If carbon and hydrogen atoms are fully ionized, the electron number density is around 1.51×10^{23} cm^{-3}. The FEL pulse can propagate in the liquid target as the normalized plasma density is $n_e/n_c = 8.47 \times 10^{-4}$, where $n_c \approx 1.78 \times 10^{26}$ cm^{-3} is the critical density for reflection.

2. Results

2.1. *Refluxing electron injection*

Due to the ultrafast photoionization process, the cryogenic liquid target can be ionized and isochoric heated by the X-ray pulse to form very uniform plasma. [1] Sharp vacuum-plasma transition can be formed and electron dynamics in density

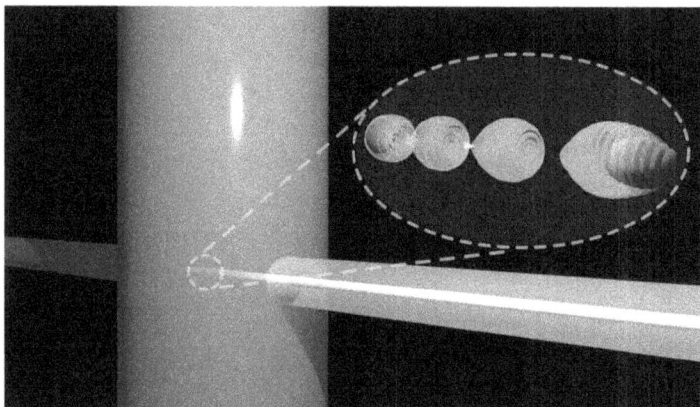

Fig. 1. Schematic of an intense attosecond free electron laser pulse (red) propagating through liquid methane jet (green) and generation of ultrahigh brightness electron beams (white). The inset shows the spatial distribution of X-ray pulse (red), plasma acceleration cavity (green) and injected electrons (white) from a 3D PIC simulation.

discontinuities can leads to electron injection.[24] For coherent X-ray laser drivers, the longitudinal equations of motion averaged by laser cycles can be given as[10]

$$\frac{dp_x}{dt} = -E + F_p, \quad \frac{dx}{dt} = p_x/\gamma. \tag{2}$$

t is time normalized by ω_p^{-1}, where $\omega_p = \sqrt{n_e e^2/m_e \varepsilon_0}$ is the plasma frequency, n_e is plasma density, e is elementary charge, m_e is electron mass at rest and ε_0 is vacuum permittivity. p_x is the longitudinal momentum and is normalized by $m_e c$, where c is the light velocity in vacuum. x is longitudinal position normalized by k_p^{-1}, where $k_p = \omega_p/c$ is the plasma wavenumber. E is the longitudinal electric field and is normalized by $m_e \omega_p c/e$. $F_p = -\frac{1}{4\gamma}\frac{\partial a^2}{\partial x}$ is the ponderomotive force of a linearly polarized laser driver, normalized by $m_e \omega_p c$, a is the laser vector potential normalized by $m_e c/e$ and $\gamma = \sqrt{1 + a^2/2 + p_x^2}$ is the Lorentz factor. All the physical quantities are in SI units. Before sheet crossing happens, the order of electron sheets is not disturbed and the electric force can be obtained from Poisson's equation as $E = \int_{-\infty}^{x}[n_i(x) - n_e(x)]dx = \int_{x_0}^{x} n_0(x)dx$, where $n_i(x)$ is the positive charge density, $n_e(x)$ is electron density and $n_0(x)$ is the initial undisturbed electron density.[24] Here ions are assumed to be fixed and all the density values are normalized using $n_0(x = 0)$. For simplicity without losing generality, the density profile of a vacuum-plasma transition (locates at $x = 0$) can be written as $n_0(x) = H(x)$, where $H(x)$ is the unit step function, and the electric force for electrons initially at x_0 can be written as $E = xH(x) - x_0$. Equations (2) can be integrated numerically for a linearly polarized gaussian laser pulse with $a = a_y = a_0 \exp\left[-(t - x/v_d - t_{\text{delay}})^2/\tau^2\right]$, where a_0 is the peak vector potential, v_d is the velocity of the driver, t_{delay} is the time delay of the pulse peak and τ is the 1/e half pulse duration. For drivers with small amplitudes ($a_0 \sim 1$), $v_d \approx \sqrt{1 - n_0/n_c}$, where n_c is the critical plasma density.[10] Figure 2(a) shows the numerical solutions of equations (2), where the rainbow-colored wavy lines are particle trajectories moving in the laser and charge separation fields, with colors being the initial positions of the particles, x_0, given by the color bar on top. Some particles (like red ones in Fig. 2(a)) oscillate and remain approximately the same region in the plasma while the laser moves forward into the plasma. Electrons near the boundary (the blue ones in the back) are ejected into the vacuum, and sheet crossing happens shortly after they re-enter the plasma. Electrons initially deep inside the plasma are oscillating with constant amplitudes and by neglecting the ponderomotive force term in Eqs. (2), one can obtain a constant of motion for the oscillation, i.e. $\mathcal{H} = \sqrt{1 + p_x^2} - \Psi$, where Ψ is the electrostatic potential and $\partial \Psi / \partial x = -E$. An approximate solution for small amplitude oscillation can be obtained as

$$p_x(t, x_0) = -p_m \cos [\omega_\gamma (t - x_0/v_d - t_0)],$$
$$x(t, x_0) = x_0 - \delta_m \sin[\omega_\gamma (t - x_0/v_d - t_0)]. \tag{3}$$

p_m is the maximum oscillation momentum, $\gamma_m = \sqrt{1 + p_m^2}$ is the maximum Lorentz factor after the driver and $\delta_m = \sqrt{2(\gamma_m - 1)}$ is the maximum displacement of the oscillation. t_0 is the time electron with $x_0 = 0$ enters the vacuum. For gaussian drivers with $a_0 \sim 1$, $\tau \approx \pi/2$ and $v_d \sim 1$, we found $\gamma_m \approx 1.005 + 0.166(a_0 - 0.5)^2$. ω_γ is the relativistic oscillation frequency, and can be fitted as $\omega_\gamma \approx \gamma_m^{-0.44}$. For electrons have initial positions $0 < x_0 < \delta_m$, they follow the same oscillation motion until they reach the plasma boundary and then enter the vacuum ($x < 0$), where they experience constant positive electric forces as $E(x) = -x_0$. By this force, they will return to the plasma with a maximum momentum equal to p_m, as shown in Fig. 2(a). The refluxing time, which is defined as the time from the refluxing electron leaving its initial position with negative velocity to its returning, can be written as $t_{re} = 2 \arccos(p_0/p_m)/\omega_\gamma + 2p_0/x_0$, where $p_0 = \sqrt{(\gamma_m - x_0^2/2)^2 - 1}$ is the electron momentum at $x = 0$. As one can see, the refluxing time t_{re} is a function of the initial position x_0 and its value ranging from π/ω_γ to $+\infty$. For refluxing electrons with proper refluxing time, their trajectories will intersect with the trajectories of oscillating electrons (Fig. 2(a)). The trajectories of refluxing electrons inside the plasma and the trajectory of the outermost oscillating electrons ($x_0 = \delta_m$) will cross on condition that

$$x_0 + \delta_m \sin[\omega_\gamma(t_{sc} - x_0/v_d - t_{re} - t_0)]$$
$$= \delta_m - \delta_m \sin[\omega_\gamma(t_{sc} - \delta_m/v_d - t_0)]. \tag{4}$$

Solving Eq. (4), one can obtain the sheet crossing time $t_{sc} = t_0 + x_0/v_d + t_{re} + \arcsin[A/\sqrt{2 + 2\sin(B)} - C]/\omega_\gamma$, where $A = 1 - x_0/\delta_m$, $B = \omega_\gamma[(\delta_m - x_0)/v_d - t_{re}]$, $C = \arcsin[\cos(B)/\sqrt{2 + 2\sin(B)}]$.

Sheet crossing will always happen near the vacuum-plasma transition, but only when the plasma waves are strong enough can refluxing electrons be trapped. After sheet crossing, due to the disordering of the electrons, it will be difficult to obtain the electric fields analytically. To find out the injection condition and injection threshold, we neglect the beam loading effects of the refluxing electrons after they cross with the outermost oscillating electron, i.e. they do not contribute to the electric field and disturb the motions of oscillating electrons. We can make this assumption because near the injection threshold, the injected charge is small and the beam loading effects are not significant. With this assumption, the electric force of the refluxing electrons can be obtained using the quasistatic theory[10] and their trajectories can be computed after sheet crossing as shown in Fig. 2(b). As a reference, the trajectories obtained from a 1D PIC simulations with the same parameters are shown in Fig. 2(c), and one can see the dynamics of the injected electrons are much alike despite the approximations we have made. Refluxing electrons can be injected into plasma waves if they are above the separatrix of injection when they cross with the outermost oscillating electron.[10,25] The momentum of the outermost oscillating electron at sheet crossing moment is $p_{os} = -p_m \cos[\omega_\gamma(t_{sc} - \delta_m/v_d - t_0)]$, the separatrix of injection can be written as $p_{sp} = \left[v_d D - \sqrt{D^2 - (1 - v_d^2)}\right]/(1 - v_d^2)$, where $D = \sqrt{1 - v_d^2} - \sqrt{1 + p_m^2} + \sqrt{1 + p_{os}^2} + v_d(p_m - p_{os})$. The momentum of refluxing electron at sheet crossing time t_{re} is $p_{re} = p_m \cos[\omega_\gamma(t_{sc} - x_0/v_d - t_{re} - t_0)]$, and the injection condition can be thus written as

$$p_{re} - p_{sp} > 0. \tag{5}$$

The sheet crossing time t_{sc} and momentum difference $p_{re} - p_{sp}$ against different initial position x_0 are plotted in Fig. 2(d). As one can see, there are several separated injection areas with $p_{re} - p_{sp}$ above zero, and according to the corresponding t_{sc}, we can tell that electrons from these injection area are loaded into different acceleration buckets of the wakefields. In 1D theory, REI can not load electrons into the first acceleration bucket. The second acceleration bucket has the longest injection length and highest injected charge. Theoretical injection length of the second bucket as a function of normalized laser amplitude a_0 is shown in Fig. 2(f) together with numerical results from 1D PIC simulations. The injection threshold, or the minimum a_0 required for injection, is about 1.05 according to the theory and 1D PIC simulations. The 1D PIC injection length is smaller than the theory for a_0 larger than the injection threshold.

2.2. *Electron beam evolution*

Injected electrons are trapped by the quasilinear plasma wakefields, in which they experience longitudinal acceleration and transverse focusing. Electrons exhibit transverse oscillations in the focusing electric and magnetic forces, known as betatron motions. The amplitudes of betatron motions decrease as $\gamma^{-1/4}$ during acceleration, while the betatron frequency decreases as $\gamma^{-1/2}$, where γ is the Lorentz factor of electrons.[23] The transverse bunch sizes in y and z directions are oscillating and gradually decreasing over time due to the acceleration as shown in Fig. 3(a). The acceleration fields experienced by the electron bunch are not uniform. The positive longitudinal acceleration gradient imprints a positive energy chirp on the electron

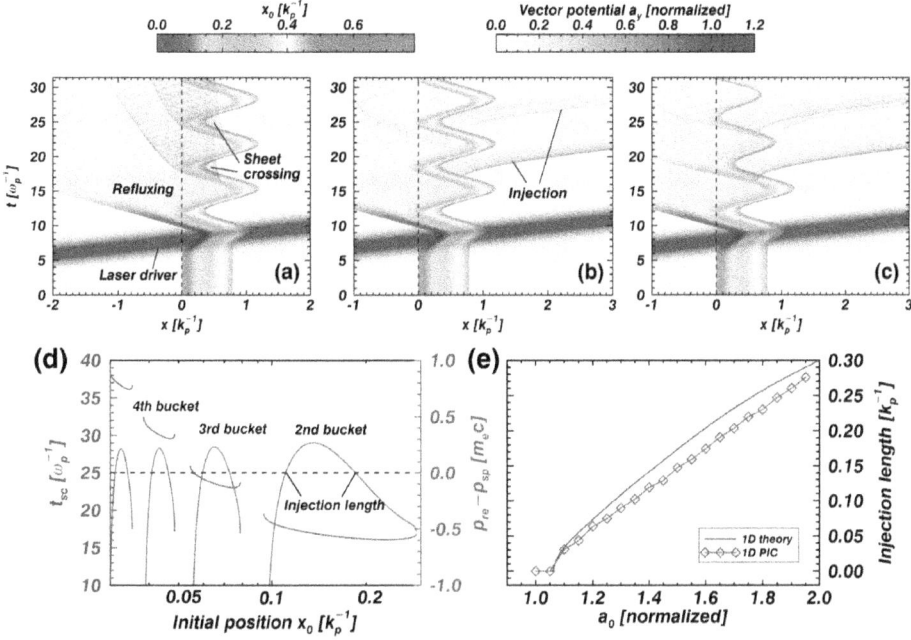

Fig. 2. Trajectories of electrons with different initial positions marked with different colors. (a) Numerical solutions of Eqs. (2) without modifications after sheet crossing. (b) Numerical solutions including quasistatic theory after sheet crossing. (c) Trajectories obtained from a PIC simulation with the same parameters. (d) Refluxing time t_{sc} and momentum difference $p_{re} - p_{sp}$ as a function of electron initial position x_0. (e) Injection length of second bucket for different laser peak amplitude a_0. The filled contours in subfigures (a)–(c) show the spatiotemporal profiles of the driving laser and the dashed lines indicate the initial plasma boundary. The laser is incident from left to right, and plasma initially locates in the $x > 0$ area. $a_0 = 1.2$ are used in subfigures (a)–(d) and $n_0/n_c = 0.004$ are used in all subfigures.

bunch. The head of bunch moves slightly faster than the tail, and the longitudinal bunch size is slowly stretching over time as shown in Fig. 3(a). The root-mean-squared (RMS) emittances in y and z directions are increasing initially and saturated around 5 nm as shown in Fig. 3(b). The main cause of the emittance growth is transverse phase mixing induced by the energy chirp.[26] The electron bunch initially have relatively large transverse size as shown in Fig. 3(c). In the focusing fields of wakes, the transverse size decreases as shown in Fig. 3(d). The betatron frequencies are different for electrons with different energies, and the betatron phases begin to diverge after injection. The emittance is proportional to phase space area occupied by the electron beam, formally defined as $\varepsilon_n = \sqrt{\langle \Delta x^2 \rangle \langle \Delta p_x^2 \rangle - \langle \Delta x \Delta p_x \rangle^2}/m_e c$, where $\Delta x = x - \langle x \rangle$, $\Delta p_x = p_x - \langle p_x \rangle$, $\langle \cdot \rangle$ represents the ensemble average over the electron phase space distribution and x can be x, y or z for each direction. After mixing of the betatron phases, the electron beam occupies the whole phase ellipse as shown in Fig. 3(e) and the emittance saturates as shown in Fig. 3(b). Electrons with larger radii (marked by red color) have larger betatron amplitudes and larger transverse momenta as depicted in Fig. 3(e). It indicates that reducing the initial size of the electron bunch can reduce the saturated emittance. The longitudinal phase space distribution is shown in Fig. 3(f). The slice energy spread of the bunch head is much smaller than the bunch tail. The increasing of slice energy spread in the bunch tail is due to the longitudinal phase mixing. The velocities of electrons are all close to light speed, electrons with larger transverse momenta will have smaller longitudinal velocities. Figure 3(c) shows the electron real space distribution before phase mixing, electrons with different radii are marked with different colors. After propagation about 1.8 μm in plasma, electrons initially with larger radii slip backwards with respect to bunch head as shown in Fig. 3(d). Due to the longitudinal gradient of the acceleration field, electrons with larger initial radii have slightly higher energy than those initially behind them with smaller radii.

After propagating for about 6.0 μm in plasma, the electron beam parameters are stabilized after transverse and longitudinal phase mixings. The detailed parameters that are of interest in applications of the electron beams are shown in Table 1. The nominal case has simulation parameters described above, notably with $a_0 = 1.6$ and FEL beam waist $w = 100$ nm. The electron beam energy is about 100 MeV and the acceleration gradient is about 16.7 TV/m, which is two orders higher than that in gas density plasma. The 5D-brightness $B_{5D} \approx 2I_p/(\epsilon_{ny}\epsilon_{nz}) = 5.42 \times 10^{20}$ A $\cdot$ m^{-2} $\cdot$ rad^{-2}, which is several orders higher than the state-of-art linac sources. The averaged transverse emittance $\varepsilon_n = \sqrt{\varepsilon_{nx}\varepsilon_{ny}}$ is 5.03 nm. The RMS bunch duration is only 17.4 as. The energy spread of the accelerated beam is changing during the acceleration. The longitudinal gradient of the acceleration field changes the energy chirp of the injected electron beam.[27,28] Electrons are trapped near the end of the acceleration bucket where the acceleration field gradient is positive. Electron beam becomes positively chirped and the absolute energy spread increases during acceleration. As electrons are faster than wakefields, they advance to the acceleration region with negative acceleration gradient. The energy spread decreases in negative acceleration gradient and reaches minimum after propagation of about 10 microns in plasma. If the diameter of the liquid methane jet can be controlled around 10 microns, electron beams with percent level energy spread can be generated.

3. Discussion

Electron beams generated by X-ray laser pulses interacting with liquid or solid density plasmas are unprecedented radiation sources combining ultrahigh brightness, ultralow emittance and ultrashort duration. The generated electrons beams can be used to probe microscopic electric and/or magnetic fields with ultrahigh spatial and temporal resolution. Such electron beams can also be injected to conventional linac or plasma based accelerators[29,30] and accelerated to higher energies for different applications like Thomson/Compton scattering sources,[23] FELs[31] and colliders.[32,33] With shorter and brighter FELs driving acceleration in higher density plasmas, even

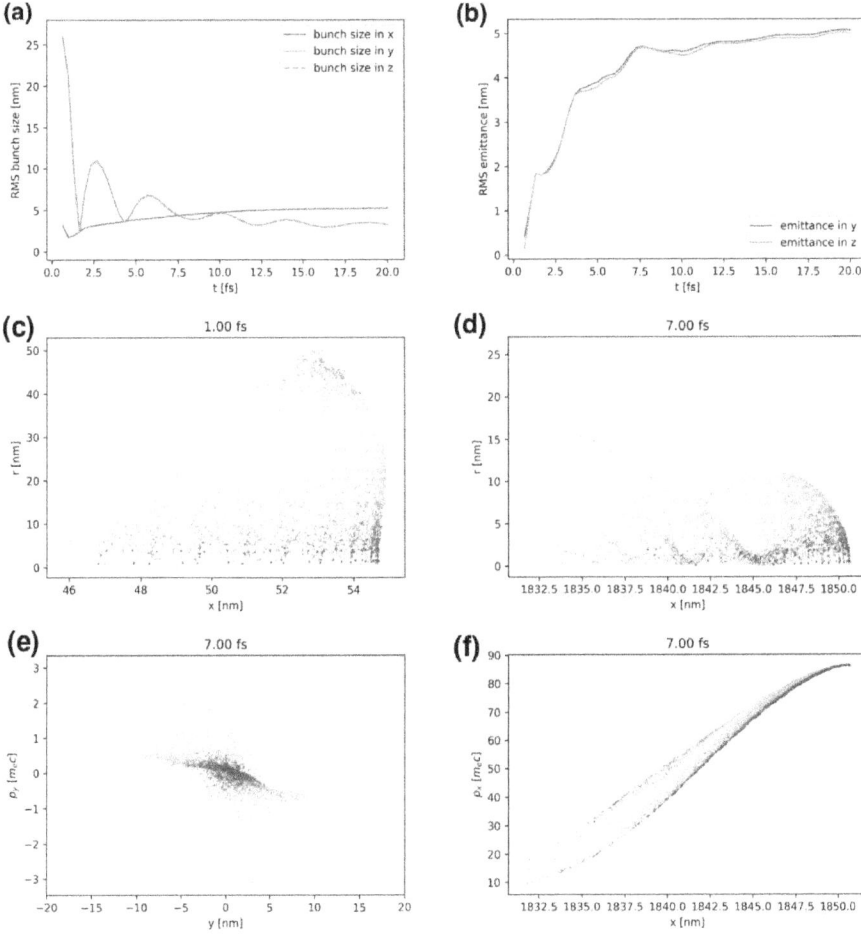

Fig. 3. (a) Electron bunch size evolution. (b) Electron beam transverse emittances as a function of time. (c) Electron real space distribution at $t = 1.00$ fs. (d) Electron real space distribution at $t = 7.00$ fs. (e) Electron transverse phase-space at $t = 7.00$ fs. (f) Electron longitudinal phase-space at $t = 7.00$ fs. Electrons with different initial radii are marked with different colors in (c), (d), (e) and (f).

shorter and brighter electron beams can be generated and zeptosecond physics can be explored with such unique sources.[34,35]

4. Methods

4.1. *Particle-in-cell simulations*

Quasi-cylindrical 3D particle-in-cell code FBPIC are used to investigate the dynamical propagation of X-ray pulse, generation of wakefields and relativistic dynamics of electrons. The computational domain has a length of 0.4 μm in longitudinal direction and a width of 0.2 μm in radial direction. The computational grid consists of 4000 cells in longitudinal direction and 50 cells in radial direction. The time step for field solver and particle pusher is 0.334 as. Two azimuthal modes are used to model cylindrical asymmetry. The total macroparticle number is 1.6×10^7 for electrons, and ions are assumed to be uniform positive charge background. The vacuum-plasma transition is numerically modeled with step function.

Acknowledgments

We would like to acknowledge the open-source projects FBPIC and OpenPMD.

References

1. C. Bostedt, S. Boutet, D. M. Fritz, Z. Huang and G. J. Williams, Linac coherent light source: The first five years, *Rev. Mod. Phys.* **88**, 015007 (2016).
2. G. Mourou and T. Tajima, More intense, shorter pulses, *Science* **331**, 41 (2011).
3. T. Tanaka, Proposal for a pulse-compression scheme in x-ray free-electron lasers to generate a multiterawatt, attosecond x-ray pulse, *Phys. Rev. Lett.* **110**, 084801 (2013).
4. E. Prat and S. Reiche, Simple method to generate terawatt-attosecond x-ray free-electron-laser pulses, *Phys. Rev. Lett.* **114**, 244801 (2015).
5. Z. Wang, C. Feng and Z. Zhao, Generating isolated terawatt-attosecond x-ray pulses via a chirped-laser-enhanced high-gain free-electron laser, *Phys. Rev. Accel. Beams* **20**, 040701 (2017).
6. C. H. Shim, Y. W. Parc, S. Kumar, I. S. Ko and D. E. Kim, Isolated terawatt attosecond hard x-ray pulse generated from single current spike, *Sci. Rep.* **8**, 7463 (2018).
7. T. Tajima and J. M. Dawson, Laser electron accelerator, *Phys. Rev. Lett.* **43**, 267 (1979).
8. X. Zhang, T. Tajima, D. Farinella, Y. Shin, G. Mourou, J. Wheeler, P. Taborek, P. Chen, F. Dollar and B. Shen, Particle-in-cell simulation of x-ray wakefield acceleration and betatron radiation in nanotubes, *Phys. Rev. Accel. Beams* **19**, 101004 (2016).
9. V. Malka, J. Faure, Y. A. Gauduel, E. Lefebvre, A. Rousse and K. T. Phuoc, Principles and applications of compact laser plasma accelerators, *Nature Phys.* **4**, 447 (2008).
10. E. Esarey, C. B. Schroeder and W. P. Leemans, Physics of laser-driven plasma-based electron accelerators, *Rev. Mod. Phys.* **81**, 1229 (2009).
11. A. J. Gonsalves, K. Nakamura, J. Daniels, C. Benedetti, C. Pieronek, T. C. H. de Raadt, S. Steinke, J. H. Bin, S. S. Bulanov, J. van Tilborg, C. G. R. Geddes, C. B. Schroeder, C. Tóth, E. Esarey, K. Swanson, L. Fan-Chiang, G. Bagdasarov, N. Bobrova, V. Gasilov, G. Korn, P. Sasorov and W. P. Leemans, Petawatt laser guiding and electron beam acceleration to 8 gev in a laser-heated capillary discharge waveguide, *Phys. Rev. Lett.* **122**, 084801 (2019).
12. C. J. Zhang, J. F. Hua, X. L. Xu, F. Li, C. H. Pai, Y. Wan, Y. P. Wu, Y. Q. Gu, W. B. Mori and C. Joshi, Capturing relativistic wakefield structures in plasmas using ultrashort high-energy electrons as a probe, *Sci. Rep.* **6**, 29485 (2016).
13. C. J. Zhang, J. F. Hua, Y. Wan, C.-H. Pai, B. Guo, J. Zhang, Y. Ma, F. Li, Y. P. Wu, H.-H. Chu, Y. Q. Gu, X. L. Xu, W. B. Mori, C. Joshi, J. Wang and W. Lu, Femtosecond probing of plasma wakefields and observation of the plasma wake reversal using a relativistic electron bunch, *Phys. Rev. Lett.* **119**, 064801 (2017).
14. O. Lundh, J. Lim, C. Rechatin, L. Ammoura, A. Benismal, X. Davoine, G. Gallot,

J. P. Goddet, E. Lefebvre and V. Malka, Few femtosecond, few kiloampere electron bunch produced by a laser-plasma accelerator, *Nature Phys.* **7**, 219 (2011).

15. C. J. Zhang, J. F. Hua, Y. Wan, B. Guo, C.-H. Pai, Y. P. Wu, F. Li, H.-H. Chu, Y. Q. Gu, W. B. Mori, C. Joshi, J. Wang and W. Lu, Temporal characterization of ultrashort linearly chirped electron bunches generated from a laser wakefield accelerator, *Phys. Rev. Accel. Beams* **19**, 062802 (2016).

16. S. K. Barber, J. van Tilborg, C. B. Schroeder, R. Lehe, H.-E. Tsai, K. K. Swanson, S. Steinke, K. Nakamura, C. G. R. Geddes, C. Benedetti, E. Esarey and W. P. Leemans, Measured emittance dependence on the injection method in laser plasma accelerators, *Phys. Rev. Lett.* **119**, 104801 (2017).

17. F. Li, Z. Nie, Y. Wu, B. Guo and W. B. Mori, Transverse phase space diagnostics for ionization injection in laser plasma acceleration using permanent magnetic quadrupoles, *Plasma Phys. Control. Fusion* **60**, 044007 (2018).

18. E. Brunetti, R. P. Shanks, G. G. Manahan, M. R. Islam, B. Ersfeld, M. P. Anania, S. Cipiccia, R. C. Issac, G. Raj, G. Vieux, G. H. Welsh, S. M. Wiggins and D. A. Jaroszynski, Low emittance, high brilliance relativistic electron beams from a laser-plasma accelerator, *Phys. Rev. Lett.* **105**, 215007 (2010).

19. W. T. Wang, W. T. Li, J. S. Liu, Z. J. Zhang, R. Qi, C. H. Yu, J. Q. Liu, M. Fang, Z. Y. Qin, C. Wang, Y. Xu, F. X. Wu, Y. X. Leng, R. X. Li and Z. Z. Xu, High-brightness high-energy electron beams from a laser wakefield accelerator via energy chirp control, *Phys. Rev. Lett.* **117**, 124801 (2016).

20. Y. Shou, H. Lu, R. Hu, L. Chen, H. Wang, M. Zhou, X. He, E. C. Jia and X. Yan, Near-diffraction-limited laser focusing with a near-critical density plasma lens, *Opt. Lett.* **41**, 139 (2015).

21. S. Boutet and G. J. Williams, The coherent x-ray imaging (CXI) instrument at the linac coherent light source (LCLS), *New J. Phys.* **12**, 035024 (2010).

22. F. Siewert, J. Buchheim, S. Boutet, G. J. Williams, P. A. Montanez, J. Krzywinski and R. Signorato, Ultra-precise characterization of LCLS hard x-ray focusing mirrors by high resolution slope measuring deflectometry, *Opt. Express* **20**, 4525 (2012).

23. S. Corde, K. Ta Phuoc, G. Lambert, R. Fitour, V. Malka, A. Rousse, A. Beck and E. Lefebvre, Femtosecond x rays from laser-plasma accelerators, *Rev. Mod. Phys.* **85**, 1 (2013).

24. J. M. Dawson, Nonlinear electron oscillations in a cold plasma, *Phys. Rev.* **113**, 383 (1959).

25. J. Faure, Plasma injection schemes for laser–plasma accelerators, *CERN Yellow Reports* **1**, 143 (2016).

26. X. L. Xu, J. F. Hua, F. Li, C. J. Zhang, L. X. Yan, Y. C. Du, W. H. Huang, H. B. Chen, C. X. Tang, W. Lu, P. Yu, W. An, C. Joshi and W. B. Mori, Phase-space dynamics of ionization injection in plasma-based accelerators, *Phys. Rev. Lett.* **112**, 035003 (2014).

27. R. Hu, H. Lu, Y. Shou, C. Lin, H. Zhuo, C.-e. Chen and X. Yan, Brilliant GeV electron beam with narrow energy spread generated by a laser plasma accelerator, *Phys. Rev. Accel. Beams* **19**, 091301 (2016).

28. A. Döpp, C. Thaury, E. Guillaume, F. Massimo, A. Lifschitz, I. Andriyash, J.-P. Goddet, A. Tazfi, K. Ta Phuoc and V. Malka, Energy-chirp compensation in a laser wakefield accelerator, *Phys. Rev. Lett.* **121**, 074802 (2018).

29. S. Steinke, J. van Tilborg, C. Benedetti, C. G. R. Geddes, C. B. Schroeder, J. Daniels, K. K. Swanson, A. J. Gonsalves, K. Nakamura and N. H. Matlis, Multistage coupling of independent laser-plasma accelerators, *Nature* **530**, 190 (2016).

30. X. L. Xu, J. F. Hua, Y. P. Wu, C. J. Zhang, F. Li, Y. Wan, C.-H. Pai, W. Lu, W. An,

P. Yu, M. J. Hogan, C. Joshi and W. B. Mori, Physics of phase space matching for staging plasma and traditional accelerator components using longitudinally tailored plasma profiles, *Phys. Rev. Lett.* **116**, 124801 (2016).

31. K. Nakajima, Compact x-ray sources: Towards a table-top free-electron laser, *Nature Phys.* **4**, 92 (2008).

32. C. B. Schroeder, E. Esarey, C. G. R. Geddes, C. Benedetti and W. P. Leemans, Physics considerations for laser-plasma linear colliders, *Phys. Rev. ST Accel. Beams* **13**, 101301 (2010).

33. J. Q. Yu, H. Y. Lu, T. Takahashi, R. H. Hu, Z. Gong, W. J. Ma, Y. S. Huang, C. E. Chen and X. Q. Yan, Creation of electron-positron pairs in photon-photon collisions driven by 10-pw laser pulses, *Phys. Rev. Lett.* **122**, 014802 (2019).

34. D. J. Dunning, B. W. McNeil and N. R. Thompson, Towards zeptosecond-scale pulses from x-ray free-electron lasers, *Phys. Procedia* **52**, 62 (2014).

35. G. Mourou, S. Mironov, E. Khazanov and A. Sergeev, Single cycle thin film compressor opening the door to zeptosecond-exawatt physics, *Eur. Phys. J. Spec. Top.* **223**, 1181 (2014).

Coherent Optical Transition Radiation Imaging for Compact Accelerator Electron-Beam Diagnostics[*]

A. H. Lumpkin

Accelerator Division, Fermi National Accelerator Laboratory,
Batavia, IL 60510, USA
lumpkin@fnal.gov

Application of coherent optical transition radiation (COTR) diagnostics to compact accelerators has been demonstrated for the laser-driven plasma accelerator case recently. It is proposed that such diagnostics for beam size, beam divergence, microbunching fraction, spectral content, and bunch length would be useful before and after any subsequent acceleration in crystals or nanostructures. In addition, there are indications that under some scenarios a microbunched beam could resonantly excite wake fields in nanostructures that might lead to an increased acceleration gradient.

Keywords: Compact accelerators; electron diagnostics; microbunching.

1. Introduction

Interest in the development of compact electron accelerators for a variety of applications including as the driver for free-electron lasers (FELs) has increased in the laser-driven plasma accelerator (LPA)[1] community in the last few years.[2,3] A recent development in this field has been reported involving coherent optical transition radiation interferometry (COTRI) for beam divergence measurements as well as near field imaging for beam size measurements just at the exit of the LPA.[4,5,6] These diagnostic techniques reveal few-micron size transverse structures that have ~1% microbunching fraction longitudinally at beam energies of 215 MeV. It is suggested here that the techniques could be used to evaluate beams before entering, and after leaving, crystalline- or nanostructure-based accelerator configurations presented in this Workshop. The microbunched electrons may also resonantly excite wake fields if matched to the plasma wavelength in this kind of compact accelerator. At our present state the microbunching is at visible wavelengths in the LPA so the nanostructures may be more relevant than the crystals. A brief discussion of the LPA diagnostics and a COTRI model will be presented with the possible extension to other compact accelerators described in this Workshop.

---

[*]This manuscript has been authored by Fermi Research Alliance, LLC under Contract No. DE-AC02-07CH11359 with the U.S. Department of Energy, Office of Science, Office of High Energy Physics.

2. Imaging Techniques for LPAs and COTR Basics

2.1. *Example LPA diagnostics*

A recent example of imaging diagnostics for an LPA at Helmholtz-Zentrum Dresden-Rossendorf (HZDR)[4,5] is shown in Fig. 1. In this case a 150-TW laser interacts with a He gas jet (with 3% N_2) with a plasma density $n_e = 3 \times 10^{18}/cm^3$ [7] resulting in electrons at about 215 MeV in the quasi-monoenergetic peak. Due to the LPA process transversely localized (few-micron size) electron distributions are microbunched in the visible wavelength regime resulting in COTR enhancements of $\sim 10^5$. The optical signal is then split over several paths to cameras to provide single-shot, near-field (NF) and far-field (FF) images of the beam at the exit of the LPA as shown in Fig. 1b. It is suggested that similar diagnostics would be applicable to crystalline- or nanostructure-based accelerator configurations, particularly if microbunched beams are employed.

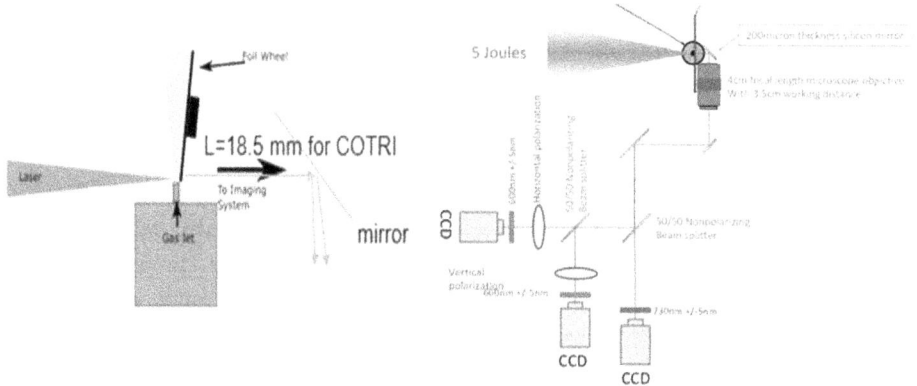

Fig. 1. Schematic of LPA with laser, gas jet, foils and mirror (L) and optical diagnostics configuration with NF and FF optical setups (R).[4]

2.2 *COTR basics*

Optical transition radiation (OTR) is generated with the transit of a charged particle beam through the boundary of two media with different dielectric constants as schematically shown in Fig. 2. For a vacuum-metal interface the approaching charge induces surface currents in the metal that result in the emission of OTR. There are forward and backward radiation cones emitted at the two interfaces with opening angle of $1/\gamma$ (where γ is the relativistic Lorentz factor) around the angle of specular reflection for the backward OTR and around the beam direction for the forward OTR.

The number W_1 of OTR photons that a single electron generates per unit frequency ω per unit solid angle Ω is

$$\frac{d^2W_1}{d\omega d\Omega} = \frac{e^2}{\hbar c}\frac{1}{\pi^2\omega}\frac{(\theta_x^2+\theta_y^2)}{(\gamma^{-2}+\theta_x^2+\theta_y^2)^2}, \tag{1}$$

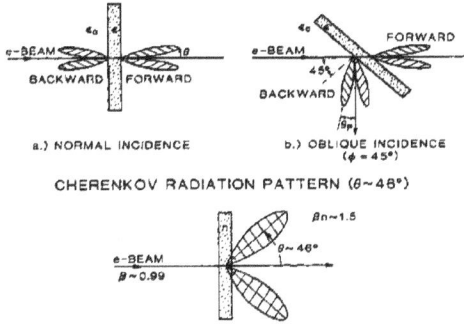

a.) NORMAL INCIDENCE   b.) OBLIQUE INCIDENCE (ϕ ~ 45°)

CHERENKOV RADIATION PATTERN (θ ~ 46°)

Fig. 2. Schematic of OTR generation at boundaries of vacuum and materials for a) normal incidence and b) oblique incidence.[8]

where $\hbar$ is Planck's constant$/2\pi$, e is the electron charge, c is the speed of light, and θ_x and θ_y are radiation angles.[9] The addition of the interference term $I(k)$ and the coherence function $J(k)$ as shown in Eqs. (2)–(4) include the effects of the microbunching fraction N_B. This formalism has been described in detail previously[6,9] and thus will not be given in detail here. The base COTRI equations are:

$$\frac{d^2W}{d\omega d\Omega} = |r_{\parallel,\perp}|^2 \frac{d^2W_1}{d\omega d\Omega} I(k)J(k), \tag{2}$$

$$I(k) = 4\sin^2\left[\frac{kL}{4}(\gamma^{-2} + \theta_x^2 + \theta_y^2)\right], \tag{3}$$

$$J(k) = N + N_B(N_B - 1)|H(k)|^2, \tag{4}$$

where k is the wave number, L is the foil separation, and $H(k)$ is the Fourier transform of the charge form factors given in Ref. 9. For a nominal case of 215 MeV, a foil separation $L = 1.85$ cm, a wavelength of 633 nm, and a beam divergence of 0.2 mrad, one obtains the single foil (black curve) and two-foil angular distribution patterns as shown in Fig. 3. The fringe visibility decreases with larger divergence values, and this dependence then can be used for a divergence measurement. The coherence function is angle dependent for $N_B > 0$.

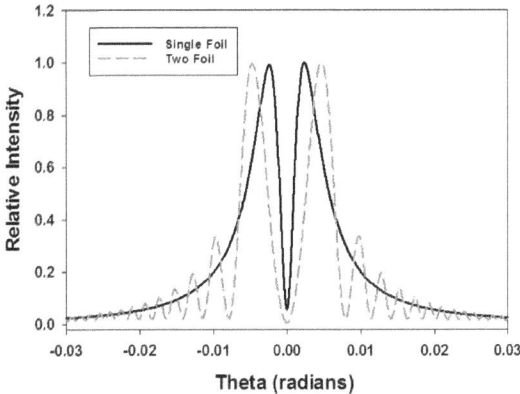

Fig. 3. Calculated angular distribution patterns for single foil and a two-foil interferometer.

3. Example Images of an LPA Beam in a Compact Configuration

As an example of the imaging of LPA generated electrons, Fig. 4 shows the NF and FF images for $E = 215$ MeV and $\lambda = 600$ and 633 nm, respectively. In the NF vertically polarized image, the 2.5-μm beam size was dominated by the coherent point spread function lobe structure. The two beamlets are about 6 μm apart along the x axis (laser polarization axis). In the FF image at the right, the number of interference fringes observed are indicative of sub-mrad divergence. Comparison to the COTRI model curves for different divergences from 0.1 mrad to 1.0 mrad leads to an estimated value of $\sigma_\theta = 0.5 \pm 0.2$ mrad.

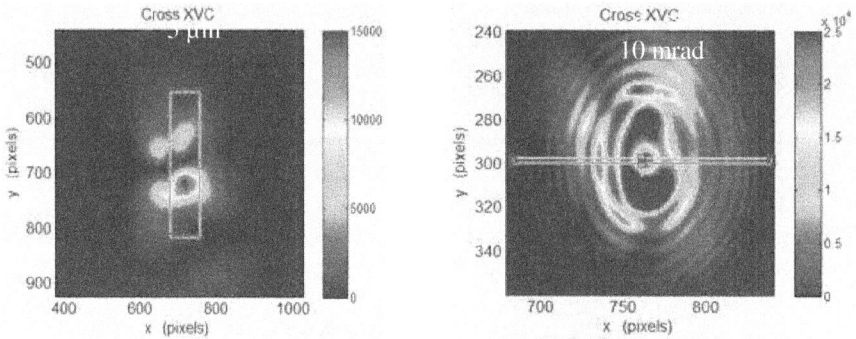

Fig. 4. Examples of NF and FF images from the same shot on an LPA at a beam energy of 215 MeV.[6]

4. Possible Extension to Crystalline- or Nanostructure-based Accelerators

The LPA diagnostics case for its compact accelerator configuration appears to lend itself to potential other compact configurations based on crystals or nanostructures as presented at this Workshop. The spatial resolution at the few-micron level using the coherent point spread function signature in the NF images, and the sub-mrad divergence information carried in the fringes observed in the FF seem sufficient for anticipated beams in the next several years. The 1% scale of microbunching in the LPA beam localized transversely to a few microns suggests potential for a short transport to a crystal, or more likely a specially designed nanostructure since such microbunching might resonantly excite wake fields.

5. Summary

In summary, the demonstration of high-resolution, COTR-based electron beam diagnostics on existing LPAs should make them relevant to future compact accelerator configurations based on crystals or nanostructures. In one simple extension, one might use the microbunched LPA beam to excite wake fields in appropriately designed nanostructures. Experimental demonstrations are needed to help benchmark present and future simulations.

Acknowledgments

The author acknowledges the collaborations with M. LaBerge and Prof. M. Downer of the University of Texas-Austin in the   realization of these COTR diagnostics initially proposed for the LPA in 2016. He also acknowledges recent discussions with T. Tajima (UC-Irving) and support from C. Drennan and M. Lindgren of Fermilab.

References

1. T. Tajima and J. M. Dawson, Laser electron accelerator, *Phys. Rev. Lett.* **43**, 267 (1979).
2. C. Lin *et al.*, Long-range persistence of femtosecond modulations on laser-plasma-accelerated electron beams, *Phys. Rev. Lett.* **108**, 094801 (2012).
3. Also see for example, *Proc. of AAC18, Breckenridge, Colorado* (IEEE, 2019).
4. M. LaBerge *et al.*, Talk presented at *EAAC17, September 2017*, Elba, Italy.
5. A. H. Lumpkin *et al.*, Observations of OTR interference fringes generated by LPA electron beamlets, *Proc. of AAC18* (IEEE, 2019).
6. A. H. Lumpkin *et al.*, Interferometric optical signature of electron microbunches from laser-driven plasma accelerators, submitted to *Phys. Rev. Lett.* July 12, 2019.
7. M. Mirzaie *et al.*, *Sci. Rep.* **5**, 14659 (2015).
8. A. H. Lumpkin, Advanced, time-resolved imaging techniques for electron-beam characterizations. Accelerator Instrumentation Workshop 1990, Batavia, *AIP Conf. Proc.* **229**, 151 (1991).
9. D. W. Rule *et al.*, Analysis of coherent optical transition radiation interference patterns produced by SASE-induced microbunches, in *Proc. of the 2001 Particle Accelerator Conference, Chicago* (IEEE, 2001), pp. 1288–1290, www.JACoW.org.

A Survey of Fiber Laser Technology in Light of Laser Particle Accelerator

Weijian Sha

2500 Walsh Ave., Santa Clara, CA 95051, USA

weijian.sha@commscope.com

Fiber laser technology and related performance and maturity are surveyed with the perspectives of laser acceleration. Advantageous fiber attributes allow high efficiencies in energy-conversion and heat removal and superior performance in high repetition. Wide utilization in the industries has led to high reliability and cost effectiveness in fiber laser technology. With coherent combination of fiber amplifiers, average-power and peak-power levels are targeted for practical laser-driven plasma acceleration applications.

Keywords: Fiber lasers; coherent combination; laser acceleration.

1. Introduction

This survey and study on fiber laser technology is intended to provide perspectives to the particle accelerator community in applying fiber lasers within the laser acceleration scheme. Reviews on laser-driven plasma accelerator development since the genesis of the concept[1] can be found in literature.[2] Intensive laser pulses are used to excite wake field in plasmas for acceleration of electrons, protons and charged ions. Therefore high peak-power laser is a foundation of the laser acceleration field.

As the push continues for ever higher peak-power in ultrafast laser systems, a necessary additional dimension — significantly increasing repetition of the intensive pulses which is associated with higher average power — is proposed for more practical laser acceleration application[3] and development is underway.[4,5,7,9–13] High repetition rate can deliver beam current in laser plasma accelerator to approach levels of conventional RF accelerator. In Ref. 3, proposed targets of such systems are 100 TW peak power with >100 kW of average power at >10 kHz. These targets lay out the goals and stimulate plans and actions to realize the dual requirements of high average power and high peak power.

Fiber laser is identified as the candidate for this new dimension, to overcome the limits of laser-crystal based systems due to thermal dissipation constrain and low wall-plug efficiency. It is worthwhile to survey the general real-world applications, as fiber lasers are becoming the mainstream of laser technology today thanks to their advantageous characteristics. In a number of ways fiber lasers have surpassed or are surpassing laser-crystal or solid-state lasers in performance, power efficiency and cost efficiency. This study will attempt to review the current state of fiber laser industry as a measure of technology progress and maturity, and also to survey basic attributes of fiber lasers to identify its advantages in performance as well as the limits in ultrafast amplification. With the understanding of the constraints but at the same time recognizing its scalability, we can view individual fiber amplifier as the building element for a larger, coherently combined system. We then can remark the new window of development for scaling up peak power and average power towards the targets of practical laser accelerators.

The angle of study in this paper is more from an industrial standpoint. The survey is mostly aimed at examining the fiber laser/amplifier element regarding the practical factors leading to high efficiencies, the underlying technology maturity, considerations of reliability and modular-ability. Coherent combination is the next frontier of fiber laser or even laser-in-general technologies and the subject is covered by another paper in this workshop. Here it is briefly described so that we come to a more complete picture on using fiber for laser acceleration.

2. Industrial Base of Fiber Lasers

2.1. *Fiber amplifiers in telecom*

A major drive for optical fiber development, passive or active fibers, has been in the telecom industry. Today's internet infrastructure employs erbium-doped fiber amplifiers as one of the basic building blocks and owns the largest base of active fiber components. The emphasis of engineering in the telecom is to transmit data signal with best quality and photon economy. Fibers are operated in the linear regime, engineered with low noise and dispersion compensation. One watt of power in a fiber can deliver huge data speed almost uncomprehend-able. Reliability is another key factor. In trans-oceanic fiber systems, fiber and pump laser components have mean-time-between-failures over 100 years.

2.2. *Industrial fiber lasers for metal processing*

While telco fiber design focuses on the delicate side, industrial fiber lasers for metal processing thrive on sheer light power. Fiber laser systems on roller wheels, including all its elements such as laser diode pumping, cooler and power supply can deliver continuous-wave (CW) power up to 100 kilo-watt. With breakthrough engineering in component damage control and heat removal, these fiber lasers are replacing or expanding what other lasers can do and cannot do. A fiber laser with a few kilo-watts can cut sheet metals in high processing speed unparalleled by conventional tools.

2.3. *Industrial ultrafast fiber lasers*

A third class of fiber lasers in industrial use produces ultrafast pulses at high repetition rates. This class of fiber laser or amplifier is the element for the coherently combined system aiming at laser acceleration application. Ultrafast (referred to picosecond, sub-picosecond and femtosecond) fiber laser is a very active research field since 1990's and continuing on at present. Concurrently we are seeing commercialization from academic research in a time span from 5 to 10 years. Within the last five years or so, many companies had offerings in ultrafast fiber lasers with 100 micro-joule pulse energy at

MHz repetition rate. With single transverse-mode beam quality which facilitates tight focusing, these pulses have electric field strength greater than molecular bonding force, causing (cold) ablation even in the hardest material such as glass or diamond. A new field of precision micro-machining has found many industrial usages. Developments from commercial companies make ultrafast fiber lasers turn-key systems, ruggedized under wide environmental conditions and highly reliable over tens of thousands of hours of continuous operation.

2.4. *Current state of commercial fiber lasers*

A summary of current commercial CW and ultrafast fiber lasers can be further illustrated in Fig. 1. Axis parameters are in pulse energy, repetition and average power (diagonal lines, CW operation only in average power).

1) CW fiber laser operating at kW levels and above is a common place. These lasers require large-core fibers to stay below damage intensity thresholds and operate in high-order spatial modes. The resulting beam quality is adequate for bulk metal processing, but rather poor compared to single transverse mode (or single mode or SM in short). Beam quality is measured by the M^2 parameter, which can be viewed as focus-ability of the beam. Single mode has $M^2 = 1$ at diffraction limit. The data point of 1 kW at SM with 15-μm core speaks of the high sustaining intensity of this fiber laser under continuous operation.

2) Nano-second (ns) pulse lasers can operate from hundreds of watts to 1 kW. With ns pulses, fiber medium starts to see nonlinear effects. Peak power is still relatively low compared to picosecond (ps) and femtosecond (fs) pulses. Dispersion effect is tolerable given narrow bandwidth and the relatively long duration of ns pulses. This class of laser can be viewed as the intermediate regime between CW and ps/fs operation in average power, peak power and beam quality.

3) Ultrafast lasers are seen to have a ceiling in pulse energy, regardless the repetition rate. Pulses with ps and sub-ps duration have 3–4 orders of magnitude higher in peak power over ns pulses, causing large nonlinear detrimental effects. Pulse broadening is subject to large bandwidth and the relative effect of shorter pulse. Overcoming nonlinearity and managing dispersion is the central engineering design in ultrafast fiber lasers. Pulse energy ceiling is related to nonlinear effects but it is also affected by the choice of fiber due to the trade-off between mode quality, manufacture-ability, and assembly handling.

Fiber Laser Systems at 1 μm (M² < 1.3 except for > 2 kW)

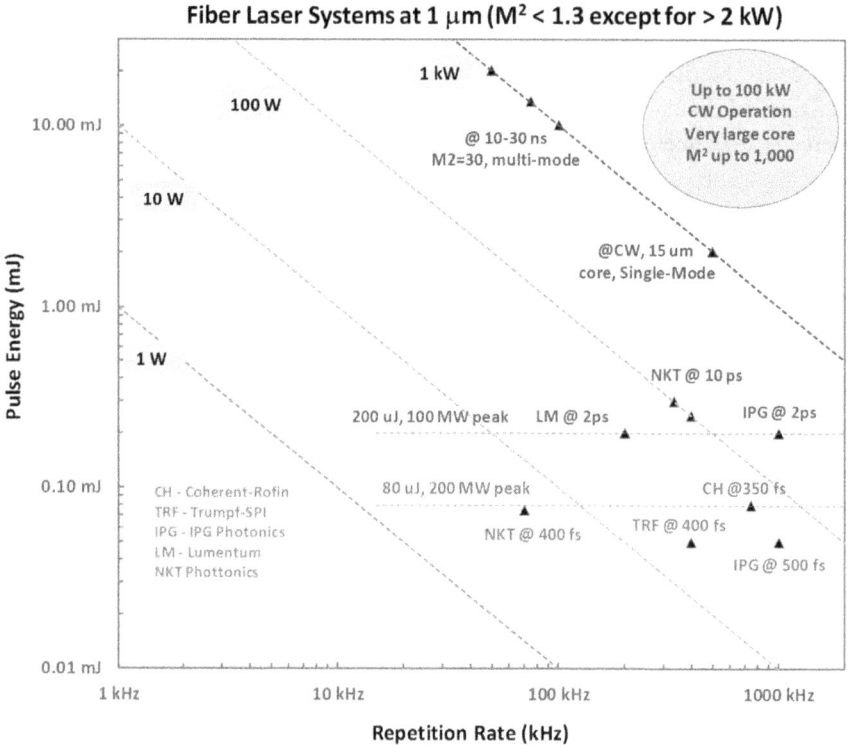

Fig. 1. Survey of commercial fiber laser performance operating in continuous wave, nanosecond, picosecond and femtosecond regimes.

3. Basic Fiber Attributes

The purpose here is to provide a quick overview of fiber attributes. Fiber technology is a large-body engineering field with broad subjects, thus brief discussions here can only touch on some practical aspects of optical fiber and fiber laser regarding mode guiding, nonlinearity, thermal management, electrical efficiency. Laser diode as the pump source is discussed since it is an integral part of the fiber laser.

3.1. *Mode guiding*

Light guiding in optical fiber can be viewed as total internal reflection using Snell's law. Even with this simple view, one can come to recognize the underlying design accessibility of light guiding in the optical index structure in Figs. 2(a) and 2(b). The sensitivity of guiding angle to index contrast gives rise to many design varieties towards diverse applications. Index control in fiber glass material is precise but also limited to a few percent of index contrast within compatible materials.

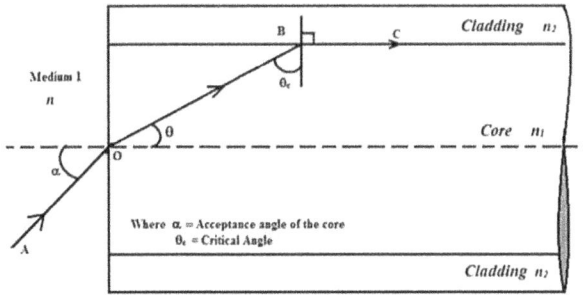

Fig. 2(a). The optical fiber is defined by the core with a higher refractive index n_1 and the cladding with n_2. The working mechanism of total internal reflection is quantified by critical angle θ_c while light ray guiding is defined by guiding angle $\theta = 90° - \theta_c$.

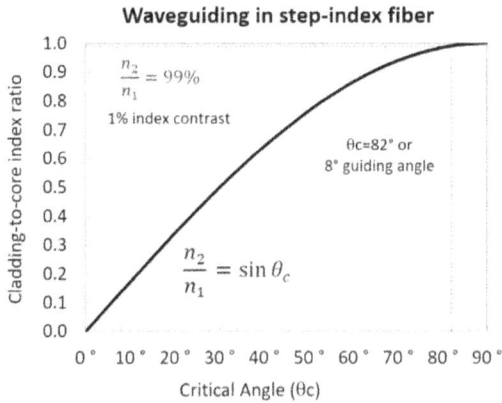

Fig. 2(b). Guiding angle of light ray vs. index contrast using Snell's law. Curvature of sinusoidal function towards 90° produces leverage from small index contrast to significant guiding angle. An index contrast of 1% between the core and the cladding results in 8° guiding angle, enough for single-mode guiding in most transmission fibers.

A major design merit for ultrafast fiber laser is to make the fiber core as large as possible to minimize nonlinear effect while maintaining single transverse mode. Multi-modes are associated with different wave propagation constants resulting in modal dispersion that cannot be compensated. Coherent combination of fiber amplifiers, whether in CW or short pulse operation, requires uniform phase front of each element thus single-mode is necessary.

Single-mode fiber designs with step-index or graded-index profiles are limited to 15 to 20 μm. Photonic crystal fibers[6] (PCF) is an important breakthrough in recent fiber technology and become one of the key components for short pulse lasers. PCF employs special design using periodic index modulation in the fiber to achieve single mode at much larger core sizes. Commercial PCF fibers up to 50 to 60 μm core diameters are now available. Note scaling of fiber power (more so on peak power since fiber is more limited by peak rather than average power) is proportional to core area so for example increase of diameter from 10 μm to 50 μm results in 25-fold increase of area. Core diameter of PCF as large as 105 μm is reported and is a key element for the record peak power above GW.[7]

3.2. *Double clad fiber*

The invention of double-clad fiber enables high output power and high efficiency in fiber-based lasers and amplifiers. Figure 3 illustrates geometric structure of a double-clad fiber and exchange between pump light and signal light. Signal is amplified in the single-mode core surrounded by the inner cladding. This inner clad to the signal is in-turn the core of the pump light which is confined with the outer clad. Both signal and pump are guided along the entire length of this double-clad structure, allowing long interaction length for signal to extract pump power at a high fraction from 80% to 90% or even above.

The larger pump core, 100 μm and above, is significant in collecting broad-area laser diodes for pump power up to hundreds of watts. The double clad fiber is seen as a brightness up-converter that the core delivers single-mode diffraction limited output from lower intensity pump light. A fiber laser can further scale power by utilizing multiple of such fiber-coupled pump modules with pump combiner.

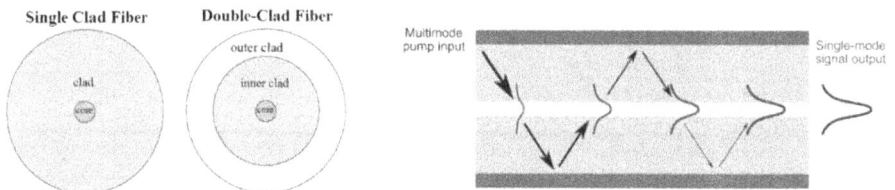

Fig. 3. Illustration of double clad fiber and pump-signal exchange.

Another source of high efficiency for fiber laser is the high quantum efficiency of active laser ion in the host material. Most of the high power fiber lasers operate at 1030 nm or 1050 nm using ytterbium-doped glass fiber. Yb3+ ion absorbs pump light at 940 nm yielding quantum efficiency at 90% level.

3.3. *Laser diode as pump source, fiber laser integration*

The overall high wall-plug efficiency in fiber lasers are also attributed to semiconductor laser technology. The technology has matured that an active emitter of a few micron wide and less than a micron thick can produce a few watts of power, reliably for tens of thousands of hours. Reliability can be prolonged at de-rated power levels. Due to monolithic fabrication process, elements can be aggregated for power output can be fabricated economically. Volume production has allowed a cost factor around $5 per watt or less for multi-mode pumps which are used as most fiber laser sources. Bandgap engineering of semiconductor laser has yielded electrical to optical efficiency at 60% to 70% level, including fiber coupling loss.

Fiber laser completely removes free-space optics thanks to the availability of all other fiber components such as fiber pump combiners, fiber Bragg gratings, dispersion compensators, etc. Assembly employs only fiber splicing procedure. This allows stable, compact operation in modular form which can then be used for next level scaling, the necessary arrangement of the proposed large system by coherent combination.

3.4. *Fiber laser efficiency of power conversion and heat dissipation*

It is important to note that high extraction and quantum efficiency in active fiber greatly reduce the thermal generation in the laser amplification process. The unit-length thermal load is further reduced by the long fiber length that is typically 1 meter to 10 meters. Heat dissipation is facilitated by the large surface area to volume ratio with fiber. Fiber diameters are typically 100 μm to 500 μm, compared to laser crystal of millimeters or centimeters. Favorable geometrical factors for heat remove in fiber allows fiber laser to operate at much higher average power than laser crystal counterpart.

Overall wall-plug power efficiency of fiber laser is around 40%, the composite of the various parts as discussed above.

3.5. *Nonlinearity in ultrafast fiber lasers*

Fiber is a double-sided sword as its application in short pulse amplification is concerned. Guided structure allows concentration of intensity and long interaction length between the signal and pump to drive high efficiency and favorable thermal management. But ps and fs laser pulses peak up power many orders of magnitude and so is intensity. Ultrashort pulses will accumulate nonlinear phase quickly even with small nonlinearity index n_2 in silica glass ($\sim 3 \times 10^{-16}$ m^2/W).

The dominate nonlinear effect here is the phase accumulation φ_{nl} due to self-phase modulation (SPM) which can be expressed as,

$$\varphi_{nl} = \gamma P L ,$$

where γ relates to fiber parameters including n_2 and mode area A_{eff} by $\gamma = 2\pi n_2 / \lambda A_{eff}$, and P is instantaneous power or peak power and L is fiber length.

Certain fiber length, typically from 1 meter to 10 meters, is needed to take the advantages of the fiber scheme. Also there is certain doping level limit of the active ions without adverse effects. Fiber length contributes proportionally to SPM.

Peak power can be reduced by chirped pulse amplification[8] (CPA) by 3 to 4 orders of magnitude in stretching the pulses up to a few ns long.

Geometrically γ can be decreased by larger effective area A_{eff}. Large mode area fibers (LMF) are possible with the advent of photonic crystal fibers as mentioned before. Increase of A_{eff} can be 1 to 2 orders using LMF compared to standard fiber.

Figure 4 provides a snapshot of the research works on state-of-the-art ultrafast fiber lasers. The vertical axis is peak power, and the horizontal axis is effective area of fiber core. While peak power can be increased with larger effective area, data points show results are along an intensity line at 10 GW/cm². This trend indicates the nonlinear limiting factor imposed from SPM. Further increase of fiber diameter is getting more difficult at 1 μm. For example, 100 μm is already 100 times of the wavelength because it is hard to discriminate higher modes at this size. In addition, more complex design in PCF has to deal with delicate fabrication and assembly.

CPA Fiber Amplifiers (with stretched pulses in fiber)

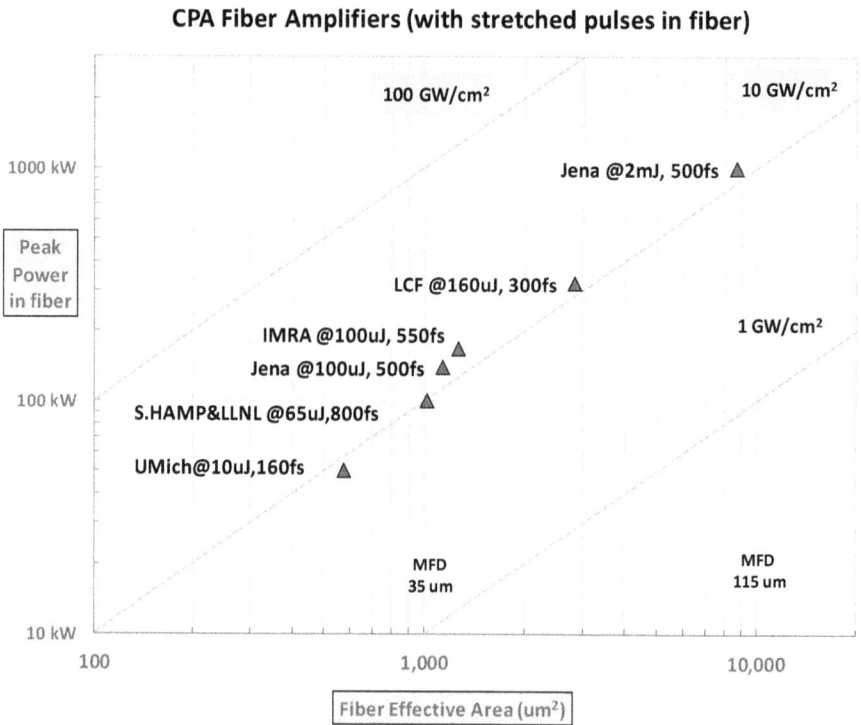

Fig. 4. Research works on femtosecond fiber lasers. While peak power can be increased with larger effective area, data points show results are along an intensity line at 10 GW/cm².

3.6. *Fiber summary*

Advantageous fiber attributes allow high efficiencies in energy-conversion and heat removal. However, ultrafast amplification with individual fiber device is more restrained due to its peak power limit than average power. This barrier to reach fiber laser potential can be overcome by combining individual elements coherently in a large assemble. The coherent combination scheme enables the next window for extending peak- and average-power performance of ultrafast fiber lasers.

4. Coherent Combination

4.1. *Scheme and progress*

An in-depth review and the state-of-the-art of the subject is provided in a presentation in this workshop.[5] A brief description is provided here. Coherent beam combination (CBC) scheme gathers a large number of elemental amplifiers spatially and synchronize each element in optical phase. Coherently combined beams act like a single laser and the constituent element is part of a large wave front which has a simple phase structure. With such property, the combined beams can focus at or near diffraction limit, enabling multi-fold increase of intensity over individual element proportionally to the number of beams.

As phase coherence is the key factor in CBC, each element must have a uniform phase front. This requires individual fiber element operate in single mode as discussed earlier.

Progress of this research frontier is impressive. The latest result was presented in this workshop on successful coherent combination of 61 lasers, successively from previous 7, 19, 34 beam combinations. Optical phase of each beam is actively lined up with the phase of a reference beam. The advanced control loop is capable of controlling optical phase variation due to fiber length fluctuation far less than operating wavelength at 1 μm.

4.2. *System engineering*

Considering engineering aspects in system integration of a (large) coherent system, there are a number of areas that can be foreseen based on known fiber laser attributes.

First, light guiding in fiber leads to fast build-up of amplified spontaneous emission (ASE) if not handled properly. ASE is accumulated in the background and fills in the lapse between pulses, decreasing contrast ratio between pulse and noise. This also deteriorate detector system for control feedback. ASE can be reduced by inserting isolator and band-pass filter between amplification stages but at the expense of system complexity.

Secondly, as CBC requires sub-optical-cycle precision, temperature has a large effect due to the long length of fiber, even though temperature coefficient of optical index is only a few ppm per °C. Optical path variation is given by $\Delta n(T) \cdot L / \lambda$ which produces a few optical cycle for 1 meter of fiber over 1°C of temperature variation. Temperature stabilization is helpful at least in easing feedback control. Given the sealed fiber structure, water-cooling of fiber does not interfere optical path and can be used with well-controlled chiller.

Last but not least, any large system has different characteristics in reliability from individual element. Convolution of reliability between element and system needs to be well understood and engineered. Fiber components and pump diodes have in high reliability only when free from fabrication defect or faulty process. Reliability engineering and control can ensure successful system integration. There are well-established industrial processes that can be utilized.

5. Towards CAN Targets for Laser Particle Acceleration

The call[3] for femtosecond laser systems of having both high peak power and average power presents opportunities and challenges. This survey study aims to provide an analysis of fiber laser as the promising candidate. To put the analysis in a larger reference frame, following diagram depicts the various steps of technology and research towards the target levels that are suitable for laser-driven plasma acceleration and to summarize this study.

The diagram is presented in peak power measures. Average power is the natural result when peak power stacks up, since fiber lasers are more constrained in peak power. Requirement of single-mode beam quality is assumed in the discussion.

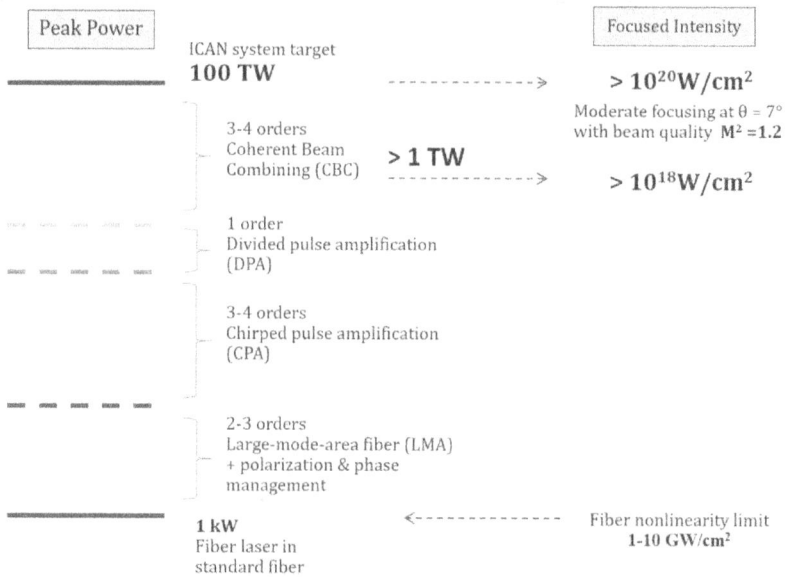

Fig. 5. Pathway to laser systems with high peak and average power.

There are 11 orders of magnitudes of peak-power increase, starting from nonlinearity limit 1 kW level when a standard single-mode fiber is used. These leaps are listed as followed:

(1) By enlarging the core size of the fiber while maintaining single-mode, 2 to 3 orders of magnitude increase can be obtained on fiber design.

(2) Chirped pulse amplification is applied to gain 3 to 4 orders.

(3) Divided pulse amplification,[8] a temporal coherent method, can advance another 1 order.

(4) Coherent beam combination[4,5,9–13] of amplifying elements is projected to lift 3 to 4 orders.

Target level 100 TW can be approached when these techniques are collectively used. Even with 1 TW of power such system can deliver intensity level greater than

10^{18} W/cm^2 that can accelerate electron to relativistic speed, at the desired higher repetition rates.

Further compressing output pulses from the system output by an order or more can increase peak power by the same magnitude.[5]

Acknowledgments

The author is grateful to Toshiki Tajima and Jonathan Wheeler for stimulating discussions on the topic.

References

1. T. Tajima and J. M. Dawson, Laser electron accelerator, *Phys. Rev. Lett.* **43**, 267–270 (1979).

2. S. M. Hooker, Developments in laser-driven plasma accelerators, *Nat. Photonics* **7**, 775 (2013) and references therein.

3. G. Mourou *et al.*, The future is fibre accelerators, *Nat. Photonics* **7**, 258 (2013) and references therein.

4. A. Heilmann *et al.*, Coherent beam combining of seven fiber chirped-pulse amplifyiers using an interferometric phase measurement, *Opt. Express* **26**, 31542–31553 (2018).

5. J. Wheeler *et al.*, Thin film compression and CAN laser experimental results, paper in this workshop publication.

6. P. Russell, Photonic crystal fibers, *Science* **299**(5605), 358–362 (2003) and references therein.

7. T. Eidam *et al.*, Fiber chirped-pulse amplification system emitting 3.8 GW peak power, *Opt. Express* **19**, 255–260 (2011).

8. D. Strickland and G. A. Mourou, Compression of amplified chirped optical pulses, *Opt. Commun.* **55**, 447–449 (1985).

9. F. Guichard *et al.*, High-energy chirped- and divided-pulse Sagnac femtosecond fiber amplifier, *Opt. Lett.* **40**, 89–92 (2015).

10. J. Le Dortz *et al.*, Highly scalable femtosecond coherent beam combining demonstrated with 19 fibers, *Opt. Lett.* **42**, 1887–1890 (2017).

11. A. Klenke *et al.*, Coherently combined 16-channel multicore fiber laser system, *Opt. Lett.* **43**, 1519–1522 (2018).

12. L. Siiman *et al.*, Coherent femtosecond pulse combining of multiple parallel chirped pulse fiber amplifiers, *Opt. Express* **20**, 18097 (2012).

13. T. Zhou, T. Sano, and R. Wilcox, Coherent combination of ultrashort pulse beams using two diffractive optics, *Opt. Lett.* **42**(21), 4422–4425 (2017).

Demonstration of Thin Film Compression for Short-Pulse X-ray Generation

D. M. Farinella,* M. Stanfield, N. Beier, T. Nguyen, S. Hakimi, T. Tajima, F. Dollar

Department of Physics and Astronomy, University of California,
Irvine, CA 92697, USA
https://uci.edu
* dfarinel@uci.edu

J. Wheeler and G. Mourou

Department of Physics, École Polytechnique,
91128 Palaisaeu Cedex, France

Thin film compression to the single-cycle regime combined with relativistic compression offers a method to transform conventional ultrafast laser pulses into attosecond X-ray laser pulses. These attosecond X-ray laser pulses are required to drive wakefields in solid density materials which can provide acceleration gradients of up to TeV/cm. Here we demonstrate a nearly 99% energy efficient compression of a 6.63 mJ, 39 fs laser pulse with a Gaussian mode to 20 fs in a single stage. Further, it is shown that as a result of Kerr-lensing, the focal spot of the system is slightly shifted on-axis and can be recovered by translating the imaging system to the new focal plane. This implies that with the help of wave-front shaping optics the focusability of laser pulses compressed in this way can be partially preserved.

Keywords: Thin film compression; nonlinear optics; wakefield acceleration; relativistic compression.

1. Introduction

In the past few decades, the Nobel Prize winning technology chirped pulse amplification (CPA)[1] has enabled the physical realization of plasma wakefield acceleration[2] at gaseous plasma densities culminating in electron acceleration gradients on the order of GeV/cm.[3] Recently, new developments in laser technology such as thin film compression (TFC)[4] and relativistic compression[5] provide a route[6] to generating the X-ray laser pulses needed for driving wakefields in solid density plasmas which have the potential to generate acceleration gradients as high as TeV/cm.[7-10] In order to generate the X-ray pulses needed for solid state wakefield accelerators, relativistic compression requires conventional laser pulses with durations in the single-cycle regime. TFC has been proposed for the compression of 25 fs laser pulses to the single-cycle limit without a restriction on pulse energy.[6]

Similar to conventional pulse compression techniques[11-14] TFC utilizes self-phase modulation (SPM) and chirp compensation to compress laser pulses. SPM arises from the Kerr effect where the index of refraction is modified by the local intensity of the laser pulse where $n(r,t) = n_0 + n_2 I(r,t)$. Here n is the index of refraction, n_0 is the linear index of refraction, n_2 is the nonlinear index of refraction, and $I(r,t)$ is the local intensity as a function of space and time. SPM broadens the laser spectrum resulting in a new Fourier transform limit (FTL) and imparts a spectral phase which is roughly quadratic in the case of a Gaussian temporal profile. Chirp compensation can then be used to remove the roughly quadratic spectral phase, bringing the laser pulses to the FTL.[15,16]

Unlike conventional pulse compression techniques such as that of gas-filled capillaries, TFC allows for the compression of pulses with energies far beyond $\sim \mathcal{O}(1\,\mathrm{mJ})$. TFC proposes using intense $\sim \mathcal{O}(1\,\mathrm{TW/cm^2})$ flat-top laser pulses which enables one to utilize a thin medium to induce uniform spectral broadening across the beam profile. Furthermore, since the process is unguided, there is no restriction on beam diameter and the nonlinear material can be placed at Brewster's angle thus allowing this technique to be very energy efficient.

In this work TFC pulse compression as well as the focusability of a Gaussian beam after nonlinear broadening is investigated. Section 2 briefly describes the experimental layout. Section 3 presents the spectral broadening and pulse compression results. Finally, Sec. 4 investigates the focusability of the compressed laser pulses.

2. Experimental Setup

A 1 kHz, 6.63 mJ, 39 fs, 800 nm laser with $1/e^2$ diameter of 11.7 mm and a Gaussian mode is used in this experiment. These laser parameters represent a peak intensity of $\sim 0.3\,\mathrm{TW/cm^2}$. The pulse duration of the laser output is continuously variable by means of a grating compressor which allows varying amounts of group delay dispersion (GDD) to be added or removed from the laser pulses. An overview of the experimental layout can be seen in Fig. 1.

The laser pulses first travel through two pieces of 4 mm thick fused silica windows on flip mounts to broaden their spectrum. Then the laser pulses are attenuated by means of a wedge where the front reflection is sent to two chirped mirrors (Ultrafast Innovations - HD58) compensating -275 fs² each. A second removable mirror then sends the laser to a second wedge for further attenuation before sending the laser to the diagnostics. The front reflection of the second wedge is sent to a home built scanning second-harmonic generation frequency resolved optical gating (SHG FROG) device to measure pulse duration, while the reflection from the back surface is sent to a focal spot camera to monitor changes in wavefront. The laser is focused by a f = 122 mm achromat and a 40× PLAN achromat objective is used to collimate the beam to re-image the focal spot onto the surface of a silicon CMOS camera with a second f = 122 mm achromat.

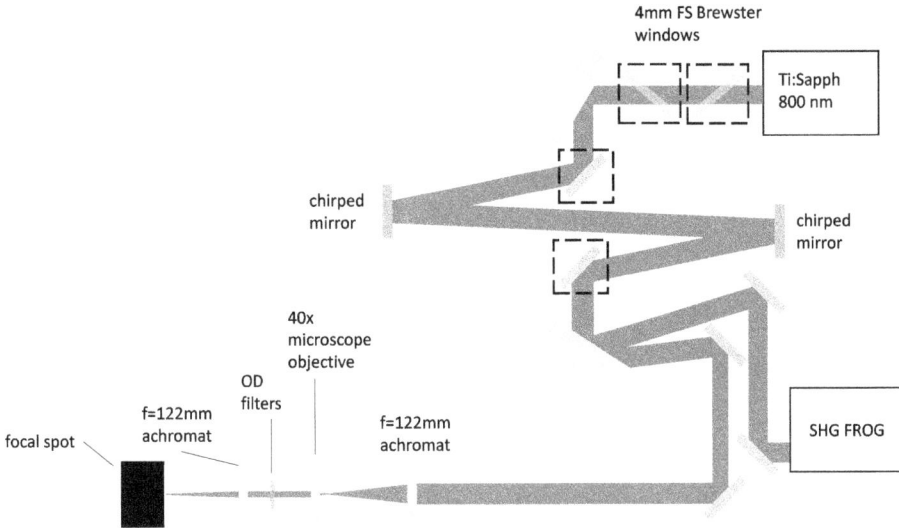

Fig. 1. Experimental layout of the pulse compression experiment and beam path leading to the laser diagnostics. Dotted lines encompass optics that are on flip mounts or kinematic bases and are therefore removable. All steering mirrors except for wedges and chirped mirrors have gold (Au) coatings.

3. Pulse Compression

Laser pulses are broadened in 2×4 mm fused silica (FS) windows oriented at Brewster's angle such that they compensate each others Snell shift, providing ~ 9.7 mm of refracted path length. In this regime the fused silica thickness required for significant SPM also leads to the accrual of non-negligible amounts of GDD due to the relatively thick material. Therefore the condition under which maximum spectral broadening occurs is found by adjusting the initial GDD to maximize the spectral broadening. With the fused silica windows in the beam, the compressor is adjusted to provide the broadest spectrum as measured by a spectrometer (Ocean Optics - Flame) shown in red in Fig. 2 (a).

The fused silica windows are then removed from the beam and the initial spectrum is measured as seen in blue in Fig. 2 (a). The spectral broadening represents a decrease in the FTL from 29 fs to 18 fs. Further, the energy throughput of the Brewster window pair is measured to be 6.55 mJ out of an initial 6.63 mJ ($\sim 99\%$) which is a significantly more efficient spectral broadening than offered by other pulse compression schemes using fibers or gas-filled capillaries.[17]

The compressed laser pulses were found to have a FWHM duration of 20 fs as seen in red in Fig. 2(b), which is very close to the FTL of 18 fs. The red-dotted line in Fig. 2(b) shows that the remaining phase is largely quadratic, implying that further compression toward the FTL is possible.

Initial laser pulses were measured by removing the FS Brewster windows and

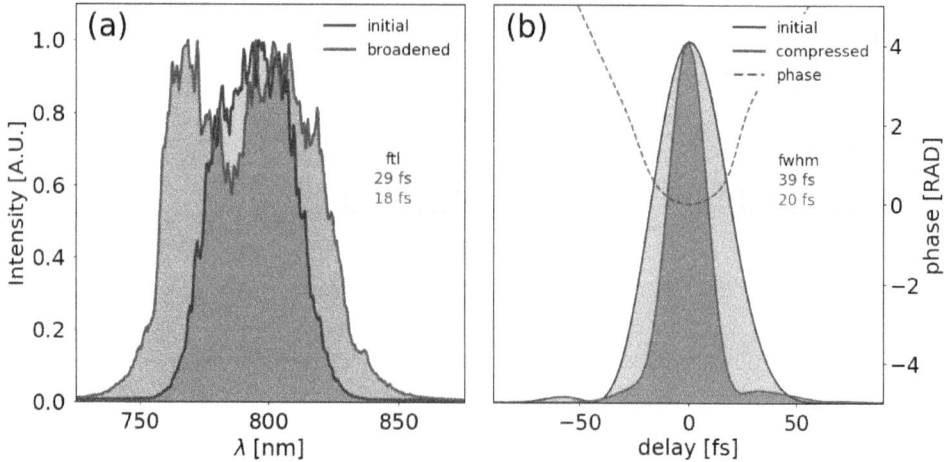

Fig. 2. The laser spectrum (a) before (blue) and after (red) nonlinear spectral broadening, and the laser pulse duration (b) before (blue) and after (red) pulse compression. The red-dotted line in (b) represents the measured phase of the compressed pulses showing that further compression is possible by tuning GDD compensation. The FTL of the laser spectra and the FWHM of the laser pulse duration is noted in (a) and (b) respectively.

removable Au mirrors. After tuning the compressor to the shortest pulse possible, the initial laser pulses are simply reflected by the first and second wedge and sent directly into the SHG FROG. The FWHM duration of the initial pulse was found to be 39 fs. An energy efficiency of $\sim 99\%$ and a laser pulse compression from 39 fs to 20 fs represents a power amplification of nearly a factor of 2 of the central part of the beam where the intensity is roughly constant.

4. Focusability

When utilizing the Kerr nonlinearity in an unguided medium for pulse compression, one must also consider wave-front effects. SPM of a laser pulse with a Gaussian temporal intensity profile leads to a broadened spectrum with a largely quadratic spectral phase. This allows for compression using optics that provide quadratic phase of the opposite sign (i.e. chirped mirrors) as demonstrated in Sec. 3. However, if there are beam-scale intensity gradients in space (i.e. a Gaussian mode) one must also contend with radial changes in the index of refraction (i.e. Kerr-lensing). This complication is avoided in the initial TFC proposal[4] by requiring that the mode be flat-top, which would minimize the effect of Kerr-lensing and induce uniform spectral broadening across the laser profile. Other recent efforts using concave optics have also demonstrated that uniform spectral broadening can be achieved with a Gaussian beam profile.[18–20] The scope of this section is to address the effects of induced wave-front deviations when using a Gaussian beam in TFC pulse compression.

In this experiment, the focal spot is measured before and after nonlinear pulse

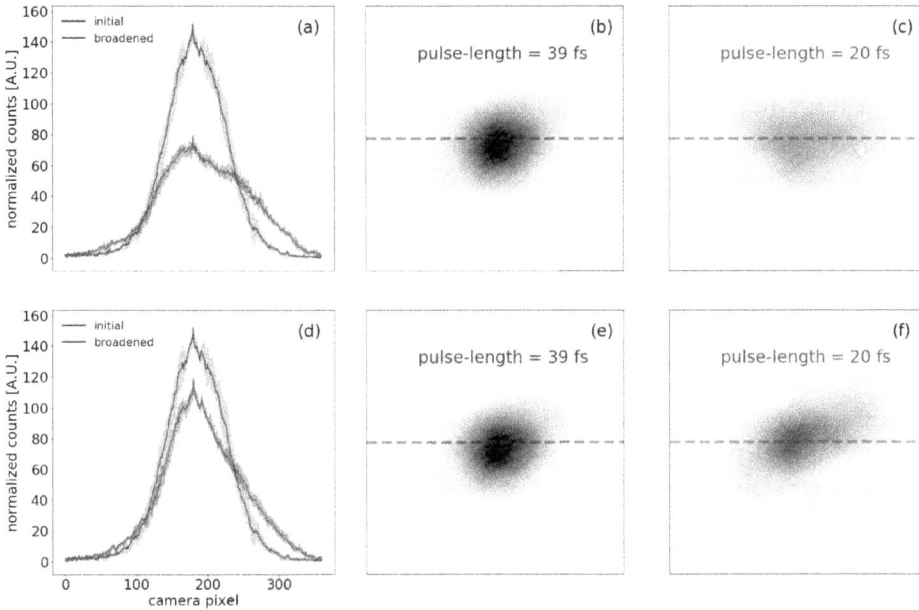

Fig. 3. Average horizontal focal spot line-out at the max pixel value (a) for initial (blue) and compressed (red) pulses and sample focal spots of the (b) initial and (c) compressed pulses in the initial focal plane. The shading in (a) represents the standard deviation of all compressed (light red) and initial (light blue) focal spot line-outs. A second set of plots (d), (e) and (f) show a similar comparison after the 40X objective was shifted on-axis to the new focal plane after Kerr lensing. The overlays of the line-out location on the initial focal spot (b), (e) and the spot after compression (c), (f) are indicated by dotted lines.

compression. To characterize the initial focal spot, the FS Brewster windows are removed from the beam and the laser pulses are sent through the rest of the optical system depicted in Fig. 1 with the removable Au mirrors in the path. The 40X PLAN achromat is aligned and translated to the focal plane of the first f = 122 mm achromat. The focal spot is then imaged onto the camera after being attenuated by an OD2 filter. Five image acquisitions were taken of the unbroadened focal spot with a gain of 1, and an integration time of 180 ms without the FS Brewster windows. These camera settings were then used for all subsequent measurements. A typical image can be seen in Fig. 3 (b). The FS Brewster windows are then inserted into the beam and the beam is probed in the same plane, showing apparent focal spot degradation which can be seen in Fig. 3(c). Figure 3(a) shows the average horizontal line-out at the maximum pixel value (as indicated by dotted lines in Figs. 3(b) and 3(c)) of the integrated beam profile before and after nonlinear pulse compression. Here the light colored shading represents the standard deviation of the 5 camera images. The peak counts are seen to decrease from ~ 152 to ~ 79 ($\sim 52\%$ of it's initial value).

However, by simply translating the objective closer to the first achromat we can image the new focal plane to the camera. Again, an example image of the focal spot in the original focal plane before the FS Brewster windows were inserted and an image of the focal spot in the new focal plane can be seen in Figs. 3(e) and 3(f) respectively. As before, a line-out of the integrated beam profile at the peak pixel value before and after nonlinear pulse compression can be seen in Fig. 3(d) showing that the peak counts of the focal spot actually only decrease from ~ 152 to ~ 119, which is $\sim 78\%$ of its initial peak counts.

5. Conclusion

TFC pulse compression of 6.63 mJ, 39 fs laser pulses to 20 fs with $\sim 99\%$ energy efficiency has been demonstrated which represents a power amplification of the central part of the Gaussian beam by nearly a factor of two. Further, it has been demonstrated that TFC on laser systems with a Gaussian mode can be optimized by taking into account the Kerr-lens induced by the nonlinear interaction. Further improvements on focal spot quality using Gaussian laser systems will likely come from the use of wave-front shaping optics such as deformable mirrors. The high efficiency pulse compression demonstrated here showcases the the utility of TFC in approaching the single-cycle regime for the generation of X-ray laser pulses needed to drive laser wakefields at solid densities.

Acknowledgments

The authors would like to thank Sergey Mironov for insightful discussions.

References

1. D. Strickland and G. Mourou, Compression of amplified chirped optical pulses, *Optics Communications* **56**, 219 (1985).
2. T. Tajima and J. Dawson, Laser electron accelerator, *Physical Review Letters* **43**, p. 267 (1979).
3. W. Leemans, B. Nagler, A. Gonsalves, C. Toth, K. Nakamura, C. Geddes, E. Esarey, C. Schroeder and S. Hooker, GeV electron beams from a centimetre-scale accelerator, *Nature Physics* **2**, 696 (2006).
4. G. Mourou, G. Cheriaux and C. Raider, Device for generating a short duration laser pulse (July 2009), US Patent 20110299152 A1.
5. N. Naumova, J. Nees, I. Sokolov, B. Hou and G. Mourou, Relativistic generation of isolated attosecond pulses in a lambda 3 focal volume, *Physical Review Letters* **92**, p. 063902 (2004).
6. G. Mourou, S. Mironov, E. Khazanov and A. Sergeev, Single cycle thin film compressor opening the door to Zeptosecond-Exawatt physics, *European Physical Journal Special Topics* **223**, 1181 (2014).
7. T. Tajima and M. Cavenago, Crystal x-ray accelerator, *Physical Review Letters* **59**, 1440 (September 1987).
8. T. Tajima, Laser acceleration in novel media, *European Physical Journal Special Topics* **223**, 1037 (May 2014).

9. X. Zhang, T. Tajima, D. Farinella, Y.-M. Shin, G. Mourou, J. Wheeler, P. Taborek, P. Chen, F. Dollar and B. Shen, Particle-in-cell simulation of x-ray wakefield acceleration and betatron radiation in nanotubes, *Physical Review Accelerators and Beams* **19**, p. 101004 (October 2016).

10. S. Hakimi, T. Nguyen, D. Farinella, C. K. Lau, H.-Y. Wang, P. Taborek, F. Dollar and T. Tajima, Wakefield in solid state plasma with the ionic lattice force, *Physics of Plasmas* **25**, p. 023112 (February 2018).

11. C. V. Shank, R. L. Fork, R. Yen, R. H. Stolen and W. J. Tomlinson, Compression of femtosecond optical pulses, *Applied Physics Letters* **40**, 761 (1982).

12. C. Rolland and P. B. Corkum, Compression of high-power optical pulses, *JOSA B* **5**, 641 (1988).

13. M. Nisoli, S. De Silvestri and O. Svelto, Generation of high energy 10 fs pulses by a new pulse compression technique, *Applied Physics Letters* **68**, 2793 (1996).

14. M. Nisoli, S. De Silvestri, O. Svelto, R. Szipcs, K. Ferencz, C. Spielmann, S. Sartania and F. Krausz, Compression of high-energy laser pulses below 5 fs, *Optics Letters* **22**, 522 (1997).

15. D. M. Farinella, J. Wheeler, A. E. Hussein, J. Nees, M. Stanfield, N. Beier, Y. Ma, G. Cojocaru, R. Ungureanu, M. Pittman, J. Demailly, E. Baynard, R. Fabbri, M. Masruri, R. Secareanu, A. Naziru, R. Dabu, A. Maksimchuk, K. Krushelnick, D. Ros, G. Mourou, T. Tajima and F. Dollar, Focusability of laser pulses at petawatt transport intensities in thin-film compression, *JOSA B* **36**, A28 (February 2019).

16. S. Y. Mironov, V. N. Ginzburg, I. V. Yakovlev, A. A. Kochetkov, A. A. Shaykin, E. A. Khazanov and G. A. Mourou, Using self-phase modulation for temporal compression of intense femtosecond laser pulses, *Quantum Electronics* **47**, p. 614 (August 2017).

17. S. Bohman, A. Suda, T. Kanai, S. Yamaguchi and K. Midorikawa, Generation of 5.0 fs, 5.0 mJ pulses at 1 kHz using hollow-fiber pulse compression, *Optics Letters* **35**, 1887 (2010).

18. S. Mironov, V. Lozhkarev, G. Luchinin, A. Shaykin and E. Khazanov, Suppression of small-scale self-focusing of high-intensity femtosecond radiation, *Applied Physics B* **113**, 147 (October 2013).

19. S. Mironov, P. Lassonde, J.-C. Kieffer, E. Khazanov and G. Mourou, Spatially-uniform temporal recompression of intense femtosecond optical pulses, *The European Physical Journal Special Topics* **223**, 1175 (May 2014).

20. P. Lassonde, S. Mironov, S. Fourmaux, S. Payeur, E. Khazanov, A. Sergeev, J.-C. Kieffer and G. Mourou, High energy femtosecond pulse compression, *Laser Physics Letters* **13**, p. 075401 (2016).

Laser-Wakefield Application to Oncology

B. S. Nicks*, T. Tajima and D. Roa

University of California, Irvine, CA 92697, USA
* bnicks@uci.edu

A. Nečas

TAE Technologies, Inc., Foothill Ranch, CA 92610, USA

G. Mourou

École Polytechnique, Palaiseau, 91128, France

Recent developments in fiber lasers and nanomaterials have allowed the possibility of using laser wakefield acceleration (LWFA) as the source of low-energy electron radiation for endoscopic and intraoperative brachytherapy, a technique in which sources of radiation for cancer treatment are brought directly to the affected tissues, avoiding collateral damage to intervening tissues. To this end, the electron dynamics of LWFA is examined in the high-density regime. In the near-critical density regime, electrons are accelerated by the ponderomotive force followed by an electron sheath formation, resulting in a flow of bulk electrons. These low-energy electrons penetrate tissue to depths typically less than 1 mm. First a typical resonant laser pulse is used, followed by lower-intensity, longer-pulse schemes, which are more amenable to a fiber-laser application.

Keywords: Brachytherapy; fiber lasers; laser wakefield; electron beams; critical density; sheath physics.

1. Introduction

The treatment of cancer remains one of the most pressing concerns of medical research. One promising avenue of cancer treatment is brachytherapy, in which a source of radiation is brought inside the body close to the tissues requiring treatment.[1] This technique localizes the radiation dose to source of radiation, limiting collateral damage to surrounding healthy tissue. In contrast, more conventional external sources of radiation can cause significant damage to intervening tissues. Typically, small quantities of radioisotopes provide the dose source for brachytherapy, but such sources suffer from decay, which lengthens treatment times, and shielding costs. The use instead of an electron beam as the radiation source would eliminate the first challenge and greatly mitigate the second.

A potential means for generating such an electron beam is Laser Wakefield Acceleration (LWFA),[2] a compact method of accelerating electrons to high energies which was first proposed by Tajima and Dawson[3] in 1979. While conventional linear accelerators (LINAC), relying on large and costly wave-guide cavities, are limited in acceleration gradient by the breakdown threshold of metals, the strong electric fields accessible in plasmas (which are already broken-down) allow much higher accelerating gradients: GeV per cm or higher. Such a high acceleration gradient reduces the necessary machine size and also increases their availability through reduction in cost. Typical electron energies needed for brachytherapy lie in the range between 100 keV and 5 MeV, and thus LWFA can accelerate electrons to these energies in micron to millimeter scale lengths. LWFA was demonstrated experimentally[4,5] after the invention of Chirped Pulse Amplification (CPA),[6] which allowed access to the necessary high intensity laser regime. Since then, many more experiments have demonstrated this technique in different regimes, and the field has been growing steadily.

Research in the use of LWFA to generate electron beams for medical applications has proceeded for at least two decades now. Initially, these efforts focused on generation of high-quality electron beams with energies roughly in the range 6–25 MeV, as would be applicable for conventional, external sources of radiation for cancer therapy.[7–13] Recent innovations in the field of fiber lasers has offered a new leap forward in this effort: the Coherent Amplification Network (CAN),[14] in which many individual micron-scale fiber lasers are coherently combined and amplified to provide both high-rep rate and high power. This innovation allows medical LWFA to proceed into a new regime of applications, though a CAN laser certainly is applicable as well to the traditional medical effort at producing an external electron beam.

As we see below, we find that even at very modest intensity, LWFA can produce electrons that are relevant for tissue penetration and delivery of beams of ionizing radiation. With this insight, in combination with the compactness afforded by the recent developments in fiber lasers, we are led to consider the new situation of *in situ* radiation sources (of electrons). In this vision, we see three chief schemes in which the wakefield electron source could be brought directly to the cancer. First, the laser-wakefield accelerator could be inserted in an intraoperative fashion,[10,11] which involves surgically opening the intervening tissues and can presently be used for LINAC sources in some instances. A surgeon may also use such an operation to remove any residual cancer or clean affected tissues by hand. Less invasive is brachytherapy, where the laser is injected discreetly into the body, such as through a blood vessel or directly through tissue. Finally, it may be possible to carry the laser into the body in a endoscope, whereby the surgeon could potentially both diagnose and treat the cancer simultaneously. It is our goal here to devise a way in which any or all of these methods might be possible with LWFA.

In each of these cases, the electrons beam need only have shallow penetrative power, as they need not traverse the body before reaching the tissues to be treated. The desirable energy for an electron beam then reduces to the order of 10^2 keV. We thus seek a means of producing low-energy electrons. The electron energy gain from LWFA is given by $\Delta \mathcal{E} = 2g(a_0)m_e c^2 (n_c/n_e)$, where a_0 is the normalized laser intensity, $g(a_0)$ represents the function dependence of energy gain on a_0, n_c is the laser critical density, and n_e is the plasma density.[3] This relation suggests a path to low-energy electrons through a high plasma density and modest laser intensity, parameters that are also favorable to fiber lasers, as will be discussed. To provide a target material near the critical density of an optical laser ($n_c = 1.11 \times 10^{21}$ cm^{-3} for a laser wavelength of 1 micron), it may be best to use a porous nanomaterial,[15–17] such as porous alumina or carbon nanotubes, as is the main focus of this workshop. Such a target material for irradiation by the laser would also avoid the presence of ionized gas inside the body.

While the low-density, high-energy regime of LWFA has been studied extensively, the high-density, low-energy regime has been less explored in detail.[18] Indeed, the expression for electron energy gain given above was established by studying the low-density regime. As a foundation for these applications, the physics of the high-density regime of LWFA has been recently studied in Nicks *et al.*, 2019.[19] This work studied first the scaling laws of the electron energy gain for n_c/n_e and a_0. It was found that $g(a_0)$ is well-represented by the ponderomotive potential $g(a_0) = \sqrt{1 + a_0^2} - 1$. Additionally, it was found that $\Delta\mathcal{E} \propto n_c/n_e$, the proportionality predicted by low-density wakefield theory, was obeyed in the regime $n_c/n_e \gtrsim 1$ for the highest-energy electrons. Next, the mechanics of electron acceleration in the regime $n_c/n_e \sim 1$ were studied, revealing sheath acceleration[20] that generates a peculiar stream-like distribution of bulk electrons in phase space. This bulk acceleration by sheath contrasted significantly with the trapping and acceleration seen in low-density wakefield. This work then attempted to quantitatively distinguish this high-density, sheath regime from the more typical wakefield physics seen at low densities. Finally, this work briefly examined the penetrative power of such an accelerated spectrum of electrons in water, approximating biological tissue.

2. Acceleration in the High-Density Regime

To study laser-wakefield physics in the high-density regime, we use the 1D particle-in-cell (PIC) code EPOCH. An optical ($\lambda = 1$ micron) laser is injected from vacuum into a uniform slab of plasma with a temperature of 100 eV. First, we seek to understand the wakefield physics at high density with an approach that may be impractical with regard to fiber lasers but more easily understood in the context of typical wakefield physics. We first use a flat-top resonant laser pulse, which has a pulse length equal to half of the wakefield wavelength, and the plasma density is taken to be the laser critical density. The laser intensity is taken to be the somewhat powerful value of $a_0 = 1$. A resonant laser pulse in a plasma near the critical density ($n_c/n_e \leq 2$) must necessarily be sub-cycle, or at most single cycle. While sub-cycle lasers have been demonstrated experimentally,[21,22] such a setup would be generally difficult to implement, particular in a fiber laser.[14] Instead, a long, self-modulating pulse would be much more practical a fiber laser application, as is discussed in Sec. 3. It will be shown that with a longer pulse the essential physics remains unchanged.

The interaction of the laser with the plasma in the two contrasting regimes of high and low density has a convenient analogy in tsunami waves in a certain property of the wave dynamics. The phase velocity of the wake field matches the laser group velocity in the plasma ($v_g = c\sqrt{1 - n_e/n_c}$), much as sea-floor depth determines the phase velocity of an ocean wave. In the open ocean, where the water depth is great, tsunami waves propagate with a fast phase velocity and thus do not couple to stationary objects. Boats in the ocean, for instance, may move slightly in the transverse direction (vertically), but are not otherwise affected. The waves in this regime also do not couple to the sediment on the sea floor, and so remain "blue". An object such as a surfer could only be trapped and accelerated by these waves by exerting a great deal of effort to approach the wave phase velocity. So it is with wakefield: in the low-density regime, the wake phase velocity is near the speed of light, and while a small number of elections may be accelerated to high energies, the wake does not couple to the bulk motion of the plasma. The wake and accelerated electrons form a clean and coherent train.

Near the shore, however, the increasingly shallow water causes the phase velocity of the wave to slow down, which leads to amplification and steepening of the wave until breaking occurs. The slow velocity of the wave near the shore then causes catastrophic "trapping" of stationary objects. Additionally the slow wave velocity couples with turbulence created by wave breaking to create anomalous transport on the sediment bed. Significant amounts of sediment quickly pass into the wave, creating a visibly "black" tsunami from the clean, "blue", off-shore starting wave and leading to momentum transport of the sediment. [19,23] Similar physics occurs in wakefields near the critical density. The laser couples strongly to the bulk motion of the plasma, creating a qualitatively distinct regime that features sheath acceleration and bulk flow of electrons.

The typical approach to LWFA, which seeks to generate high-energy electrons, is to use a low-density plasma ($n_c/n_e \gg 1$). In this regime, which we may call the "blue wave" case ($v_g \approx c$), a train of coherent wake waves follows the laser pulse. This wake then accelerates a regular train of electrons. The wakefield, which is a longitudinal electric field, reaches a saturation value on the order of the Tajima-Dawson field [3,24]: $E_{TD} = m_e \omega_p c/e$, where ω_p is the plasma frequency. This state of affairs reigns until the laser has traveled either the dephasing length, at which electrons begin to decelerate, or the pump depletion length, at which the laser has lost significant energy to the plasma and can no longer excite a robust wakefield. The high group velocity of the laser leaves the bulk plasma intact and creates a "blue" wakefield.

In contrast, the case of $n_e = n_c$, the "black wave", shown in Fig. 1, exhibits quite different behavior. Here, $v_g = 0$, restricting the laser-plasma interaction to within one plasma wavelength. The long train of trapped electrons in the low-density case becomes replaced by streams of low-energy ($\Delta \mathcal{E} \sim 100$ keV) electrons ejected from the site of laser entry roughly every plasma period. These electrons are accelerated by an oscillating sheath that is formed by the laser at the initial boundary of the domain. This behavior is somewhat reminiscent of laser interaction with a solid target and previous sheath acceleration efforts.[20] However, some diminished part of the laser is still able to propagate through the target, and the ions have essentially no response. The "blackness" of the wave derives from its strong coupling to the bulk electrons, which occurs because $v_g = 0$ for the laser, much as an ocean wave that slows down near the shore becomes turbulent and dredges sediment from the sea-floor. As n_e is decreased from the critical density, a transition between the "black" and "blue" regimes is seen, as is shown in Fig. 2 for the "grey" case of $n_c/n_e = 1.7$, which shows elements of the bulk flow of high-density wakefield as well as elements of more typical, low-density wakefield. These sheath mechanics may also be useful for the understanding of related ion acceleration. dynamics[2,25]

The maximum electron energy in the phase-space distribution shown in Fig. 1 is roughly 1 MeV, which is encouraging for the medical aims in this study. A naïve next step to increase the electron dose and alleviate the difficulty with a sub-cycle pulse might then be to increase the length of the laser pulse to produce a larger dose of electrons. Such a case is shown in Fig. 3 for $n_c/n_e = 1$ and a laser pulse length of $\lambda_{lp}/\lambda_p = 8$ for a snapshot at time $t/\tau_p = 96$, where $\tau_p = 2\pi/\omega_p$ is the plasma period. At this time the laser interaction has finished, as well as most of the electron acceleration, and thus this snapshot represents the near-final behavior. The overall interaction with the plasma is much more violent. For $x/\lambda_p \gtrsim 32$ a large body of overlapping electron streams is seen, which originate in the initial laser interaction. For $x/\lambda_p \approx 2$ a longitudinal oscillation similar to that in the resonant pulse case is formed which generates subsequent electron streams. This scheme is also more energetically efficient than that of Fig. 1, indicating that both the quantity and quality of low-energy electron production is improved. However, this case relies on the relatively strong intensity of $a_0 = 1$, which is likely untenable for fiber lasers.

3. High-Density LWFA in Fiber Lasers

Fiber lasers, particularly in coherent networks, offer the potential for high-power and high-rep rate lasers. Yet, individual fibers are also subject to a number of constraints not shared by large, conventional lasers. Perhaps most importantly, the material properties of fibers, which are typically made from silica, place a much more stringent limit on the intensity of the transmitted laser pulse.[26] Certainly for any fiber application inside the body, it is paramount that material damage to the fibers be avoided, as this damage would then harm the surrounding tissue as well, with potentially severe consequences. In addition to the requirement of avoiding ionization of the fiber, an even more restrictive condition may be the level of tolerable accumulation of nonlinear phase in the laser pulse, the severity of which grows with laser intensity. Given this background from the field of fiber optics, a acceptable order of magnitude for laser intensity that safely avoids these issues may be

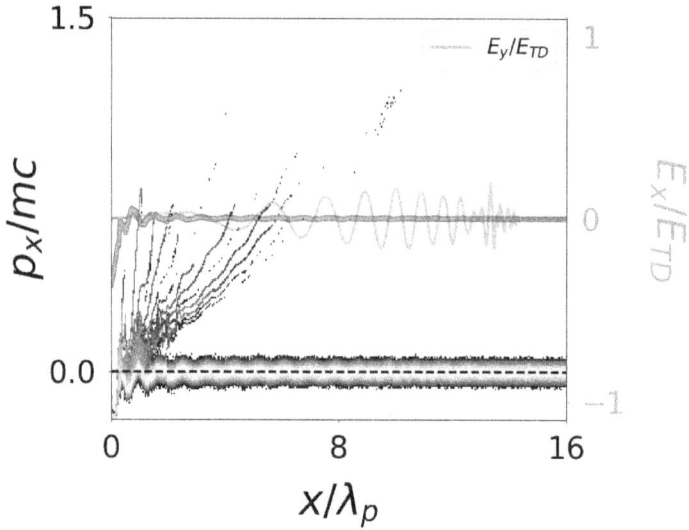

Fig. 1. A snapshot of the electron phase space p_x vs. x (heat-map, with warmer colors representing higher density) and longitudinal E_x (green) and laser E_y (translucent blue) fields for high-density ("black") case $n_c/n_e = 1$ with laser intensity $a_0 = 1$. The plasma wavelength is given by $\lambda_p = 2\pi c/\omega_p$. The time step is $t = 14.4\tau_p$, where τ_p is the plasma period. (Similar run in Nicks et al., 2019.[19])

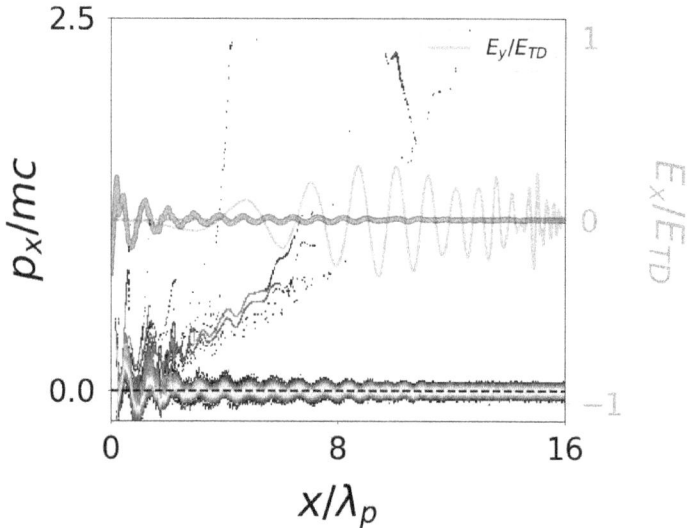

Fig. 2. The electron phase space and field structure of the intermediate ("grey") case of density $n_c/n_e = 1.7$ at the laser intensity $a_0 = 1$. The time step is $t = 16.0\tau_p$

10^{14} W cm^{-2}. We take this value as a starting point, but subsequent technical and material efforts may require a more conservative value. This intensity corresponds

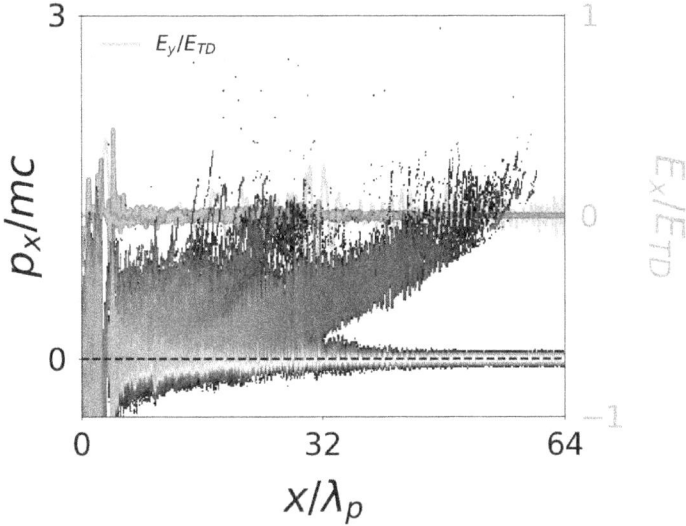

Fig. 3. The electron phase space and field structure of the critical density case $n_c/n_e = 1$ for a laser pulse of length $8\lambda_p$ at the laser intensity $a_0 = 1$, showing the "black tsunami" regime. At this snapshot ($t = 80.6\tau_p$) most of the electron acceleration is concluded and most of the laser has exited the domain. (Similar run in Nicks *et al.*, 2019.[19])

to $a_0 \approx 0.01$, which is far below the relatively powerful value of $a_0 = 1$ considered thus far. The expression for electron energy gain $\Delta\mathcal{E} = 2g(a_0)m_e c^2(n_c/n_e)$, where $g(a_0) = a_0^2/2$ for $a_0 \ll 1$ if $g(a_0)$ takes the form of the ponderomotive potential, suggests that one means of compensating for this low individual fiber laser intensity to attain 10^2 keV electrons is simply to use a coherent network of ~ 100 lasers, though in a endoscopic application this number may be difficult to achieve. Phase-matching each fiber would also be technically difficult, but may be possible making use of the partial reflection of the laser pulse at the boundary between the fiber and high-density material to be irradiated. However, by lowering the target material density, one can use fewer coherent lasers, such as with ~ 10 lasers at $n_c/n_e = 10$. Another alternative is the use of a hollow fiber, which can have a diameter of order 10^2 times larger than that of a typical fiber, thereby accommodating much more pulse energy.

Another chief concern with fiber lasers is the pulse length. While thus far the simulations in this work at $n_c/n_e \approx 1$ have used resonant laser pulse lengths of ~ 2 fs (a difficult, sub-cycle pulse in its own right), the shortest pulse practically achievable for a fiber laser is likely around 100 fs. Letting λ_l be the laser pulse length, a resonant pulse has $\lambda_l/\lambda_p = 0.5$, where in contrast at the critical density, a 100 fs pulse has $\lambda_l/\lambda_p \approx 30$. An attractive alternative is provided by the phenomenon of self-modulation[5,27–30], where a large laser pulse ($\lambda_l/\lambda_p \gg 1$) becomes spontaneously broken into units of length λ_p and reproduces the desired wakefield behavior. This regime is called self-modulated LWFA or SM-LWFA, and is likely

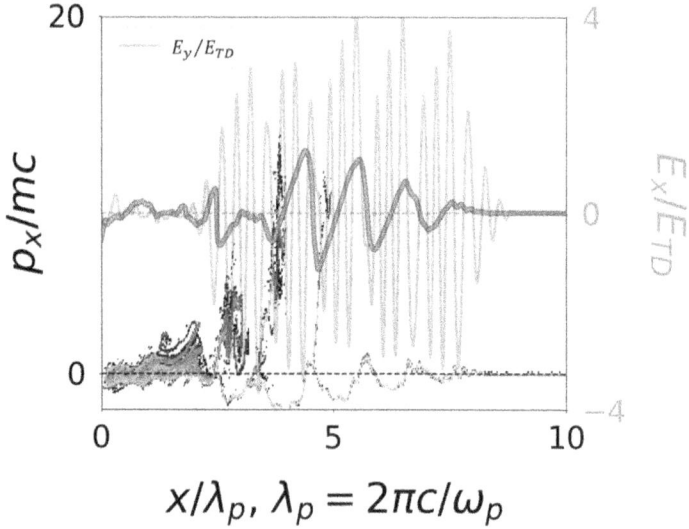

Fig. 4. A demonstration of the self-modulation of a laser pulse into resonant pieces in the "blue" regime of $n_c/n_e = 10$. Here $\lambda_{lp}/\lambda_p = 5$. The time step is $t = 9.2\tau_p$. (Similar run in Nicks et al., 2019[19]).

much more amenable to fibers than the resonant pulse case. Stated simply, fibers generally prefer a longer, low-amplitude pulse to a short, intense pulse.

To demonstrate the onset of SM-LWFA, a low-density plasma of $n_c/n_e = 10$ and $a_0 = 1$ is injected with laser pulse with length ($\lambda_l/\lambda_p = 5$), as is shown in Fig. 4.[19] As progressive peaks pass a point in the plasma, the wakefield is strengthened, and ultimately the strength of the wakefield is enhanced in the self-modulated cases compared to that of the resonant case. Electron acceleration is also seen following the wakefield.

Applying this technique to a fiber laser case, Fig. 5a shows the case of 15 coherently added fiber lasers at $a_0 = 0.01$ with $n_c/n_e = 10$. The pulse is now Gaussian, also for practical considerations, with FWHM a pulse length of 100 fs. For self-modulation, it is desirable that the laser power be greater than the critical power,[24] while in this case, without considering laser guiding conditions, this condition may not be satisfied. (A rough calculation using a laser spot size of $\pi\lambda_p^2$ gives a laser power of 7.5 GW, while the critical power is given by $P_c = 17(n_c/n_e)$ GW $= 170$ GW.) Nonetheless, substantial acceleration in Fig. 5a is seen up to nearly 1 MeV.

Many low-energy electrons are accelerated as well. This coupling to bulk electrons may be caused by the generation of lower-frequency laser components through Raman forward scattering, which is expected in the presence of a long laser pulse.[31–33] A spectral analysis of the laser field in Fig. 5a reveals the presence of down-shifted frequency components at $\omega = \omega_0 - \omega_p$ and $\omega = \omega_0 - 2\omega_p$, as well as a small up-shifted component at $\omega = \omega_0 + \omega_p$, where ω_0 is the nominal laser frequency.

(a). Self-Modulation

(b). Beat-Wave

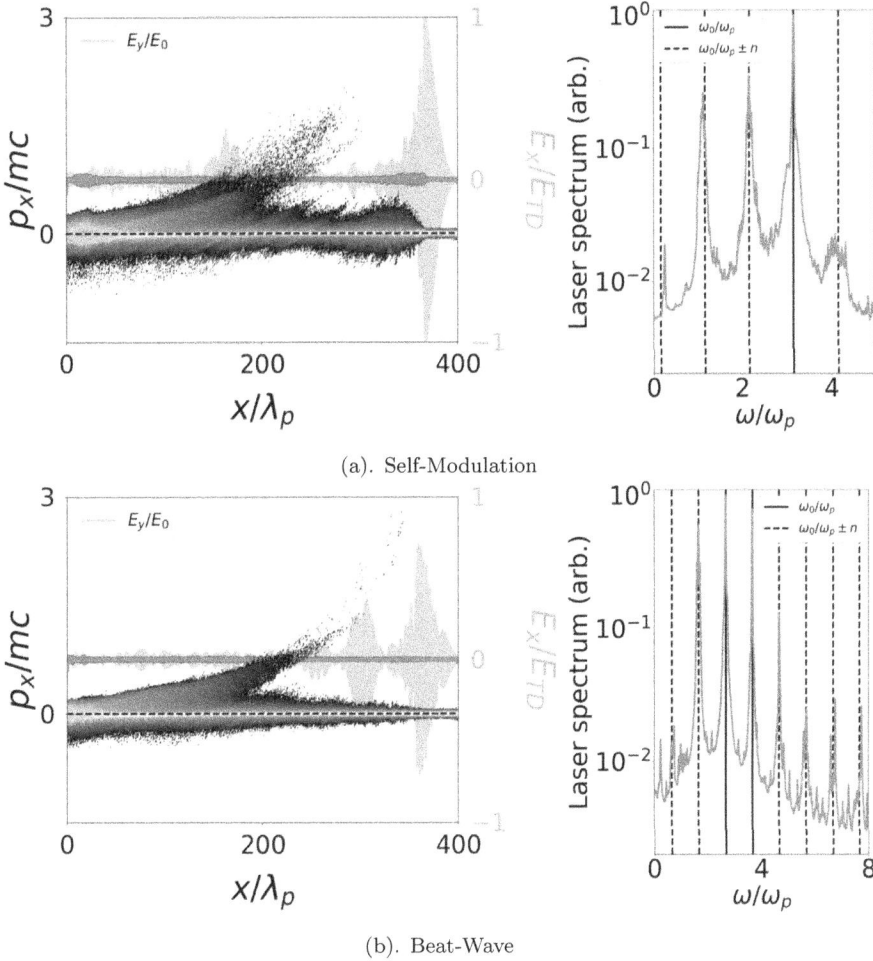

Fig. 5. The electron phase space and field structure (left) for the case of a 100 fs pulse at $n_c/n_e = 10$, as well as the laser (E_y) frequency spectrum (right). Fifteen lasers each contributing intensity $a_0 = 0.01$ are coherently added, demonstrating practical parameters for fiber laser applications. In 5a, the laser pulse undergoes self-modulation, while in 5b, each of the five laser contributions is further divided into two components that beat at ω_p, thus resonantly seeding the wakefield (beat-wave acceleration). Note that here the laser field E_y is normalized with respect to the initial combined amplitude E_0 of all the fiber contributions. For the frequency spectra, solid vertical lines indicate the nominal laser frequency (ω_0/ω_p) while dashed lines indicate harmonics, which differ from the nominal frequency by some integer multiple n of the plasma frequency ω_p.

These components arise almost immediately after the laser enters, and begins interacting with, the plasma. With $n_c/n_e = 10$ in this case, $\omega_0/\omega_p = \sqrt{10} \approx 3.2$. A frequency down-shifted from this value in multiples of ω_p thus becomes very close to resonance with ω_p; the two down-shifted components $\omega = \omega_0 - \omega_p$ and $\omega = \omega_0 - 2\omega_p$ are equivalent to a laser with nominal density ratios of $n_c/n_e = 5$ and $n_c/n_e = 1.25$, respectively. The latter of these values is firmly within the "black tsunami" regime of Figs. 1 and 3. Further down-shifting is suppressed because frequencies lower than ω_p would not be able to resonantly excite the Raman forward scattering instability and would be immediately absorbed by the plasma.

Another approach using practical fiber laser parameters is the laser beat-wave accelerator,[3,34] which was used historically in the early years of LWFA. In this scheme, two lasers with frequencies differing by ω_p create a modulation at the plasma frequency and resonantly excite the wakefield. Exploration of this possibility at $n_c/n_e = 10$ with a 100 fs Gaussian pulse and 15 $a_0 = 0.01$ lasers, each divided into two equal components separated in frequency by ω_p, shown in Fig. 5b, yielded slightly more efficient acceleration, with electrons reaching energies slightly in excess of 1 MeV. This relatively clean acceleration is expected given the seeded plasma oscillation and provides a confirmation of the physics of the pre-modulated laser field. As in the self-modulating case, here a lower laser harmonic equivalent to $n_c/n_e = 4$ is seen, as well as harmonics higher than the nominal frequency, the former perhaps aiding bulk acceleration of electrons and the latter pulling the highest-energy electrons past the energies reached in self-modulated case (Fig. 5a).

In these examples, an initially "blue" wave is converted into a "black" wave that can efficiently accelerate low-energy electrons even at very low laser intensity. Together with a variable number of coherently added fibers, this effect may provide substantial practical flexibility for a medical fiber laser application. For instance, if an optimized setup required an even lower individual laser intensity than 10^{14} W cm^{-2}, the target density could be modestly lowered, preserving the bulk flow of electrons in the desired energy range. Furthermore, if a beat-wave laser is possible, a potentially cleaner electron beam can also be produced if desired. It is remarkable that these benefits derive from the requirement of a long laser pulse, which can be considered one of the "limitations" of fiber lasers.

4. Electron Tissue Penetration

We may now consider more closely the interaction of an electron population like those in Figs. 3 and 5 with human tissue for radiation therapy. Conventional radiation therapy typically relies on exposing the body to an external source of radiation, whether X-ray, gamma-ray, protons, or electrons. In this process, the radiation passes through a significant depth of healthy tissue, causing collateral cellular damage. Three techniques to avoid this collateral damage include intraoperative radiation therapy (IORT),[10] where the source of radiation is surgically brought to the tumor, brachytherapy, in which the laser is injected into the body, or endoscopic radiation therapy (ESRT), where a small endoscope is internally brought to the tumor site. Consequently, in all of these cases, the radiation produced need only penetrate a short distance, perhaps millimeters or less. In such a scheme, the distribution in Figs. 3 and 5, possessing a large spread of low-energy electrons, may be particularly fitting. The recent development of coherent networks of fiber lasers (CAN)[14] has allowed LWFA to branch into this new and distinct application.

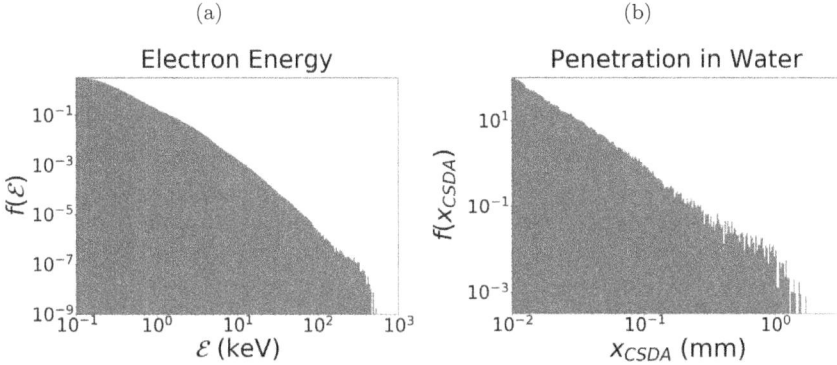

Fig. 6. Electron penetration in the high-density LWFA regime. (6a) shows the normalized electron energy distribution for setup in figure 5a, which models a bundle of 15 fiber lasers each with $a_0 = 0.01$ coherently added with plasma density $n_c/n_e = 10$ and a pulse length of 100 fs. (6b) shows the resulting normalized distribution of electron penetration depth in the continuous slowing-down approximation (CSDA).

The penetration depth in human tissue can be approximated by integrating the stopping power of electrons in water, giving the stopping distance in the continuous slowing-down approximation (CSDA).[35,36] At the critical density, the distribution of low-energy electrons in Fig. 5a has the energy distribution shown in Fig. 6a. This distribution $f(\mathcal{E})$ corresponds to a maximum penetration depth x_{CSDA} in water of about $\lesssim 1$ mm, as is shown in Fig. 6b as a function of x_{CSDA}. Tuning the plasma density allows control of the penetration depth. By changing the density of the irradiated material, the penetration depth can be tuned to the desired value; lower material density will give deeper penetration. The laser intensity a_0 or number of fibers can also be tuned for the desired electron energies produced. As an additional benefit, near the critical density, a significant acceleration of the bulk population of electrons occurs, potentially creating a far larger overall dose of radiation than would occur for more typical wakefield acceleration. This combination of a large dose and shallow, yet tunable, penetration may be ideal for intraoperative, brachytheraputic, and endoscopic medical applications.

5. Conclusions

We have studied the dynamics of LWFA in the high-density regime. In this domain, where $n_c/n_e \sim O(1)$, sheath dynamics emerges with an important role, producing a large flux of low-energy electrons (the "black tsunami" regime). Furthermore, we have found that the self-modulation or Raman forward-scattering process allows a conversion of the "blue" or "grey" regimes into the "black" regime, providing efficient generation of a bulk flow of low-energy electrons despite the presence of a density ratio n_c/n_e greater than unity. Along with the invention of the Coherent Amplification Network (CAN)[14] fiber laser technology, these dynamical characteristics in the appropriately chosen regime of operation open a pathway to creating far more compact electron radiation sources through LWFA and thereby a radically new radiotherapy using compact electron sources.

Regarding the material to be irradiated by the laser, nanomaterials with an open structure, such as carbon nanotubes[15] or porous alumina, present an attractive means for achieving the critical density of a 1-micron laser (approximately 10^{21} cm^{-3}) while avoiding the presence of ionized gas inside of the body. Such a medium would also provide the benefit of guiding the laser. It might also be possible to tailor the design of the nanomaterial to suit the desired plasma density.

One potential challenge of this approach is that the population of accelerated electrons generated by LWFA at high density is non-monoenergic and probably of high emittance. We may also strive to further increase the efficiency. Toward such a purpose we may wish to employ a graded density of plasma to control the phase gradation of the wakefield.[37,38] Nonetheless, interesting physics has already emerged from these efforts, and the richness of a new regime is evident. Within the last decade, the technology needed to realize endoscopic electron therapy through LWFA has come of age, and serious endeavors for implementation may now proceed.

Acknowledgments

The present paper arose from the term-project efforts of the students in the tri-campus (UCI, UCLA, UCSD) graduate physics course Special Topics in Plasma Physics PHY249, "Nonlinear Plasma Physics" (Winter, 2019), led by the instructor T. Tajima. We also tried to tie plasma physics with other disciplines such as medical physics and geophysics to broaden the students' experience in physics. The tri-campus plasma physics graduate course was launched in academic year 2018, and this course was one of three such courses. The materials are partially available on Google Drive upon request. This original work is found in B. S. Nicks, S. Hakimi, E. Barraza-Valdez, K. D. Chesnut, G. H. DeGrandchamp, K. R. Gage, D. B. Housley, G. Huxtable, G. Lawler, D. J. Lin, P. Manwani, E. C. Nelson, G. M. Player, M. W. L. Seggebruch, J. Sweeney, J. E. Tanner, K. Thompson, and T. Tajima, Electron Dynamics in the High-Density Laser-Wakefield Acceleration Regime, Submitted to Phys. Rev. Accel. Beams (2019)[19]. We are thankful for discussions with J. Wheeler, J. C. Chanteloup, X. Yan, and N. Beier. This work was partially supported by the Norman Rostoker Fund at UCI.

References

1. F. M. Khan, *The Physics of Radiation* (Lippincott Williams & Wilkins, Philadelphia, PA, 2010).
2. T. Tajima, K. Nakajima and G. Mourou, Laser acceleration, *La Rivista del Nuovo Cimento* **40**, 33 (Jan 2017).
3. T. Tajima and J. M. Dawson, Laser electron accelerator, *Phys. Rev. Lett.* **43**, 267 (Jul 1979).

4. K. Nakajima *et al.*, Laser wakefield accelerator experiments using 1 ps 30 TW Nd:glass laser, in *Proceedings of International Conference on Particle Accelerators*, (Washington, DC, 1993).
5. K. Nakajima *et al.*, A proof-of-principle experiment of laser wakefield acceleration, *Phys. Scr.* **T52**, 61 (Jan 1994).
6. D. Strickland and G. Mourou, Compression of amplified chirped optical pulses, *Opt. Commun.* **56**, p. 219 (1985).
7. T. Tajima, Prospect for compact medical laser accelerators, *J. Jpn. Soc. Therp. Radiat. Oncol.* **9**, 83 (1997).
8. C. Chiu, M. Fomytskyi, F. Grigsby, F. Raischel, M. C. Downer and T. Tajima, Laser electron accelerators for radiation medicine: A feasibility study, *Med. Phys.* **31**, 2042 (2004).
9. K. K. Kainz, K. R. Hogstrom, J. A. Antolak, P. R. Almond, C. D. Bloch, C. Chiu, M. Fomytskyi, F. Raischel, M. Downer and T. Tajima, Dose properties of a laser accelerated electron beam and prospects for clinical application, *Med. Phys.* **31**, 2053 (2004).
10. A. Giulietti, N. Bourgeois, T. Ceccotti, X. Davoine, S. Dobosz, P. D'Oliveira, M. Galimberti, J. Galy, A. Gamucci *et al.*, Intense γ-ray source in the giant-dipole-resonance range driven by 10-TW laser pulses, *Phys. Rev. Lett.* **101**, p. 105002 (Sep 2008).
11. A. Giulietti (ed.), *Laser-Driven Particle Acceleration Towards Radiobiology and Medicine* (Springer International Publishing, Switzerland, 2016).
12. K. Nakajima, J. Yuan, L. Chen and Z. Sheng, Laser-driven very high energy electron/photon beam radiation therapy in conjunction with a robotic system, *Applied Sciences* **5**, 1 (2015).
13. K. Nakajima, Laser-driven electron beam and radiation sources for basic, medical and industrial sciences, *Proceedings of the Japan Academy, Series B* **91**, 223 (2015).
14. G. Mourou, W. Brocklesby, T. Tajima and J. Limpert, The future is fibre accelerators, *Nat. Photonics* **7**, p. 258 (2013).
15. T. Tajima, Laser acceleration in novel media, *Eur. Phys. J. Spec. Top.* **223**, 1037 (May 2014).
16. N. V. Myung, J. Lim, J.-P. Fleurial, M. Yun, W. West and D. Choi, Alumina nanotemplate fabrication on silicon substrate, *Nanotech.* **15**, 833 (Apr 2004).
17. X. Zhang, T. Tajima, D. Farinella, Y. Shin, G. Mourou, J. Wheeler, P. Taborek, P. Chen, F. Dollar and B. Shen, Particle-in-cell simulation of x-ray wakefield acceleration and betatron radiation in nanotubes, *Phys. Rev. Accel. Beams* **19**, p. 101004 (Oct 2016).
18. F. Sylla, A. Flacco, S. Kahaly, M. Veltcheva, A. Lifschitz, V. Malka, E. d'Humières, I. Andriyash and V. Tikhonchuk, Short intense laser pulse collapse in near-critical plasma, *Phys. Rev. Lett.* **110**, p. 085001 (Feb 2013).
19. B. S. Nicks *et al.*, Electron dynamics in the high-density laser-wakefield acceleration regime, *Phys. Rev. Accel. Beams (Submitted)* .
20. F. Mako and T. Tajima, Collective ion acceleration by a reflexing electron beam: Model and scaling, *The Physics of Fluids* **27**, 1815 (1984).
21. B. Rau, T. Tajima and H. Hojo, Coherent electron acceleration by subcycle laser pulses, *Phys. Rev. Lett.* **78**, 3310 (Apr 1997).
22. M. T. Hassan *et al.*, Optical attosecond pulses and tracking the nonlinear response of bound electrons, *Nature (London)* **530**, p. 66 (2016).
23. H. Lamb, in *Hydrodynamics*, (Dover Publications, New York, 1945), New York, ch. 8.
24. E. Esarey, C. B. Schroeder and W. P. Leemans, Physics of laser-driven plasma-based electron accelerators, *Rev. Mod. Phys.* **81**, 1229 (Aug 2009).

25. E. Fourkal, B. Shahine, M. Ding, J. S. Li, T. Tajima and C.-M. Ma, Particle in cell simulation of laser-accelerated proton beams for radiation therapy, *Med. Phys.* **29**, 2788 (2002).

26. G. P. Agrawal, *Nonlinear Fiber Optics* (Oxford:Academic, 2013).

27. N. E. Andreev, L. M. Gorbunov, A. A. Pogasova, R. R. Ramazashvili and V. I. Kirsanov, Resonant excitation of wake fields by a laser pulse in a plasma, *JETP Lett.* **55**, 571 (1992), [Pisma Zh. Eksp. Teor. Fiz.55,551(1992)].

28. J. Krall, A. Ting, E. Esarey and P. Sprangle, Self-modulated laser wake field acceleration, in *Proceedings of the 1993 Particle Accelerator Conference*, (Washington, DC, 1993).

29. K. Nakajima *et al.*, Observation of ultrahigh gradient electron acceleration by a self-modulated intense short laser pulse, *Phys. Rev. Lett.* **74**, 4428 (May 1995).

30. A. Modena *et al.*, Electron acceleration from the breaking of relativistic plasma waves, *Nature (London)* **377**, 606 (1995).

31. C. Joshi, T. Tajima, J. M. Dawson, H. A. Baldis and N. A. Ebrahim, Forward raman instability and electron acceleration, *Phys. Rev. Lett.* **47**, 1285 (Nov 1981).

32. D. L. Fisher and T. Tajima, Enhanced raman forward scattering, *Phys. Rev. E* **53**, 1844 (Feb 1996).

33. W. B. Mori, The physics of the nonlinear optics of plasmas at relativistic intensities for short-pulse lasers, *IEEE Journal of Quantum Electronics* **33**, 1942 (Nov 1997).

34. M. N. Rosenbluth and C. S. Liu, Excitation of plasma waves by two laser beams, *Phys. Rev. Lett.* **29**, 701 (Sep 1972).

35. M. J. Berger, J. S. Coursey and M. A. Zucker, ESTAR, PSTAR, and ASTAR: Computer programs for calculating stopping-power and range tables for electrons, protons, and helium ions (version 1.21) (Jan 1999).

36. D. K. Brice, Stopping powers for electrons and positrons (ICRU report 37; international commission on radiation units and measurements, Bethesda, Maryland, USA, 1984), *Nuclear Instruments and Methods in Physics Research Section B: Beam Interactions with Materials and Atoms* **12**, 187 (1985).

37. R. Hu, H. Lu, Y. Shou, C. Lin, H. Zhuo, C. Chen and X. Yan, Brilliant GeV electron beam with narrow energy spread generated by a laser plasma accelerator, *Phys. Rev. Accel. Beams* **19**, p. 091301 (Sep 2016).

38. A. Döpp, C. Thaury, E. Guillaume, F. Massimo, A. Lifschitz, I. Andriyash, J.-P. Goddet, A. Tazfi, K. Ta Phuoc and V. Malka, Energy-chirp compensation in a laser wakefield accelerator, *Phys. Rev. Lett.* **121**, p. 074802 (Aug 2018).

Paradigm of Experimental High Energy Physics: A Personal Perspective

S. Chattopadhyay

Fermi National Accelerator Laboratory
Kirk Road and Pine Street, Batavia, Illinois 60510-5011, USA
and
Northern Illinois University
1425 W. Lincoln Hwy., DeKalb, Illinois 60115-2828, USA
swapan@fnal.gov

Keywords: Acceleration; colliders; energy; luminosity; early universe; quantum sensors; coherent quantum vacuum.

1. Prelude

The field of high energy particle physics started with the discovery of very high energy particles and radiation from 'cosmic rays', explored via balloon and mountaineering expeditions, exposing specially designed emulsion plates to the cosmic rays in the first decades of the twentieth century. The discovery of the positron, an antiparticle of the electron, owes much to cosmic ray explorations in the 1920s and 1930s. In parallel, with the discovery of the atomic nucleus by Rutherford at the turn of the twentieth century, establishment of the quantum structure of the Bohr Atom, invention of the 'Cyclotron' by Ernest Orlando Lawrence in Berkeley in 1920s and the 'splitting of the atom' by Cockcroft and Walton in 1932,[1] it became increasingly attractive to consider creating in the experimental laboratories terrestrial scale experiments that could produce accelerated high energy particles for detailed fundamental investigations of subatomic physics, without relying on costly and challenging cosmic ray explorations — and so started the present paradigm of charged particle accelerator-driven high energy particle physics. This was further advanced by the possibility of colliding beams in storage rings in 1950s and 1960s, leading to GeV- to TeV-scale electron-positron and proton-proton collisions in the laboratory.

The paradigm of laboratory accelerator-based high energy physics depends on creating the necessary 'high energy' and 'high energy density', luminous collisions in the laboratory itself, that could penetrate ever deeper into the shorter and shorter distance scales of ordinary matter, the resolving power limited only by the energy of the colliding particles. In the conventional paradigm, the experimental approach depends critically on two important concepts: (i) the penetration and resolving power into subatomic structures depends on the energy reach of the particle accelerator; (ii) while the rate 'R' of production of relevant and interesting subatomic processes depends on the associated geometric cross-section 'a' of the process under investigation, enhanced by the 'luminosity, L' of the collisions, which is a measure of how luminous are the colliding beams to start with:

$$R = L \cdot a .$$

There are two fundamental limitations of such an approach: as the energy goes up, it is nature's revenge that the corresponding cross-sections tend to either go down or go up much slower with energy with a certain scaling, barring occasional most interesting 'resonances', which of course we want to observe. This often means that along with increasing energy, one must increase the luminosity to reach the physics potential. Currently, the energy and luminosity reach of state-of-the-art particle colliders (electron-positron or proton-proton) are limited to scales of TeVs in energy and 10^{34} cm^{-2}s^{-1} in luminosities.[2,3] The beam dynamics and 'radiative' physics of the luminous collisions have been studied in detail in earlier studies,[4,5] as well as in a paper by Nakajima *et al.* in these proceedings. Extending beyond these values is proving to be technically limited by conventional technologies demanding significant resources and real estate (e.g. tens of billions of dollars and hundreds of kilometers, as envisioned for the 100 km-scale, 100 TeV colliders in the FCC and CepC programs at CERN, Switzerland and China). The challenges of future high-energy colliders are captured also in Ref. 6 and a paper by Shiltsev in these proceedings.

I propose below two alternate paradigms of experimental research below.

A. COSMIC ARCHAEOLOGY: The Energy Reach: Exploit Signals from the Early Universe via Precision Sensing: Quantum Sensors in the Laboratory

Our energy reach, in the first instance, can never compete with mother Nature's original design — the highest energies and energy densities were already created 13.8 billion years ago during the beginning of the 'hot big-bang' universe, the presently accepted cosmological model of the universe. So why bother working so hard, spending such resources on creating such extreme energies in the laboratory rather than exploring the relics, fossils and left-over signals — no matter how "weak" they may be — by development of laboratory-scale specially sensitive detectors that can 'see', 'sense' and 'listen' to very weak signals from the big-bang? Such detectors can be based upon novel techniques exploiting quantum mechanical 'entanglement' process, involving "quantum squeezing and interference". Many promising techniques have emerged in the last decades using cavity electrodynamics, superconducting 'Qubits', atomic beam interferometers, novel NMR (nuclear magnetic resonance) materials, quantum 'correlated' or 'entangled' states in solids (e.g. nitrogen vacancies in diamond) and topological states in condensed matter (e.g. 'Dirac' and 'Weyl' semi-metals and semi-conductors), that offer us laboratory investigations of the high energy frontier — especially its 'dark' sector involving 'dark' matter and 'dark' energy.[7,8] Crystal channeling and acceleration may play a role here.

B. COHERENT QUANTUM VACUUM vs Luminosity: Exploit 'Coherent', 'Condensate' Nature of Quantum Vacuum vs Incoherent Scattering

The Luminosity argument on the other hand is based upon the geometric and kinematic concept of an 'incoherent' albeit deep and inelastic scattering of the constituents of matter. There is no reason to believe the 'Standard Model Vacuum' — that is whatever is left over when one removes the Standard Model constituents of quarks, leptons, force carrier Bosons and even the 'Higgs' Boson — is an 'incoherent' state, amenable to penetration via incoherent scattering and not a 'condensate', coherent state, that can simply be penetrated by individual highly energetic single particles. It is quite possible that a PeV or higher energy particle or beam, even with very low luminosity, can penetrate and polarize the 'Standard Model Vacuum' effectively. Hence, one must consider available techniques of reaching very high energies alone, independent of luminous colliding beams. It is in this context that current advanced accelerator techniques using laser-driven wakefield acceleration, nanostructures and metamaterials and crystals/carbon nanotubes — many of which are discussed in these proceedings are relevant topics, offering alternates to current paradigm of experimental approaches to high energy particle physics. Table 1 below summarizes some of the strengths and parameters of experimental high energy physics explorations.

Table 1

Physical entity	Associated strength	Units
Particle energy (real or 'dark')	1	meV–TeV
Accelerating gradient	GV–TV	m^{-1}
Collider luminosity	10^{34}	$cm^{-2}s^{-1}$
Collider center-of-mass energy	10–100	TeV
Equivalent dark matter field	10^{4} [mass density = energy density/c^2]	V/m
Equivalent dark energy field	10	V/m
Quantum sensor sensitivity	10^{-25}	Probability of a signal in the 'dark' sector
Early universe space-time cosmic background tremor	10^{-22}	Dimensionless relative strain of space-time, dl/l, in the band 100 mHz to 1 Hz

2. Single Particle Acceleration and Focusing Limits in Structures: Laser-driven Plasma Wakefields, Structured Nanomaterials and Crystals

Typically, highest achievable accelerating fields in plasmas, nanostructures and crystals, depend on the available 'polarized' participating electric charge densities, n_0, roughly according to the following scaling law:

$$E \ (\text{GV/m}) = 100 \ [n_0 \ (10^{18} \ /\text{cm}^{-3})]^{1/2} \ ,$$

while available strong electrostatic electric field between crystal atoms are:

$$E = 100 \ \text{V/Angstrom} = 1 \ \text{GV/cm} \ .$$

Various innovative techniques using lasers, plasmas and particle beams [9,10,11,12,13,14,15] have been progressed and continue to be developed that have the promise of generating beams reaching up to PeV energies, the significant energy where the 'knee' in the 'cosmic ray' spectrum occurs. Even without a luminous beam, such tenuous beams or even a few energetic particles should be able to penetrate Standard Model Vacuum and tell us something about nature at extreme high energies. These proceedings contain much material on reaching higher energies with structured materials under various settings, including lasers, plasmas, particle beams, nanostructures and crystal structures.

3. Quantum Sensors of the "Dark" Universe and Weak Processes

In the emerging 'quantum science' initiative around the world and in the US in particular[7,8] in terms of developing innovative quantum sensors to probe 'dark matter', 'dark energy' and primordial cosmic gravitational wave background, one senses fields and their fluctuations with extreme sensitivities. For example, superconducting Qubits, embedded in superconducting microwave cavities, can sense low-mass 'axion-like' dark matter particles in the meV mass range with sensitivities of one oart in 10^{25}. Similarly, atomic interferometers over long baselines of 100 meters to kilometers scale, including eventually space-based experiments, will reveal much information about the fluctuations and energy-matter-space-time emergence in the early universe with unprecedented sensitivity. Such techniques will complement accelerator, energetic charged particle- or beam-based investigations, in addressing both the 'discovery' and 'precision' frontiers.

4. Outlook

As I tried to argue qualitatively, with appropriate adjustment of our paradigm of experimental high energy physics research, there is plenty of room of gaining information about the physical world and early universe at extreme energies and energy densities via either penetrating the 'coherent quantum vacuum' by one single or a few energetic single high energy particles without having recourse to luminous collision or sense extremely weak fundamental signals via ultra-sensitive quantum sensors. The former aspect can be easily addressed by various structure-driven particle acceleration schemes, e.g., laser-plasma, structured materials including crystals discussed in these proceedings, while the latter will be addressed by development innovative detectors and quantum sensors using exotic topological materials and exploiting engineered control of 'entangled' quantum states.

Acknowledgments

This work was supported by the University President's office via a grant to the author for research in his position as President's Professor of Research, Scholarship and Artistry and Director of Accelerator Research at Northern Illinois University.

References

1. M. Poole, J. Dainton and S. Chattopadhyay, Cockcroft's subatomic legacy, *CERN Courier* **October 2007** (2007).
2. S. Chattopadhyay, "Advances in beam physics and technology: Colliders of the future", Invited paper presented at the *Tamura Symposium on Accelerator Physics*, Austin, TX, USA, November 14–16, 1994, LBL-37966.
3. O. Bruning, V. Chohan and S. Chattopadhyay, Accelerator physics challenges of the Large Hadron Collider, in *Proc. Indian National Academy of Sciences* (Springer Verlag, 2009).
4. S. Chattopadhyay, D. Whittum and J. Wurtele, Advanced accelerator technologies — A Snowmass '96 subgroup summary, in *Proc. of Snowmass Workshop, June 1996, SNOWMASS 1996 – ACC 060*, pp. 356–370, LBL-39655, SLAC-PUB-9914.
5. S. Chattopadhyay and R. M. Jones, Radiative regime of linear colliders, high repetition rate free electron lasers and associated accelerator structures, *Nucl. Instrum. Methods Phys. Res. A* **657**(1), 168–176 (November 2011).
6. S. Chattopadhyay and M. Syphers, Perspectives on future particle accelerators at the energy and intensity frontiers, in *Proc. of ICHEPP 2016, Chicago*, published in *Proceedings in Physics* (2016).
7. S. Chattopadhyay, Viewpoint: In search of hidden light, *CERN Courier* **April 2015** (2015).
8. S. Chattopadhyay, R. Walsworth and R. Falcone, Quantum sensors for fundamental science, quantum information science and advanced computing, *Report of the US DOE Round table on Quantum Sensors*, OSTI report, February 2016.
9. S. Chattopadhyay, Role of lasers in linear accelerators, in *Proc. of LINAC '96*, August 1996, CERN 96-07.
10. W. Leemans *et al.*, Interactions of relativistic electrons with ultrashort laser pulses: generation of femtosecond X-rays and microprobing of electron beams, *IEEE J. Quantum Electron.* **33**, 1925 (May 1997).
11. S. Chattopadhyay, Alight a beam and beaming light: A theme with variations, *Phys. Plasmas*, **5**(5), 2081 (May, 1998).
12. G. Xia *et al.*, Ultracold and high brightness electron source for next generation particle accelerators, in *Proc. of IPAC 2013, Shanghai, China*, pp. 452–454 (2013).
13. A. Caldwell *et al.*, The AWAKE experiment at CERN: Possibilities of reaching TeV electron energies via proton-driven plasma wakefield acceleration, in *Proc. of Controlled Thermonuclear Fusion 2014*.
14. G. Xia *et al.*, Collider design issues based on proton-driven plasma wakefield acceleration, *Nucl. Instrum. Methods Phys. Res. A* **740**, 173–179 (2014).
15. A. Caldwell *et al.*, Path to AWAKE. *Nucl. Instrum. Methods Phys. Res. A* **829**, 3–16 (2016), arXiv: 1511.09032 [physics.plasm-ph], FERMILAB-CONF-15-650.

Wakefield Acceleration Towards ZeV from a Black Hole Emanating Astrophysical Jets

T. Ebisuzaki

RIKEN Cluster for Pioneering Research,
2-1 Hirosawa, Wako 351-1198, Japan
ebisu@postman.riken.jp

T. Tajima

Department of Physics and Astronomy,
University of California,
Irvine, CA 92679, USA
ttajima@uci.edu

We consider that electromagnetic pulses produced in the jets of this innermost part of the accretion disk accelerate charged particles (protons, ions, electrons) to very high energies via wakefield acceleration, including energies above 10^{20} eV for the case of protons and nucleus and 10^{12-15} eV for electrons by electromagnetic wave-particle interaction. Thereby, the wakefield acceleration mechanism supplements the pervasive Fermi's stochastic acceleration mechanism (and overcomes its difficulties in the highest energy cosmic ray generation). The episodic eruptive accretion in the disk by the magneto-rotational instability gives rise to the strong Alfvenic pulses, which acts as the driver of the collective accelerating pondermotive force. This pondermotive force drives the wakes. The accelerated hadrons (protons and nuclei) are released to the intergalactic space to be ultra-high energy cosmic rays. The high-energy electrons, on the other hand, emit photons to produce various non-thermal emissions (radio, IR, visible, UV, and gamma-rays) of active galactic nuclei in an episodic manner, giving observational telltale signatures.

Keywords: Wakefield acceleration; blackhole; accretion disk; jet; cosmic-rays; gamma-rays; neutrinos.

1. Introduction

In space physics and astrophysics the established known mechanism of high energy particles (hadrons in particular and often electrons as well) has been the nearly almighty (in astrophysics) and venerable Fermi's stochastic acceleration ever since his epoch making paper [1]. (He was also visionary to suggest his PeV accelerator, which encircles the entire Earth by the accelerating rf tube). This can explain many observed features, while there emerge several difficulties (as discussed below in more details), in particular the tail of the highest energy cosmic rays beyond 10^{19} eV, at which energy even protons would begin to lose energy by synchrotron radiations through bending their orbits that are inevitable in the stochastic multiple magnetic bendings that necessitate in Fermi's mechanism. On the other hand, the wakefield acceleration is basically straight acceleration, readily alleviating this severest difficulty.

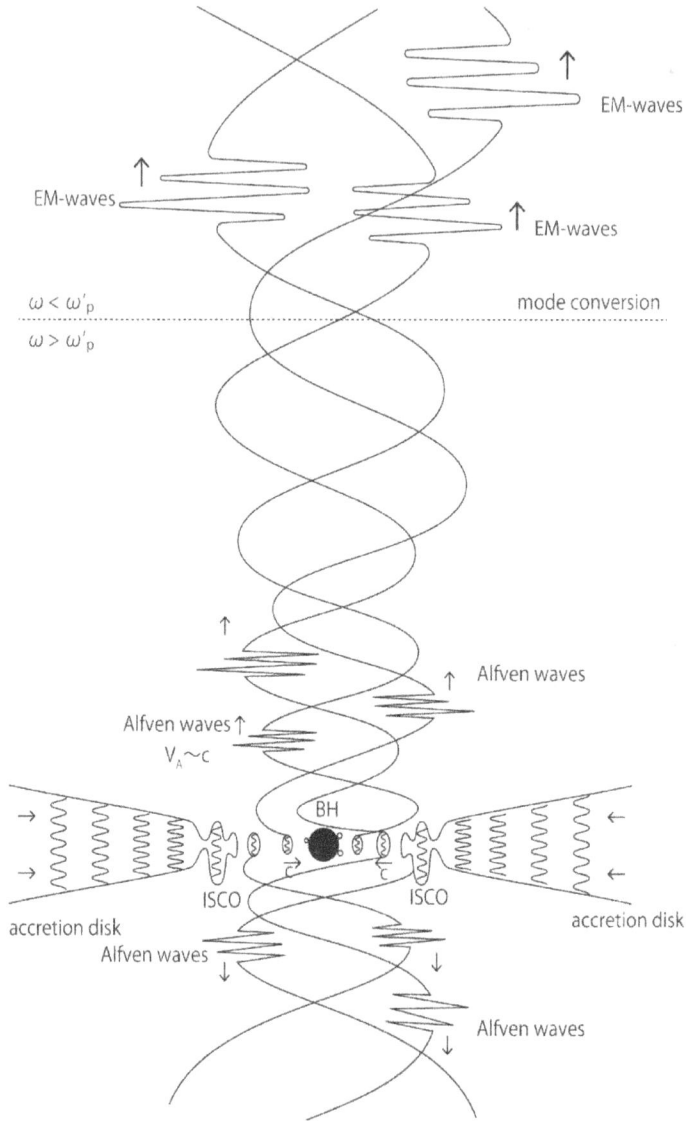

Fig. 1. Gas clumps are formed around the inner edge of the accretion disk. When they fall down to the blackhole during its transition, magnetic field penetrating jets are strongly shaken and electromagnetic disturbances propagates along the jets as bursts of the Alfven/whistler waves.

In the present paper, we consider that electromagnetic pulses produced in the jets of the innermost part of the accretion disk accelerate charged particles (protons, ions, electrons) to very high energies including energies above 10^{20} eV for the case of protons [2,3] and nucleus and 10^{12-15} eV for electrons by electromagnetic wave-particle interaction (Figs. 1, 2, and 3). The episodic eruptive accretion in the disk by the magneto-rotational instability [4,5] gives rise to the strong (super-relativistic, i.e., the electron momentum far exceeding $m_e c$, where m_e is the mass of electrons) Alfvenic pulses, which acts as the driver of the

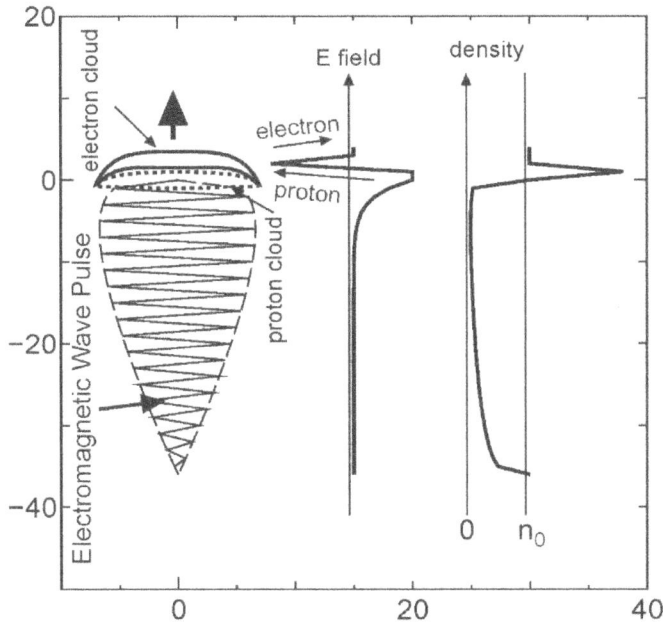

Fig. 2. The structure of bow wake. An electron cloud is formed at the front top of the wave pule and a proton cloud follows. The resultant electric field accelerates protons in the back side and electrons in front side of the bow wake (taken from [2]).

collective accelerating pondermotive force [6]. This pondermotive force drives the wakes (both bow and stern wakes [2,3]). As investigated in the present super-relativistic pulse case, the bow wake far exceeds the conventional (stern) wake acceleration (see the comparison of the bow and stern wakes in such as [7]; Fig. 2). The accelerated hadrons (protons and nuclei) are released to the intergalactic space to be ultra-high energy cosmic rays. The high-energy electrons, on the other hand, emit photons in the collisions of electromagnetic perturbances to produce various non-thermal emissions (radio, IR, visible, UV, and gamma-rays) of active galactic nuclei.

The conditions for this strong acceleration by wakefield are:

(i) the acceleration structure (wave) is very close to the relativistic propagation velocity (phase velocity), i.e. the speed of light; and

(ii) the wave has a relativistic amplitude (i.e. the particles in the wave have a relativistic momentum $e_j E/\omega > m_j c$ in one photon cycle). Where e_j and m_j are the charge and the mass of the particle j, and E and ω are the electric field and angular frequency of the wave.

The condition (b) came from the fact that the significant acceleration by a pondermotive force of the electromagnetic wave takes place only in the nonlinear case: This term becomes significant only if the amplitudes of the electromagnetic waves become relativistic [6,8,9]. If these two conditions are satisfied, as shown by a number of ground experiments, [10], this acceleration mechanism overcomes some of the difficulties of

Fermi acceleration [1], in which charged particles gradually gain energy as they are scattered many times by magnetic clouds. for the following reasons.

(1) The pondermotive force provides a very high acceleration field.

(2) No deflection of the particles is required which causes energy loss due to synchrotron radiation.

(3) The acceleration field is parallel to the direction of movement of the particles and has the same velocity (speed of light). The result is a robust built-in coherence in acceleration systems, which is called as relativistic coherence [11]. In such cases, the energy spectrum takes the form of a power function of the exponent -2 (E^{-2}; see [12,13]) due to the phase difference between the particle and the acceleration field. On the other hand, Fermi acceleration based on a large number of scattering is incoherent, it is an extended, and has no specific temporal structure.

(4) There are no escape problems [14].

(5) The acceleration field dissipates spontaneously so that the particles are free to escape from the field after the acceleration. The wave has such coherent dynamics as long as it is excited, with a sufficiently high frequency. On the other hand, in a mechanism that accelerates with an electrostatic structure [22], the acceleration of the charged particles is not possible unless there is a special reason to overcome the plasma and maintain the structure since the plasma itself has a strong tendency to destroy the electrostatic field. Also, there is no reason to have a low spectrum power.

This new acceleration mechanism, therefore, seams to solve a long-standing enigma of the origin of ultra-high energy cosmic rays (Ultra-High Energy Cosmic rays: UHECR) with energy of 10^{20} eV (e.g. [14]), in which have been discussed primarily in the framework of the Fermi acceleration [1]. The candidate astronomical objects of ULHECR were neutron stars, active galactic nuclei, gamma-ray bursts, and cosmological accretion shock waves in the intergalactic space. However, even in these candidates, the acceleration to 10^{20} eV by the Fermi mechanism was difficult because of: (1) very large number of scatterings is required to reach high energy; (2) energy loss due to the synchrotron radiation is not negligible at the time of scattering; and (3) adiabatic energy loss takes place when particles escape from the acceleration region.

The idea that electromagnetic waves accelerate charged particles is not new. For example, Takahashi *et al.* [15] and Chen *et al.* [13] showed that strong Alfven waves, produced by a neutron star collision were able to accelerate charged particles to energies above 10^{20} eV. Although it is believed to be associated with such short gamma bursts [16], the direct collisions of the two neutron stars are not very likely. This requires exactly the same masses, for the binary neutron stars because otherwise the more massive neutron star destroys the lighter stars into an accretion disk. Chang *et al.* [17] performed one-dimensional numerical calculations of the propagation of whistler wave emitted from AGNs and found that a wakefield accelerates particles to UHECR.

An accreting supermassive blackhole is the main engine of AGNs, in which Ebisuzaki and Tajima [2,3] considered wakefield acceleration to take place. Accretion disk has been shown to repeated transitions between a strongly magnetized (low β) state and a weakly

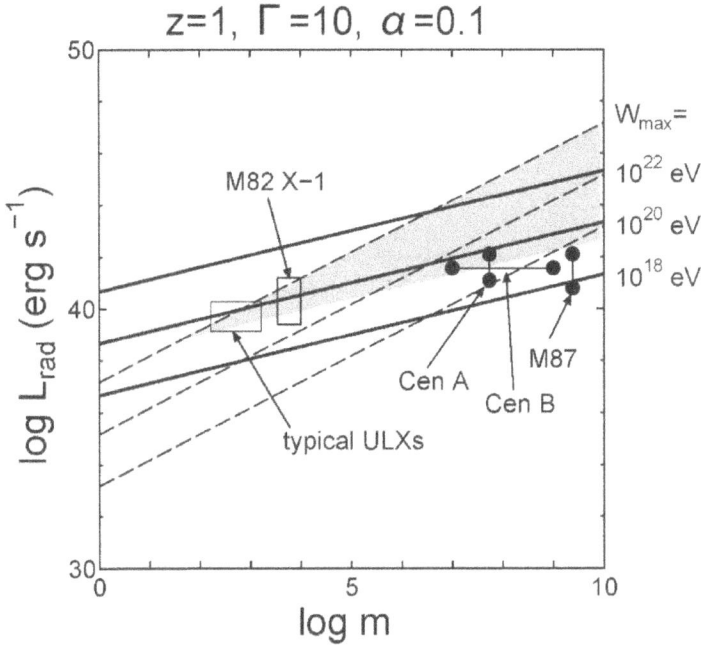

Fig. 3. The lines for the maximum energy gain, $W_{max} = 10^{18}$, 10^{20}, 10^{22} eV, are plotted in $m - L_{rad}$ diagram. M82 X-1 is located well above the line of 10^{20} eV so that be a good candidate for northern hot spot [29]. Other nearby AGNs, such as M87, Cen A and B, and typical ULXs are also possible candidates of UHECRs (> 3.9×10^{19} eV) Grey area ($W_{max} > 3.9 \times 10^{19}$ eV and $\dot{m} < 0.1$) represents possible acceleration region of UHECR. Three dashed lines are for $\dot{m} = 10^{-5}$, 10^{-3}, and 10^{-1}.

magnetized (high β) state [18]. In fact, O'Neil et al. [19] shows that the transition takes place every 10–20 orbital periods in the three-dimensional simulation. The amplitude of the distortion in the magnetic field becomes resulting in a very large amplitude at the innermost portion of the disk. At this transition from the strongly magnetized state to the weakly magnetized state, strong pulses of electromagnetic disturbance are excited in the accretion disk. This disturbance converted into strong pondermotive field by nonlinear effects in jets made of plasmas ejected from accretion disk with relativistic velocities. It is shown that this pondermotive force can spontaneously accelerate protons and nuclei to ultimate energies exceeding ZeV (10^{21} eV; Fig. 3 and Table 1). Mizuta et al. [20] performed three-dimensional MHD simulations of accretion disk and found that accretion disk exhibited strong fluctuations and that intermittently produced strong electromagnetic pulses and matter out of equilibrium was injected toward the rotational axes. The pointing fluxes agreed with those assumed by Ebisuzaki and Tajima [2,3].

On the other hand, the Telescope Array (TA) team suggested that there is a hotspot in the northern sky where the directions of arrival of the UHECRs above 57 EeV [21]. In fact, some anisotropy is expected to be observed at energies of 10^{20} eV because protons deflect only a few degrees even in the galactic magnetic field [14], and only about 100 Mpc can propagate as a result of collisions with microwave background radiation (GZK-effect [22,23]). In fact, the Pierre Auger Observatory (PAO) team also reported that a hotspot in the southern sky above 57 EeV, although it is less statistically significant [24,25]. These are signs of the anisotropy of charged particles and are key clues to reveal the origin of cosmic rays.

Table 1. Time Scales, Maximum Energy, and Luminosities [35]

Quantities	Scaling law	Units
$2\pi/\omega$	$8.2 \times 10^{-5} \, \alpha^{1/2} \dot{m} m$	s
$1/\nu$	$7.3 \times 10^{-5} \, \alpha^{-1/2} \, m$	s
D_3/c	$1.7 \times 10^2 \, \alpha^{5/6} \dot{m}^{5/3} m^{4/3}$	s
W_{max}	$2.4 \times 10^8 \, [\text{erg}] \, z\Gamma\alpha^{2/3}\dot{m}^{4/3}m^{2/3}$ $1.5 \times 10^{20} \, [\text{eV}] \, z\Gamma\alpha^{2/3}\dot{m}^{4/3}m^{2/3}$	erg eV
L_{rad}	$1.5 \times 10^{38} \, \dot{m} m$	erg s^{-1}
L_{w}	$2.8\alpha^{1/2}L_{rad}$	erg s^{-1}
L_{TCR}	$2.8\kappa\alpha^{1/2}L_{rad} = L_{T\gamma}$	erg s^{-1}
L_{UHECR}	$2.8\kappa\zeta\alpha^{1/2}L_{rad}$	erg s^{-1}
$L_{T\gamma}$	$2.8\kappa\alpha^{1/2}L_{rad} = L_{TCR}$	erg s^{-1}

He *et al.* [26] divided the events belonging to the northern hotspot into two by energy, and found that there was a systematic deviation between them. Assuming that this is due to the deflection by the magnetic field, the position of the true source was estimated. While the estimated position, though extended to 10 degrees, included several high-energy celestial objects such as M82 and Mrk 180, only M82 was located within the GZK-horizon (~100 Mpc) that the UHECR could reach. M82 is a starburst galaxy and contains many supernova remnants and high-energy objects. It also emits intense gamma rays (1–100 GeV) [27].

Furthermore, PAO team re-analyze the data in a bit lower energy and found additional enhancements of $4\,\sigma$ above 39 EeV toward two starburst galaxies (NGC253 and NGC4945) in the southern sky, too [28]. This is confirmed by Telescope Array team [29]. These starburst (and Seyfert) galaxies with only less massive blackholes, $10^3 - 10^6 \, M_\odot$ but not supermassive $> 10^7 \, M_\odot$), however, have a difficulty to accelerate up $\sim 10^{20}$ eV in the framework of Fermi acceleration. For example, Anchordoqui and Soriano [30] found that acceleration is limited to at most 10^{19} eV for proton in M82 even if they took into account of extensive shock and strong magnetic field in the nuclear region of starburst galaxies. It is not enough for northern hot spot, in which proton dominant composition was obtained by TA team [31].

Unlike Fermi acceleration, the bow wake acceleration theory [2,3] can explain that even less massive black holes ($10^3 - 10^6 \, M_\odot$) in the starburst (Seyfert) galaxies can accelerate protons to 10^{20} eV and be possible sources of high energy neutrinos (10^{15}–10^{20} eV) as well as nearby radio galaxies with supermassive ($> 10^7 \, M_\odot$) blackholes in the local supercluster.

Furthermore, the bow wake acceleration theory can explain the astronomical neutrinos in the TeV to PeV energy range, observed the IceCube Neutrino Observatory at the South Pole, which is presently the most sensitive [32–34]. The recent data set consisting 37 events corresponding spectral excess with respect to the atmospheric background with a significance more the 5σ [34]. The spectral excess is consistent with the diffuse neutrino flux with W_ν^{-2} spectrum and isotropic distribution. The observed flux for each neutrino species is well described with $W_\nu^2 F_\nu = (0.95 \pm 0.3) \times 10^{-8}$ GeV cm^{-2} sr^{-1} between 30 TeV and 2 PeV.

In the present paper, we will discuss how the bow wake acceleration theory explains the multi-messenger observations, including UHECRs, gamma-rays, and neutrinos, based on the scaling equations (Table 1) of the bow wake acceleration theory deduced by Tajima, Yan, and Ebisuzaki [35] in Section 2. We also discuss the possibility of a similar wakefield acceleration, which accompanies the formation of accretion disk and jets, even in the coalescence process of neutron star detected by gravitational wave detectors. They can clarify the origin of cosmic rays by multi-messenger astronomy as well as neutrino detectors (IceCube and POEMMA), gamma-ray detectors (Fermi gamma-ray astronomy in Section 3, Cherenkov Telescope Array), visible-infrared ray telescopes, and radio telescopes. In Section 4, we conclude the emergence of new mechanism of charged particle acceleration beyond the conventional framework of Fermi acceleration.

2. Possible UHECR Sources: Multi-Messenger Approach

The Telescope Array (TA) team suggested that there is a hotspot in the northern sky where the directions of arrival of the UHECRs above 60 EeV [23]. In addition, the Pierre Auger Observatory (PAO) team also reported that a hotspot in the southern sky above 60 EeV, although it is less statistically significant [24,25]. These are signs of the anisotropy of charged particles and are key clues to reveal the origin of cosmic rays. He *et al.* [26] divided the events belonging to the northern hotspot into two by energy, and found that there was a systematic deviation between them. Assuming that this is due to the deflection by the magnetic field, the position of the true source was estimated. The estimated position, though extended to 10 degrees, included several high-energy celestial objects such as M82 and Mrk 180, only M82 was located within the GZK-horizon (~100 Mpc) that the UHECR could reach. M82 is a starburst galaxy and contains many supernova remnants and high-energy objects. It also emits intense gamma rays (1–100 GeV) [27]. Furthermore, PAO team found additional enhancement of 4σ above 39 EeV toward two starburst galaxies (NGC253 and NGC4945) in the southern sky, too [28,29].

2.1. *Starburst galaxy M82*

M82 is a nearby (approximately 3.6 Mpc; [36]) edge-on galaxy. M82 is in a starburst state in which a large number of stars are formed at a same time as a result of strong disturbance of the low-temperature gas in the galaxy by the event of the collision with a M81 galaxy. As a result of this starburst, the supernova rate is at least one order of magnitude high compared with a normal galaxy; it is as high as once every 5–10 years [37–40]. In addition, massive and compact young star clusters are seen [41]. Furthermore, many Ultra-Luminous X-ray Sources (ULXS), which have very high luminosity ($L_X > 10^{39}$ erg/s) been discovered as in other starburst galaxy.

The M82 X-1 is the brightest ULXS in M82, located about 200 pc away from the dynamic center of the galaxy [42–45]. In general, the maximum luminosity of a star bound by gravity is limited by the Eddington luminosity, defined by:

$$L_{Edd} = 1.26 \times 10^{38} \ [\text{erg s}^{-1}]m. \tag{1}$$

Thus, it can be concluded that M82 X-1 with an X-ray flux of 2×10^{41} erg/s or more [42,44,45] must have a mass of at least around $10^3 \ M_\odot$ and most likely $\sim 10^4 \ M_\odot$. In addition, the mass of the M82 X-1 estimated from the frequency of the Quasi Periodic Oscillation (QPO) observed in the X-ray luminous intensity as to be $100–1300 \ M_\odot$ [46–50]. Such a blackhole with mass of $10^2 - 10^5 \ M_\odot$ is called an intermediate mass blackhole [51], and is considered to be an important key for solving the mystery of formation of the super massive by connecting a stellar mass blackhole ($\sim 10 \ M_\odot$) and a supermassive blackhole ($\geq 10^6 \ M_\odot$) in a galactic center [51].

2.1.1. *UHECR flux*

Applying the bow wake acceleration theory to the M82 X-1, it has been shown that acceleration to 10^{20} eV is well feasible (Fig. 3) in the accreting blockhole system. The UHECR flux, $F_{UHECR: M82}$, of M82 at the Earth is estimated as:

$$F_{UHECR: M82} = 6.7 \times 10^{-1} \left[\frac{UHECRs}{100 \ \text{km}^2 \ \text{yr}} \right] \left(\frac{\kappa}{0.1} \right) \left(\frac{\alpha}{0.1} \right)^{\frac{1}{2}} \left(\frac{\ln(W_{max}/W_{min})}{30} \right)^{-1} \times$$

$$\times \left(\frac{d}{3.6 \ \text{Mpc}} \right)^{-2} \left(\frac{L_{rad}}{10^{42} \ \text{erg s}^{-1}} \right), \tag{2}$$

where we assume isotropic emission for UHECRs. On the other hand, the TA team acquired 72 cosmic rays of 57 EeV in 5 years, and observed 19 events (4.5 events of which were expected from uniform arrival) within the hotspot [23]. Since the effective area of TA is 700 km^2, the observed excess flux in the hot spot direction is about 0.4 UHECRs/100 km^2/yr, which is consistent with Eq. (2).

Two other nearby starburst galaxies, NGC253 and NGC4945 are also suggested to be possible UHECR sources [28,29]. It is worth noting that long gamma ray bursts are shown to be related to the explosions of Walf-Rayet stars undergoing Ic supernovae, which take place in high frequency in starburst galaxies [52].

2.1.2. Gamma-ray flux

The Fermi Gamma Ray Observatory detects gamma rays of 0.1–100 GeV from M82. The flux is measured as $(15.4 \pm 1.9) \times 10^{-9}$ ph s^{-1} cm^{-2} [27,53,54], which corresponds to the gamma-ray luminosity of 1.2×10^{40} erg s^{-1}. In bow wakefield acceleration theory, on the other hand, electrons are accelerated as well as protons and nuclei. The energies obtained by electron acceleration is converted into photons (gamma-rays) by some electromagnetic interaction. Its luminosity, L_γ, is considered to be equal to be L_{TCR} at the highest cases. In other words,

$$L_\gamma \sim L_{TCR} = 8.7 \times 10^{-2} \left(\frac{\kappa}{0.1}\right) \left(\frac{\alpha}{0.1}\right)^{\frac{1}{2}} L_{rad}$$

(3)

The expected gamma-ray flux from M82 X-1 at the Earth can be calculated as $(0.4 - 1.8) \times 10^{-9}$ erg s^{-1} cm^{-2} depending on the variation of L_{rad} $(5$-$20) \times 10^{40}$ erg s^{-1}, as shown in Table 1. This expected luminosity is consistent to the observation by Fermi Gamma-ray Observatory, if we take into account of the contributions from the numerous numbers of supernova remnants in the galaxy [55]. It is worth noting the gamma-ray flux from M82 may show a significant enhancement at the luminous phase of M82 X-1, though any significant variation in gamma-ray flux have not observed yet [27,53,54].

2.1.3. Deflection by the magnetic field

If the hot spot in the northern sky comes from M82, the total deflecting angle is 17.4 degrees [26]. If it is due to the deflection by intergalactic magnetic field, we get:

$$\theta = 0.5°z \left(\frac{d}{Mpc}\right) \left(\frac{B}{nG}\right),$$

(4)

which takes the value of 17.4°. M 82 ($d = 3.6$ Mpc) is the source of a proton ($z = 1$), each requires a magnetic field of the order of $B = 9.7$ nG. In ordinal intergalactic space, the magnetic field strength is expected as low as 0.1 nG [56]. Therefore, the large deflection angle given by Eq. (4) cannot be explained.

However, Ryu et al. [57] carried out the simulation of the local large-scale structure of the universe and found that the magnetic field of about 10 nG can be expected in the filament structure. The distribution of galaxies represents the network (filament structure) of the local supercluster to which our galaxies belong. Our Milky Way Galaxy and M82 are in the same filament structure. Therefore, the magnetic field of the UHECR propagating path from M82 to the Milky Way Galaxy can be expected to be about 10 nG, which is higher than that of the ordinary intergalactic space [57]. It is expected that the UHECR propagating through at distances of 3.6 Mpc is deflected by nearly about 18 degrees.

2.2. Nearby AGNs in the local supercluster (M87 and Cen A and B) and background components of UHECR

Nearby AGN in the local supercluster, such as M87, Cen A and B can be considered possible sources of UHECRs. First, M87 is difficult to accelerate UHECR above 3.9×10^{19} eV, since W_{max} is less than 3×10^{19} eV even for the flaring phase. Furthermore, a branch of the filaments different from that toward M82 extends towards M87 from the Milky Way Galaxy. Therefore, the path of UHECR from M87 to our galaxy is also considered to be within the filament structure. In fact, galaxies in this direction, and our galaxy, is "falling" towards M87 [58]. The distance to M87 is 16.7 Mpc [59], which is about four times larger than that of M82, the deflection due to the magnetic field would exceed 80 degrees and the spot would spread at least several tens of degrees. Thus, any excess flux from M 87, even if it exists, is difficult to detect with current ground-based detectors without all-sky observable ability.

Cen A and B are capable to accelerate UHECR during flaring phase. Another filament extends toward the Centaurus direction (direction of Cen A, B and NGC4945). The distance to Cen A is about 3.4–3.8 Mpc [60–62], which is comparable to M82, so that Cen A, as well as NGC4945 can contribute to the excess flux. In fact, the Pierre Auger Observatory teams reported that there may be excessive fluxes towards the Centaurus region [24,25]. However, the contribution from Cen B, which is much far [56 Mpc; 63] compared to the Cen A, is difficult to form a confined hot spot. The components from Cen B would contribute background component of UHECRs.

2.3. Background (isotropic) fluxes of UHECR and neutrinos

The cosmic ray spectrum averaging the contributions of such numerous numbers of UHECR sources (starburst galaxies and AGNs) over the entire sky is estimated as follows. First, the cosmic-ray flux J_{CR} per unit energy, area, time, and solid angle of cosmic-ray particle is calculated as:

$$J_{CR} = \frac{c l_\gamma \tau_{CR}}{4\pi} W^{-2} = 1.8 \times 10^{-28} \ [\text{particles}/(\text{GeV cm}^2 \text{ s sr})]$$

$$\times \left(\frac{W}{10^{19} \text{ eV}}\right)^{-2} \left(\frac{l_\gamma}{10^{38} \text{ erg s}^{-1} \text{ (Mpc)}^{-3}}\right) \left(\frac{\tau(W_{CR})}{3.4 \times 10^9 \text{ yr}}\right), \tag{5}$$

where l_γ is the local gamma-ray luminosity function of blazars, estimated as $10^{37} - 10^{38}$ erg s^{-1} (Mpc)$^{-3}$, taking into account of the beaming effect of the relativistic jets [64,65] and τ_{CR} the lifetime of the cosmic-ray particles, which is consistent to the observational flux of UHECR. For the case of UHECRs, it is determined by GZK process [21,22].

On the other hand, the IceCube Neutrino Observatory at the South Pole, which is presently the most sensitive instrument in the TeV to PeV energy range, reported evidence for the extraterrestrial neutrinos, after the observation of three PeV neutrino cascades with three-year operation [32–34]. The recent data set consisting 37 events corresponding spectral excess with respect to the atmospheric background with a significance more the 5σ [34]. The spectral excess is consistent with the diffuse neutrino flux with W_ν^{-2} spectrum and isotropic distribution. The observed flux for each neutrino species is well described with $W_\nu^2 F_\nu = (0.95 \pm 0.3) \times 10^{-8}$ GeV cm^{-1} sr^{-1} between 30 TeV and 2 PeV. This neutrino can be explained by the bow wake acceleration. Furthermore, they reported that one 290-TeV neutrino event towards Blazer TXS 0506 + 056 [66]. At that same period, the Blazer was flaring gamma rays (1–400 GeV).

At the center of active galactic nuclei, there is a supermassive blackhole ($m = 10^{8-9}$). It is thought that the surrounding accretion disk undergoes transitions from a strongly magnetized state to a weak magnetic field state (on a time-scale of 1 month to 1 year) [18], and the associated electromagnetic disturbances propagate through the jets to form a bow wakefield. This bow wakefield accelerates electrons and protons simultaneously. Both have E^{-2} spectra (see [11,12]). The electrons interact with the local magnetic field to emit synchrotron photons, which collide with the electrons to form an inverse Compton peak. That produces, a typical double-peak spectrum of the synchrotron self-Compton emission process [66]. In addition, they have found that the spectral index of 100 GeV gamma rays and the high degree of inverse correlation seen in many Blazars can be explained. At the same time, the accelerated protons become UHECR and collide with the protons in the surrounding material to release neutrinos. If a fraction f of the cosmic rays lose their energies through p-p interaction in the galaxies, resultant pions produce neutrinos through their decays. According to Waxman and Bahcall [68], the total background accumulated flux of neutrinos through the entire history of the universe can be calculated as:

$$W_\nu^2 J_\nu = \frac{f\tau_\nu}{\tau_{CR}} W_{CR}^2 J_{CR} = \frac{c f l_\gamma \tau_\nu}{4\pi}$$

$$= 2.7 \times 10^{-7} \; [\text{neutrinos GeV}/(\text{cm}^2 \text{ s sr})] \times$$

$$\times \left(\frac{f}{0.5}\right) \left(\frac{l_\gamma}{10^{38} \text{ erg s}^{-1} \text{ (Mpc)}^{-3}}\right) \left(\frac{\tau_\nu}{1.5 \times 10^{10} \text{ yr}}\right)$$

(6)

where τ_ν is the neutrino accumulation time, which is thye order of Hubble time ($\sim 10^{10}$ yr). The background flux of neutrino is consistent to the observation by IceCube in the energy range of $10^4 - 10^6$ GeV (IceCube Collaboration, [32–34]). Here, we assumed neutrinos are produced with the average energy of $W_\nu = 0.05 W_p$ by one p-p collision. In other words, the bow wakefield acceleration model can explain the IceCube Collaboration observations that flares of gamma rays and neutrino radiations occurred nearly simultaneously, because acceleration of electrons and protons inevitably takes place simultaneously. In the future, the accumulation of similar cases will be awaited.

3. Astrophysical Implications

In the present paper, we clarify that an accretion disk around a blackhole emits into electromagnetic bursts propagating along jets, and charged particles are accelerated to energies exceeding ZeV (10^{21} eV) in the electric field in a bow wakefield propagating through the jets. It was shown that the wave luminosity, L_w, is comparable to the radiation luminosity, L_{rad} as shown in Table 1 [35]. The results are applied to the intermediate blackhole M82 X-1 in starburst galaxy M82. This object can explain the excessive fluxes of hot spots in the northern sky discovered by the Telescope Array-team [21].

In addition, since the space connecting the M82 and the Milky Way Galaxy was almost within a filamentary structure of the local supercluster, the magnetic field could be strong enough to explain the deflection of 20 degrees and even in the case that the main component of the chemical composition of the UHECR was protons. Other nearby starburst galaxies (NGC253 and 4945) are also the origins of UHECRs of the arrival direction enhancement obtained by PAO collaboration in the southern sky [28,29].

In the rest part of the section, we discuss astrophysical implications, such as roles of intermediate mass blackholes, such as M87 X-1 and gravitational wave burst from the merging neutrons star event.

3.1 *Intermediate blackholes as the building blocks of the supermassive blackholes*

The ULXS is considered to be an intermediate mass blackhole of 100−10000 M. Intermediate mass blackholes (IMBH) cannot be made by stellar evolution (e.g. [69]). However, Portegies-Zwart *et al.* [70,71] numerically showed that in a dense stellar cluster, massive stars of tens of solar mass successively fall to the center of the cluster in the timescale of millions of years due to the of frictional effects acting between stars called dynamical friction, and grow to 100−1000 solar masses by later coalescence with each other. The resultant supermassive star becomes an intermediate mass blackhole through a supernova explosion. Massive stars in the cluster continue to fall down to the center to accrete to the IMBH. It is considered that the ULXS including the M82 X-1 was made in the high-density star clusters in this way.

This is consistent with the fact that starburst galaxies are rich in both dense star cluster and ULXS [72−74] is consistent with the fact that ULXS and super clusters in starburst galaxies. In fact, many ULXS have been found in super clusters.

Also, such super clusters with intermediate mass blackholes fall towards the center of the galaxy over hundreds of millions of years due to dynamical friction in the galaxy. The blackhole components coalesce with each other and grow, eventually, to a central blackhole of about 1 million solar masses, and evolve into the central core of the Seyfert galaxy. Indeed, many Seyfert galaxies are often difficult to distinguish from starburst galaxies (e.g. [75]). Conversely, the star ingredient is discharged to the outside of the star clusters by the reaction and tidal action of the sedimentation of the blackhole component.

They are believed to form the so-called galaxy bulge [76]. The idea that a galaxy bulge can be a byproduct of the formation of a central blackhole can explain the correlation between the size of the galaxy central blackhole and the size of the galaxy bulge [77–79] is successfully explained.

Because of the presence of a blackhole of 4×10^6 $M_{\odot}$ in the center of the Milky Way Galaxy, the same process from Starburst to Seyfert Galaxy in the past seems to take place in the Milky Way Galaxy. At present, there is still star clusters in the central region, which are the remnant of the activity, and the presence of the intermediate mass blackhole is suggested in a star cluster [80]. Thus, intermediate mass blackholes such as M82 X-1 and UHECR acceleration are an inevitable process of space evolution of the universe, i.e., the growth of galaxies and the formation of central blackholes.

3.2   Neutron star merging event

Takahashi et al. [15] first discussed neutron star merging event as an origin of UHECRs in terms of quark-gluon plasma phase-transition (Fig. 4). Takahashi et al. [15] and Chen et al. [13] showed that strong Alfven waves, produced by a neutron star collision were able to accelerate charged particles to energies above 10^{20} eV.

On August 17, 2017, for the first time, gravitational waves (GW170817), which seem to be due to neutron-coalescing events, was observed by Advanced LIGO and VIRGO detectors, and at the same time, the Fermi Gamma Ray Observatory detected the GRB170817A of gamma-ray bursts (Fig. 5) [81]. The latter was about 1.74 seconds behind the former. It demonstrated that the coalescence neutron star was associated with short gamma-ray bursts (e.g. [16]). The neutron stars gradually approach each other with the emission of gravitational waves, and when the lighter stars reach their Roche limits, they are destroyed by the tidal force of the heavier star, to form an accretion disk. This accretion disk becomes a neutrino-cooled accretion disk, NDAF (Neutrino Dominated Accretion Flow [82], rather than electromagnetic wave because of the high temperature ($\sim 10^{9-10}$ K). Such accretion disk is present for about 1–3 seconds until the heavier neutron star plus the lighter mass is absorbed into the center blackhole of the 2–3 solar mass. If a mechanism similar to the bow wake acceleration does work in $L_w \sim 10^{44-45}$ erg s^{-1}, which is 6–7 orders of magnitude higher than that of radiation-cooled accretion disk (Table 1), the UHECR can be expected to be formed by acceleration by stronger wakefield. On the other hand, the coalescence of neutron star should be accompanied by the dispersal of a large amount of materials processed in the star interior. When a beam of accelerated particles enters the materials, ultra-high energy (10^{20-22} eV) neutrinos are emitted by the interaction of the beam and the materials. Since most of the energy used to accelerate electrons is emitted as gamma rays, there is also a possibility that a burst of gamma rays is produced. The delay of the gamma rays to gravitational waves is consistent with the time required for the NDAF disk and jets to form. The formation of the NDAF disk and the formation of the acceleration field due to the coalescence of the neutron stars will be discussed in a separate paper.

GRB including high energy particles

(Free Space)

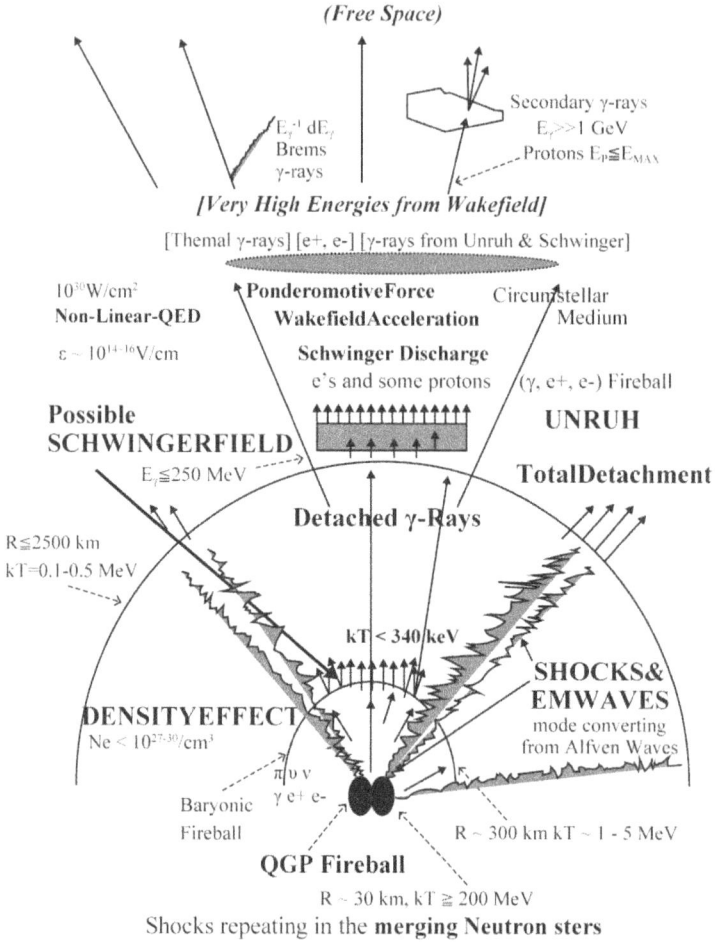

Fig. 4. Wakefield acceleration produced by electromagnetic pulses by neutron star merging events are first discussed by Takahashi *et al.* [23] in terms of quark-gluon plasma (QGP) phase transition (taken from [23] and modified by here).

4. Conclusions

As discussed in the present paper, the bow wake acceleration theory well explains the observational facts related to UHECRs so far. The further verification of the existence of acceleration of ultra-high-energy cosmic rays by bow wake in the universe should be conducted by multi-messenger observations including gravitational wave detectors, such as Advanced LIGO, the VIRGO, and KAGRA [83] and neutrino detectors, such as IceCube and POEMMA [84].

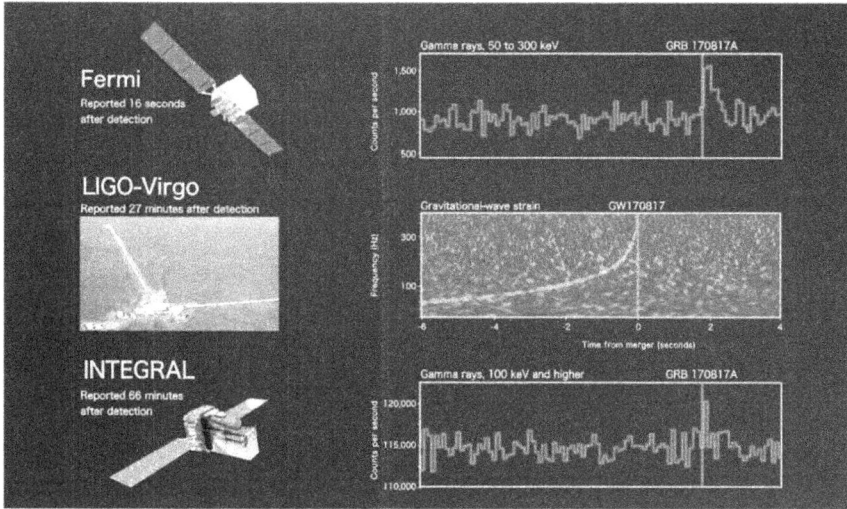

Fig. 5. Gamma-ray emission detected by Fermi and Integral satellites from the neutron star merging event (GW178017) delayed by 1.7 seconds compared with gravitational wave burst [79]. This time difference may be explained by the time to build-up the system for the acceleration of charged particles, described in the present paper, in other words, accretion disk and jets. Credit: LIGO; Virgo; Fermi; INTEGRAL; NASA/DOE; NSF; EGO; ESA (https://heasarc.gsfc.nasa.gov/docs/objects/heapow/archive/transients/gw170817.html).

There are two camps of acceleration mechanisms. The first camp is the conventional acceleration of particles by the individual force, including the Fermi stochastic acceleration [1]. The second is the collective force of the plasma through the coherent force ($\propto N^2$, where N is the number of participating particles, as opposed to the individual force [7–9]). Via using the plasma collective acceleration, we find a new path of particle acceleration that may be more intense and more compact that may suit for variety of astrophysical circumstances, some of which have been discussed above.

The origin of ultra-high energy cosmic rays (Ultra-High Energy Cosmic rays: UHECR) with energy of 10^{20} eV remain an important puzzle of astronomy, though they are commonly referred to as extragalactic origin, (e.g. [14]). So far, their acceleration mechanisms have been discussed primarily in the framework of the Fermi acceleration mechanism [1]. In the Fermi acceleration mechanisms, it is assumed that charged particles gradually retain energy as they are scattered many times by magnetic clouds. One of the necessary conditions for Fermi acceleration is magnetic confinement [85]. The candidate astronomical objects are neutron stars, active galactic nuclei, gamma-ray bursts, and cosmological accretion shock waves in the intergalactic space. However, even in these candidates, the acceleration to 10^{20} eV by the Fermi mechanism has the following problems: (1) very large number of scattering is required to reach high energy; (2) energy loss due to the synchrotron radiation is not negligible at the time of scattering; and (3) adiabatic energy loss takes place when particles escape from the acceleration region. Some of these issues may be addressed by the present mechanism.

In this corridor of the Fermilab we are standing on the shoulder of Enrico Fermi, who had started both the cosmic accelerating mechanism of the Fermi's stochastic acceleration [1] and his Erath-strapping PeV vision. It is also remarkable to observe that the wakefield acceleration may be both a key mechanism that helps alleviate some of the difficulties that Fermi's stochastic acceleration encounters in cosmic acceleration, while the wakefield accelerators now based on thousands of aligned chips, the "TeV on a chip" × 1000, (which have been discussed also here at this Workshop as seen in [Zhang, 2016; Hakimi, 2018; Hakimi in this book) might provide an alternative realization to reach PeV that Fermi dreamed of (see the picture in [9]). Wakefields may become relevant also to muons and neutrinos here at Fermilab.

Acknowledgments

The work was supported in part by the Norman Rostoker Fund. We would like to thank discussions with Prof. K. Abazajian, Dr. N. Canac, Prof. M. Teshima, Prof. S. Barwick, Dr. A. Mizuta, Prof. B. Barish, the late Prof. G. Yodh (the last Fermi's student), Prof. H. Sobel, Prof. V. Trimble, Prof. T. Tait, Dr. C.K. Lau, Ms. S. Hakimi, Dr. A. Sahai, Prof. V. Shiltsev, Prof. G. Mourou, Prof. X. M. Zhang, Prof. K. Shibata, Prof. R. Matsumoto, and Prof. H. Sagawa.

References

1. E. Fermi, Galactic magnetic fields and the origin of cosmic radiation, *Astrophys. J.* **119**, 1 (1954).
2. T. Ebisuzaki and T. Tajima, Astrophysical ZeV acceleration in the relativistic jet from an accreting supermassive blackhole, *Astropart. Phys.* **56**, 9 (2014).
3. T. Ebisuzaki and T. Tajima, Pondermotive acceleration of charged particles along the relativistic jets of an accreting blackhole, *Eur. Phys. J. Special Topics* **223**, 1113 (2014).
4. S. A. Balbus and J. F. Hawley, A powerful shear instability in weakly magnetized disks. I. linear analysis, *Astrophys. J.* **376**, 214 (1991).
5. R. Matsumoto and T. Tajima, Magnetic viscosity by localized shear flow instability in magnetized accretion disks, *Astrophys. J.* **445**, 767–779 (1995).
6. T. Tajima and J. M. Dawson, Laser electron accelerator, *Phys. Rev. Lett.* **43**, 267 (1979).
7. V. Veksler, in *Proc. of the CERN Symposium on High-energy Accelerators and Pion Physics,* Vol .1, ed. R. Edouard (CERN, 1956).
8. N. Rosstoker and M. Reiser, *Collective Methods of Particle Acceleration* (Harwood, 1978).
9. T. Tajima, K. Nakajima, and G. Mourou, Laser acceleration, *Riv. del Nuovo Cim.* **40**, 33 (2017).
10. E. Esarey, C. B. Schroeder and W. P. Leemans, Physcs of laser-driven plasma-based electron accelerators, *Rev. Mod. Phys.* **81**, 1229 (2009).
11. T. Tajima, Laser acceleration and its future, *Proc. Jpn. Acad. Sci.* **B86**, 147 (2010).
12. K. Mima, W. Horton, T. Tajima and A. Hasegawa, in *Nonlinear Dynamics and Particle Acceleration,* eds. Y. Ichikawa and T. Tajima (AIP, NY, 1991), p. 27.

13. P. C. Chen, T. Tajima and Y. Takahashi, Plasma wakefield acceleration for ultrahigh-energy cosmic rays, *Phys. Rev. Lett.* **89**, 161101 (2002).
14. K. Kotera and A. Olinto, The astrophysics of ultrahigh-energy cosmic rays, *Ann. Rev. Astron. Astrophys.* **49**, 119 (2011).
15. Y. Takahashi, T. Tajima and L. Hillman, Relativistic lasers and high energy astrophysics gamma ray bursts and highest energy acceleration, in *High Field Science*, eds. T. Tajima, K. Mima and H. Baldis (Kluwer, 2000), p. 171.
16. E. Nakar, Short-hard gamma-ray bursts, *Phys. Rep.* **442**, 166 (2007).
17. F. Y. Chang *et al.*, Magnetowave induced plasma wakefield acceleration for ultrahigh energy cosmic rays, *Phys. Rev. Lett.* **102**, 111101 (2009).
18. K. Shibata, T. Tajima and R. Matsumoto, Magnetic accretion disks fall into two types, *Astrophys. J.* **350**, 295 (1990).
19. S. M. O'Neill *et al.*, Low-frequency oscillations in global simulations of black hole accretion, *Astrophys J.* **736**, 107 (2011).
20. A. Mizuta, T. Ebisuzaki, T. Tajima and S. Nagataki, production of intense episodic Alfven pulses: GRMHD simulation of black hole accretion discs, *Mon. Not. Roy. Astron. Soc.* **479**, 2534–2546 (2018).
21. R. U. Abbasi *et al.*, Indications of intermediate-sacle anisotropy of cosmic rays with energy greater than 57EeV in the northern sky measured with the surface detector of the telescope array experiment, *Astrophys. J. Lett.* **790**, L21 (2014).
22. K. Greisen, End to the cosmic-ray spectrum? *Phys. Rev. Lett.* **16**, 748 (1966).
23. G. T. Zatsepin and V. A. Kuzmin, Upper limit of spectrum of cosmic rays, *JETP Lett.* **4**, 78 (1966).
24. J. Abraham *et al.* (Pierre Auger Collaboration), Upper limit on the cosmic-ray photon flux above 10(19) eV using the surface detector of the Pierre Auger Observatory, *Astropart. Phys.* **29**, 188 (2008).
25. P. Abreu *et al.*, Update on the correlation of the highest energy cosmic rays with nearby extragalactic matter, *Astropart. Phys.* **34**, 314 (2010).
26. H.-N. He *et al.*, Monte Carlo Bayesian search for the plausible source of the Telescope Array hotspot, *Phys. Rev. D* **93**, 043011 (3016).
27. M. Ackermann *et al.*, GeV observation of star-forming galaxies with the *FERMI* large area telescope, *Astrophys. J.* **755**, 164 (2012).
28. A. Aab *et al.* (The Pierre Auger Collaboration), An indication of anisotropy in arrival directions of ultra-high-energy cosmic rays through comparison to the flux pattern of extragalactic gamma-ray sources, *Astrophys. J. lett.* **853**, L29 (2018).
29. D. Matteo *et al.* (Telescope Array Collaboration), Investigating an angular correlation between nearby starburst galaxies and UHECRs with the Telescope Array experiment, arXiv: 1905.07994v1 (2019).
30. L. A. Anchordoqui and J. F. Soriano, Evidence for UHECR origin in starburst galaxies, arXiv:1905.13243v1 (2019).
31. R. U. Abbasi *et al*, Study of Ultra-High Energy Cosmic Ray composition using Telescope Array's Middle Drum detector and surface array in hybrid mode, *Astropart. Phys.* **64**, 49 (2015).
32. M. G. Aartsen *et al.*, First observation of PeV-energy neutrinos with IceCube, *Phys. Rev. Lett.* **111**, 021103 (2013).

33. M. G. Aartsen *et al.*, Evidence for high-energy extraterrestrial neutrinos at the IceCube detector, *Science* **342**, 6161 (2013).

34. M. G. Aartsen *et al.* Observation of high energy astrophysical neutrinos in three years of IceCube data, *Phys. Rev. Lett.* **113**, 101101 (2014).

35. T. Tajima, X. Q. Yan and T. Ebisuzaki, Wakefield acceleration, in preparation (2019).

36. W. L. Freedman *et al.*, The Hubble Space Telescope Extragalactic distance scale key project. I. The discovery of cephids and a new distance to M81, *Astrophys. J.* **427**, 628 (1994).

37. M. S. Yun *et al.*, A high resolution image of atomic hydrogen in the M81 group of galaxies, *Nature* **372**, 530 (1984).

38. P. P. Kronberg and P. N. Wikinson, High-resolution multifrequency radio observations of M82, *Astrophys. J.* **200**, 430 (1975).

39. G. H. Riek *et al.*, the nature of the nuclear sources in M82 and NCC 253, *Astrophys. J.* **238**, 24 (1980).

40. P. P. Kronberg *et al.*, The nucleus of M82 at radio and X-ray bands: discovery of new radio population of supernova candidates, *Astrophys. J.* **291**, 693–707 (1985).

41. S. Barker *et al.*, Star cluster versus field star formation in the nucleus of the prototype starburst galaxy M82, *Astron. Astrophys.* **484**, 711-720 (2008).

42. H. Matsumoto and T. G. Tsuru, X-ray evidence of an AGN in M82, *Publ. Astron. Soc. Jpn.* **51**, 321 (1999).

43. E. J. M. Colbert and R. F. Mushotzky, The nature of accreting black holes in nearby galaxy nuclei, *Astrophys. J.* **519**, 89-107 (1999).

44. T. G. Tsuru *et al.*, M82 X-1 — The hyper luminous X-ray source, *Prog. Theor. Phys. Suppl.*, **155**, 59-66 (2004).

45. A. Patruno, S. Portegies-Zwart, J. Dewi, and C. Hopman, The ultraluminous X-ray source in M82: an intermediate-mass black hole with a giant companion, *Mon. Not. R. Astron. Soc.* **370**, L6 (2006).

46. G. C. Dewangan, L. Titarchuk and R. E. Griffiths, Black hole mass of the ultraluminous X-ray source M82 X-1, *Astrophys. J.* **637**, L21 (2006).

47. H. Feng and P. Kaaret, Origin of the X-ray quasi-periodic oscillations and identification of transient ultraluminous x-ray sources in M82, *Astrophys. J.* **668**, 941-948 (2007).

48. X. L. Zhou *et al.* Calibrating the correlation between black hole mass and X-ray variability amplitude: X-ray only black hole mass estimatimates for active galactic nuclei and ultra-luminous X-ray sources, *Astrophys. J.* **710**, 16 (2010).

49. P. Casella *et al.*, Weigning the black hole in ultraluminous X-ray sources through timing, *Mon. Not. R. Astron. Soc.* **387**, 1707-1711 (2008).

50. D. Pasham, T. E. Strohmayer and R. F. Mushotzky, A 400-solar-mass black hole in the galaxy M82, *Nature* **513**, 74 (2014).

51. T. Ebisuzaki *et al.*, Missing link found? The runaway path to supermassive black holes, *Astrophys. J.* **562**, L19-L22 (2001).

52. J. K. Becker, High-energy neutrinos in the context of multimessenger astrophysics, *Phys. Rep.* **458**, 173 (2008).

53. A. A. Abdo *et al.*, Detection of gamma-ray emission from the starburst galaxies M82 and NGC253 with the Large Area Telescope on *FERMI*, *Astrophys. J. Lett.* **709**, L152 (2010).

54. B. Lacki *et al.*, On the GeV and TeV detections of the starburst galaxies M82 and NGC 253, *Astrophys. J.* **734**, 107-121 (2011).

55. B. Eichmann and J. B. Tjus, The radio-gamma correlation in starburst galaxies, *Astrophys. J.* **821**, 87 (2016).

56. K. Dolag *et al.*, Constrained simulations of the magnetic field in the local Universe and the propagation of ultrahigh energy cosmic rays, *J. Cosmol. Astropart. Phys.* **1**, 9 (2005).

57. D. Ryu, S. Das and H. Kang, Intergalactic magnetic field and arrival direction of ultra-high-energy protons, *Astrophys. J.* **710**, 1422-1431 (2010).

58. M. Aaronson *et al.*, The velocity field in the super cluster, *Astrophys. J.* **258**, 64 (1982).

59. S. Bird *et al.*, The inner halo of M87: a first direct view of red-giant population, *Astron. Astrophys.* **524**, A71 (2010).

60. L. Ferrarese *et al.* The discovery of cepheids and a distance to NGC5128, *Astrophys. J.* **654**, 186-218 (2007).

61. G. L. H. Harris, M. Rejkuba and W. E. Harris, The distance to NGC5128 (Centaurus A), *Publication of the Astronomical Society of Australia* **27**, 457-462 (2010).

62. D. Majaess, The cephids of Centaurus A (NGC5128) and implications for H0, *Acta Astronomica* **60**, 121-136 (2010).

63. P. A. Jones *et al.*, The radio galaxy Centaurus B, *Mon. Not. Roy. Astron. Soc.* **325**, 817-825 (2001).

64. M. Ajello *et al.*, The origin of the extragalactic gamma-ray background and implications for dark matter annihilation, *Astrophys. J.* **751**, 108 (2012).

65. A. E. Broderick, The cosmological impact of luminous TeV Blazers. I. Implications of plasma instabilities for the intergalactic magnetic field and extragalactic gamma-ray background, *Astrophys. J.* **752**, 22 (2012).

66. IceCube Collaboration, Neutrino emission from the direction of the blazar TXS 0506+056 prior to IceCube-170922 Alert, *Science* **361**, 147 (2018).

67. N. E. Canac *et al.*, submitted to *Astrophys. J.* (2019).

68. E. Waxman and J. Bahcall, High energy neutrinos from astrophysical sources: An upper bound, *Phys. Rev. D* **59**, 023002 (1999).

69. G. Fanbbiano, Populations of X-Ray Sources in Galaxies, *Ann. Rev. Astron. Astrophys.* **44**, 323 (2006).

70. S. Portegies-Zwart *et al.*, Star cluster ecology II. Runaway collisions in young compact clusters, *Astron. Astrophys.* **348**, 117-126 (1999).

71. S. Portgies-Zwart and S. L. W. McMillan, *Astrophys. J.* **576**, 899-907 (2002).

72. D. A. Swartz *et al.*, The ultraluminous X-ray source population from the *CHANDRA* archive of galaxies, *Astrophys. J.* **154**, 519-539 (2004).

73. A. R. Basu-Zych *et al.*, Exoring the overabundance of ULXs in metal- and dust-poor local Lyman break analogs, *Astrophys. J.* **818**, 140 (2016).

74. M. W. Pakull, F. Grise and C. Motch, Ultraluminous X-ray sources: Bubbles and optical counterparts, in *Population of High Energy Source in Galaxies. Proc. of IAU Symposium, Volume 1, No. 230*, eds. E. J. A. Meurs and G. Fabbiano (2005), pp. 293-297, doi:10.1017/S1743921306008489.

75. G. Kauffmann *et al.*, The host galaxy of active galactic nuclei, *Mon. Not. Roy. Astron. Soc.* **346**, 1055-1077 (2003).

76. S. Portgies-Zwart, H. Baumgardt and S. L. W. McMillan, The ecology of star clusters and intermediate-mass black holes in the galactic bulge, *Astrophys. J.* **641**, 319 (2006).

77. J. Kormendy and D. Richstone, Inward bound- the search for supermassive black-holes in galactic nuclei, *Ann. Rev. Astron. Astrophys.* **33**, 581 (1995).

78. J. Magorrian *et al.*, The demography of massive dark objects in galaxy centers, *Astron. J.* **115**, 2285 (1998).

79. D. Merritt and L. Ferrarese, Black hole demographics from the M-center dot-sigma relation, *Mon. Not. Roy. Astron. Soc.* **320**, L30 (2001).

80. S. Portgies-Zwart, S. L. W. McMillan and O. Gerhard, The origin of IRS 16: dynamically driven in-spiral of a dense cluster to the galactic center?, *Astrophys. J.* **593**, 352-357 (2003).

81. E. P. Abbot *et al.* (LIGO Scientific Collaboration and Virgo Collaboration, Fermi Gamma-ray Burst Monitor, and INTEGRAL), Gravitational waves and gamma-rays from a binary neutron star merger: GW 170817 and GRB 170817A, *Astrophys. J.* **848**, L13 (2017).

82. R. Narayan, T. Piran and P. Kumar, Accretion models of gamma-ray bursts, *Astrophys. J.* **557**, 949-957 (2001).

83. T. Akutsu *et al.*, Construction of KAGRA: An underground gravitational-wave observatory, *Prog. Theor. Expt. Phys.* **2018**(1), 013F01 (2018).

84. A. V. Olinto *et al.*, POEMMA: Probe Of Extreme Multi-Messenger Astrophysics, in *ICRC2017, 10–20 July 2017, Bexco, Busan, Korea* (2017); A. V. Olinto *et al.*, POEMMA (Probe of Extreme Multi-Messenger Astrophysics) design, in *Astro2020 APC white paper: Medium-class Space Particle Astrophysics Project Science White papers on Compact Objects and Energetic Phenomena; Galaxies; Cosmology.* Relevant white papers submitted include (NAS response ID): 107, 147, 206, 239, 253, 263, 265, 272, 275, 277, 377, 439 (2019).

85. A. M. Hillas, *Ann. Rev. Astron. Astrophys.* **22**, 425 (1984).